International Series in Operations Research & Management Science

Volume 133

For other titles published in this series, go to
www.springer.com/series/6161

George Samuel Fishman

Christos Alexopoulos · David Goldsman ·
James R. Wilson
Editors

Advancing the Frontiers of Simulation

A Festschrift in Honor of George Samuel Fishman

With Contributions by
Russell Cheng, Soumyadip Ghosh, Shane G. Henderson, Peter W. Glynn, Eunji Lim, James O. Henriksen, Wolfgang Hörmann, Josef Leydold, Jack P. C. Kleijnen, Pierre L'Ecuyer, François Panneton, Andrew F. Seila, Sally Brailsford, Steffen Straßburger, Thomas Schulze, Richard Fujimoto, Shing Chih Tsai, Barry L. Nelson, Jeremy Staum, Tûba Aktaran-Kalaycı

Springer

Editors

Christos Alexopoulos
Georgia Institute of Technology
H. Milton Stewart School of Industrial and Systems Engineering
765 Ferst Drive NW
Atlanta GA 30332-0205
USA
christos@isye.gatech.edu

David Goldsman
Georgia Institute of Technology
H. Milton Stewart School of Industrial and Systems Engineering
765 Ferst Drive NW
Atlanta GA 30332-0205
USA
sman@isye.gatech.edu

James R. Wilson
Edward P. Fitts Department of Industrial and Systems Engineering
North Carolina State University
111 Lampe Drive
Raleigh NC 27695-7906
USA
jwilson@ncsu.edu

ISSN 0884-8289
ISBN 978-1-4419-0816-2 e-ISBN 978-1-4419-0817-9
DOI 10.1007/b110059

Springer Dordrecht Heidelberg London New York

Library of Congress Control Number: 2009932145

Printed on acid-free paper

Springer is part of Springer Science+Business Media (www.springer.com)

Preface

This Festschrift honors George Samuel Fishman, one of the founders of the field of computer simulation and a leader of the disciplines of operations research and the management sciences for the past five decades, on the occasion of his seventieth birthday. The papers in this volume span the theory, methodology, and application of computer simulation.

The lead article is appropriately titled "George Fishman's Professional Career." In this article we discuss George's contributions to operations research and the management sciences, with special emphasis on his role in the advancement of the field of simulation since the 1960s. We also include a brief personal biography together with comments by several individuals about the extraordinary effect that George has had on all his students, colleagues, and friends.

The second article, titled "A Conversation with George Fishman," is the transcript of an extended interview with George that we conducted in October 2007.

In the article titled "Computer Intensive Statistical Model Building," Russell Cheng studies resampling methods for building parsimonious multiple linear regression models so as to represent accurately the behavior of the dependent variable in terms of the smallest possible subset of explanatory (independent) variables. The author shows how bootstrap resampling can be used not only for rapid identification of good models but also for efficient comparison of competing models.

The next article is titled "Patchwork Distributions." In this article, Soumyadip Ghosh and Shane Henderson consider a class of multivariate probability distributions that can be used to model a finite-dimensional random vector when the user has specified all the marginal distributions of the random vector, the covariance matrix of the random vector, and the probabilities that the random vector lies in certain regions.

Peter Glynn and Eunji Lim examine the foundations of the batch-means method for steady-state simulation analysis in their article titled "Asymptotic Validity of Batch Means Steady-State Confidence Intervals." Although the large-sample validity of many implementations of the batch-means method requires that the target output process must satisfy a functional central limit theorem, in this article the authors establish the validity of the batch-means method for Harris-recurrent Markov processes that satisfy a weaker (nonfunctional) central limit theorem.

In the article titled "Efficient Modeling of Delays in Discrete-Event Simulation," Jim Henriksen considers the problem of efficiently implementing four types of delays that are commonly encountered in simulation modeling, with a detailed discussion of the solution approach provided by the delay-modeling algorithms that have been developed for the SLX simulation language.

In the article titled "Sampling from Linear Multivariate Densities," Wolfgang Hörmann and Josef Leydold develop an efficient acceptance-rejection algorithm for generating a vector of dependent random variables whose joint density is linear with a domain that is bounded and symmetric about a point; and ultimately the authors extend their algorithm to generate random vectors from concave differentiable densities over point-symmetric domains.

Jack Kleijnen's article, "Factor Screening in Simulation Experiments: Review of Sequential Bifurcation," provides an overview of factor-screening methods for simulation experiments with special emphasis on sequential bifurcation, a method that is particularly effective in certain types of large-scale simulation studies.

"$\mathbb{F}_2$-Linear Random Number Generators," by Pierre L'Ecuyer and François Panneton, is the ninth article in this volume. This article reviews various construction methods for random-number generators based on linear recurrence modulo 2, examines their theoretical properties, describes the relevant computational tools and algorithms, and ends with comparisons based on various qualitative criteria. The $\mathbb{F}_2$-linear class contains many long-period random-number generators—including the Tausworthe, generalized feedback shift register (GSFR), twisted GSFR, Mersenne twister, WELL, and xorshift generators.

The tenth article is titled "Opportunities and Challenges in Health Care Simulation." In this article Andrew F. Seila and Sally Brailsford review successful applications of simulation in health care, examine the suitability of simulation models and methods for the analysis of health-care systems, and explore the reasons for the lack of adoption of simulation as a "routine" tool for health-care systems analysis. The authors end with an insightful list of ideas aimed at promoting wider adoption of simulation in health care.

In the eleventh article, "Future Trends in Distributed Simulation and Distributed Virtual Environments," Steffen Straßburger, Thomas Schulze, and Richard Fujimoto report the main results of a peer study of current trends in distributed simulation and distributed virtual environments. The survey assesses the current state of this methodology, its relevance to simulation practice, and the research challenges that must be addressed so as to facilitate the widespread use of this methodology in industry and government as well as in research organizations.

In "Combined Screening and Selection of the Best with Control Variates," Shing Chih Tsai, Barry L. Nelson, and Jeremy Staum formulate ranking-and-selection procedures with screening that exploit point estimators based on the method of control variates to gain greater statistical efficiency. Compared with previous ranking-and-selection procedures that incorporate screening and selection of the best alternative, substantial improvements in performance are achieved by the new procedures.

In the final article, "Optimal Linear Combinations of Overlapping Variance Estimators for Steady-State Simulation," Tûba Aktaran-Kalaycı, Christos Alexopoulos, David Goldsman, and James R. Wilson seek to estimate the variance parameter of a simulation output process using optimal linear combinations of variance estimators based on the methods of overlapping batch means and standardized time series. From the latter estimators they derive asymptotically valid confidence intervals for both the mean and variance parameter of the target process.

We thank the authors who have contributed to this volume, especially for their forbearance throughout the lengthy process of completing this book. Special thanks go to Yvonne Smith (Georgia Tech), who produced the transcription of our interview with George, and to Carolyn Ford at Springer, who so ably supervised the production of this book. We are also indebted to Fred Hillier (Stanford University) and Gary Folven (Springer) for patiently guiding us through the entire process.

Atlanta, GA, USA — Christos Alexopoulos
Atlanta, GA, USA — David Goldsman
Raleigh, NC, USA — James R. Wilson
March 2009

Contents

Contributors

Tûba Aktaran-Kalaycı H. Milton Stewart School of Industrial and Systems Engineering, Georgia Institute of Technology, Atlanta, GA 30332, USA, taktaran@gmail.com

Christos Alexopoulos H. Milton Stewart School of Industrial and Systems Engineering, Georgia Institute of Technology, Atlanta, GA 30332, USA, christos@isye.gatech.edu

Sally Brailsford School of Management, University of Southampton, Highfield Campus, Southampton S017 1BJ, UK, S.C.Brailsford@soton.ac.uk

Russell Cheng School of Mathematics, University of Southampton, Southampton, SO17 1BJ, UK, R.C.H.Cheng@soton.ac.uk

Richard Fujimoto Computational Science and Engineering Division, College of Computing, Georgia Institute of Technology, Atlanta, GA 30332-0765, USA, fujimoto@cc.gatech.edu

Soumyadip Ghosh IBM T. J. Watson Research Center, Yorktown Heights, NY 10598, USA, ghoshs@us.ibm.com

Peter W. Glynn Department of Management Science and Engineering, Stanford University, Stanford, CA 94305, USA, glynn@stanford.edu

David Goldsman H. Milton Stewart School of Industrial and Systems Engineering, Georgia Institute of Technology, Atlanta, GA 30332, USA, sman@gatech.edu

Shane G. Henderson School of Operations Research and Information Engineering, Cornell University, Ithaca, NY 14853, USA, sgh9@cornell.edu

James O. Henriksen Wolverine Software Corporation, Alexandria, VA 22305, USA, mail@wolverinesoftware.com

Wolfgang Hörmann Department of Industrial Engineering, Boğaziçi University, 34342 Bebek-Istanbul, Turkey, hormannw@boun.edu.tr

Jack P. C. Kleijnen Department of Information Management, Tilburg University, 5000 LE Tilburg, Netherlands, kleijnen@uvt.nl

Josef Leydold Department of Statistics and Mathematics, WU Wien, Augasse 2–6, A-1090 Vienna, Austria, josef.leydold@wu-wien.ac.at

Pierre L'Ecuyer Département d'Informatique et de Recherche Opérationnelle (DIRO), Université de Montréal, C.P. 6128, Succ. Centre-Ville, Montréal, H3C 3J7, Canada, lecuyer@iro.umontreal.ca

Eunji Lim Department of Industrial Engineering, University of Miami, Miami, FL 33124, USA, lim@miami.edu

Barry L. Nelson Department of Industrial Engineering & Management Sciences, Northwestern University, Evanston, IL 60208, USA, nelsonb@iems.northwestern.edu

François Panneton Standard Life Investments Inc., 1001 de Maisonneuve Ouest, bureau 700, Montréal, Québec, H3A 3C8, Canada, Francois.Panneton@standardlife.ca

Thomas Schulze School of Computer Science, Otto-von-Guericke University of Magdeburg, 39016 Magdeburg, Germany, tom@iti.cs.uni-magdeburg.de

Andrew F. Seila University of Georgia, 1240 Saint Andrews Drive, Bogart, Georgia 30602, USA, aseila@gmail.com

Jeremy Staum Department of Industrial Engineering & Management Sciences, Northwestern University, Evanston, IL 60208, USA, staum@iems.northwestern.edu

Steffen Straßburger School of Economic Sciences, Ilmenau University of Technology, 98684 Ilmenau, Germany, Steffen.Strassburger@tu-ilmenau.de

Shing Chih Tsai Department of Industrial and Information Management, National Cheng Kung University, Tainan, Taiwan, str6969@mail.ncku.edu.tw

James R. Wilson Edward P. Fitts Department of Industrial and Systems Engineering, North Carolina State University, Raleigh, NC 27695-7906, USA, jwilson@ncsu.edu

George Fishman Receives the INFORMS Simulation Society's Lifetime Professional Achievement Award from R. G. Sargent (2003)

George Fishman's Professional Career

Christos Alexopoulos, David Goldsman, and James R. Wilson

Abstract In this lead article for *Advancing the Frontiers of Simulation: A Festschrift in Honor of George Samuel Fishman*, we survey briefly George's professional career, summarizing his most significant contributions to the disciplines of operations research and the management sciences. We give special emphasis to George's remarkable accomplishments in helping to lay the foundation for the field of computer simulation and advancing that field over the past five decades.

1 A Brief Biography

George Fishman was born to Louis and Gertrude Fishman on July 3, 1937, in Everett, Massachusetts. Eight months later his father died of cancer, leaving his mother to bring up George and his six-year-old sister Estelle. He spent his early years in the West End of Boston near the Longfellow Bridge. In 1950, his family moved first to Roxbury and then a year later to Chelsea. George graduated from Chelsea High School in 1955 and matriculated that fall at the Massachusetts Institute of Technology (MIT). Unsure of his commitment to engineering, he took a leave of absence in November 1956 and spent the next ten months working for the Whiting Milk Company in Charlestown, Massachusetts. The nightly experience of lifting and pouring 100 forty-quart jugs of heavy cream into a tank for bottling was enough to convince George that college was worth completing. He completed his course work in December 1959 and was graduated from MIT in 1960 with a bachelor of science degree in economics.

While awaiting decisions from graduate schools, George spent January through August 1960 working as a research assistant at MIT for Professors Morris Adelman, E. Cary Brown, and Robert Solow of the Economics Department, and also for Professor Ithiel deSola Pool, chair of the Political Science Department. In many

C. Alexopoulos (✉)
H. Milton Stewart School of Industrial and Systems Engineering, Georgia Institute of Technology, Atlanta, GA 30332, USA
e-mail: christos@isye.gatech.edu

C. Alexopoulos et al. (eds.), *Advancing the Frontiers of Simulation: A Festschrift in Honor of George Samuel Fishman*, International Series in Operations Research & Management Science 133, DOI 10.1007/b110059_1, © Springer Science+Business Media, LLC 2009

respects, George's experience as a research assistant during the winter, spring, and summer of 1960 persuaded him that research was his preferred career path.

In the fall of 1960, George enrolled in the Ph.D. program of the Department of Economics at Stanford University, serving as research assistant to Professors Kenneth Arrow and Marc Nerlove. In the spring of 1962, he was invited to be a summer intern in the Logistics Department of the RAND Corporation in Santa Monica, California. The latter opportunity turned out to be fortuitous. In the summer of 1962, George was offered a full-time position as a research analyst at RAND, and he left Stanford with a master's degree in economics.

RAND was an exceptionally stimulating place, and the decision to go there was not hard for George to make. In the early 1960s, RAND's Logistics Department could have easily staffed several university-level departments in operations research. During the 1950s, RAND had been in the forefront of new developments in mathematical programming, Monte Carlo simulation, and discrete-event simulation, topics in which George had limited experience prior to joining RAND. In December 1963, Murray Geisler, then chair of the Logistics Department, asked George to referee a RAND report on statistical aspects of simulation. After reading the review, Geisler encouraged George to examine the many outstanding statistical issues in simulation. During the next seven years at RAND, George wrote a collection of publications with a focus on estimating the statistical accuracy of simulation results. Several of these papers were coauthored with Philip Kiviat, who at that time was actively engaged with Harry Markowitz in the development of the SIMSCRIPT II simulation programming language. The collaboration with Kiviat gave George a considerably broader understanding of the component parts of simulation methodology: modeling, computer languages, and statistical considerations.

By 1967, George had decided that a university setting would work best for his research interests. He enrolled in the Biostatistics Program at the University of California, Los Angeles (UCLA), from which he received a Ph.D. degree in March 1970. Shortly thereafter he and his young wife, Sue, moved to New Haven, Connecticut, where he joined the Administrative Sciences Department at Yale University as an associate professor and associate director of the Health Services Research Training Program in Yale's Institution for Social and Policy Studies. At that time, Yale's Administrative Sciences Department had an exceptionally accomplished operations research faculty, including Gordon Bradley, Eric Denardo, Matt Sobel, Harvey Wagner, and Ward Whitt.

The RAND experience, especially the collaboration with Kiviat, had provided George with an excellent preparation for teaching a graduate-level course in simulation. The notes from this course led to his book *Concepts and Methods in Discrete Event Simulation* (Fishman 1973). George's scholarly work now broadened to encompass pseudorandom number generation, random variate generation, and a new way to analyze sample path data based on the theory of regenerative processes.

In July 1974, George, Sue, and their children, Becky and Matt, moved to Chapel Hill, North Carolina, where George had accepted a professorship at the University of North Carolina (UNC) in its recently established Curriculum in Operations Research and Systems Analysis. His work on several areas of simulation methodology continued, with special emphasis on decision rules for how to make the batch-means method of analysis statistically valid in practice.

Although Sue had grown up in a small Mississippi town, George's New England roots made the warm North Carolina summers more of a challenge for him. To the family's delight, George received and accepted an invitation from Matt Sobel to spend part of the summer of 1976 at the Ecology Center of the Marine Biological Laboratory in Woods Hole, Massachusetts, to study the harvesting of elephants in African national parks. The problem called for mathematical and statistical modeling and was of sufficient interest to Dan Botkin, a principal researcher at the Center, to warrant a return the next summer. George found that Woods Hole was a delightful place to do research and spend summers.

Grants from the North Carolina Sea Grant program allowed George and his family to repeat the experience closer to home in the summers of 1978 and 1979 at the UNC Marine Research Laboratory in Morehead City, North Carolina, and the U.S. National Marine Fisheries Laboratory in Beaufort, North Carolina. The challenge was to formulate a policy for optimally determining the opening date for shrimp fishing in North Carolina intercoastal waters.

In January 1981, George became chair of the Curriculum in Operations Research and Systems Analysis at UNC. University resources for academic program development were relatively plentiful in the early 1980s; and during George's ten-year tenure as chair, the program grew in reputation within UNC, nationally, and internationally. Its core faculty, originally consisting of David Rubin, Jon Tolle, and George, was enhanced by the recruitment of Vidyadhar Kulkarni, Scott Provan, Sandy Stidham, and Mark Hartmann. In 1987, UNC acknowledged this growth in size and reputation by elevating the Curriculum in Operations Research and Systems Analysis to the status of the Department of Operations Research within the UNC College of Arts and Sciences.

Scott Provan gives a concise summary of George's contributions to the growth of the Operations Research program and the development of its faculty.

> [George] joined UNC in 1974, having been hired specifically to form the nucleus for operations research as an academic discipline within the College of Arts and Sciences. He originally occupied the sole funded position in an interdisciplinary program whose faculty included members from mathematics, statistics, biostatistics, computer science, and business. George was tenacious in his development of the program, and worked from a crystal-clear vision of its mission within the college. UNC took a farsighted position in those years of encouraging high-quality graduate programs to flourish and draw top-flight faculty and students, and George took excellent advantage of that. I joined the program in 1982 precisely because the research atmosphere was so exciting. Under his chairmanship the program gained 3.5 more funded faculty lines, and in 1987 it became a full-fledged department. George chaired the department until 1990, all the time continuing to garner international acclaim for his own research in simulation theory.
>
> It is safe to say that George single-handedly built the OR program into a nationally-respected and intellectually vibrant department during his tenure. He was relentless in his pursuit of resources, and I was constantly surprised at what he could coax out of the college for a program of our size. Just as significantly, he was able to shield the faculty from many distractions that make it difficult to pursue high-quality research by maintaining a consistent set of expectations and boundaries. I always said that he was a great guy to be behind, and a tough one to be in front of.

Vidyadhar Kulkarni, the current chair of the Department of Statistics and Operations Research at UNC adds a personal touch.

> Professor Fishman acted as my academic mentor: I learned how to write research papers and research grants under the helpful advice of George. He served as a role model of an effective chairman that helped me later. I was truly amazed at his ability of time management: his research productivity did not diminish one bit during his tenure as the chairman. I have found myself using his administrative strategies many times during my own stint as a chairman. I am thankful to him for all that I learned from him.
>
> George is now retired, but comes to the department regularly and provides the most enjoyable and erudite conversational company during our lunches. He is still a valuable resource for research problems that I continue to mine regularly. I wish him a long and productive life.

David Rubin, a professor of operations research with a primary appointment in UNC's Kenan-Flagler Business School offers a view from a slightly different angle.

> When George Fishman arrived at UNC in 1974, the Curriculum in Operations Research and Systems Analysis had control of one full position (George's) and two half positions that were joint in the mathematics and statistics departments. In addition, there were roughly eight "affiliated" faculty, like me, whose home appointments were in business, computer science, economics, and statistics. Six years later, when George became Curriculum Chair, he had control of only 2.5 faculty positions. I remember often being asked exactly what a curriculum was at UNC. My stock answer: "It's either a nascent or stillborn Department, and only time will tell." George's almost ten years as chair were a time of great growth and success for operations research at UNC. Under his able and forceful leadership, the Curriculum became the Department of Operations Research, home to 5.5 faculty members at the end of his term.

The early 1980s saw a shift in George's focus toward the application of the Monte Carlo method to networks, especially the development of efficiency-improvement (variance-reduction) techniques for estimating reliability and the distribution of maximal flow. In 1986 he began writing the text *Monte Carlo: Concepts, Algorithms, and Applications* (Fishman 1996), followed by *Discrete-Event Simulation: Modeling, Programming, and Analysis* (Fishman 2001).

In 2001, George retired from UNC; however, thanks to the generosity of the department chair, Vidyadhar Kulkarni, George was able to maintain a shared office in the department. Fortuitously, his office mate from 2001 through the summer of 2008 was Charles Dunn, whom George had as a student at Yale over thirty years earlier. Since the fall of 2008, Sandy Stidham, a friend and colleague for at least as long, has been his office mate. It was the convenience of this space that made possible publication of his latest book, *A First Course in Monte Carlo* (Fishman 2005).

Sandy Stidham succeeded George as department chair in 1990. Sandy provides a keen insight into George's leadership style and its effect on all his colleagues.

> After I came to North Carolina, it soon became apparent to me that, under George's leadership, the Curriculum in Operations Research and Systems Analysis at UNC-CH had achieved a status that is rare in academic life: it was an interdisciplinary, interdepartmental program with an independent budget and dedicated faculty lines, which had maintained vital connections with other academic units through joint faculty positions. Just as important, and just as rare, was the collegiality of the faculty, the mutual respect among its members, and the close, congenial relations between the faculty and students.

Soon after I joined the faculty in 1986, the Curriculum attained departmental status as the Department of Operations Research. George continued to provide steady leadership as Chair of the Department until 1990. I like to call it "leadership by encouragement." One of George's favorite phrases has always been "I encourage you to . . .." More than a rhetorical device, it was an emblem of his leadership style: he encouraged his colleagues to see things the way he saw them. Because he always thought through his own decisions with logic and sensitivity, the rest of us (almost always) ended up seeing things the same way.

While devoting himself to the growth and flourishing of OR at UNC-CH, George maintained an active research program and a growing international reputation in the fields of discrete-event simulation and Monte Carlo. When these subjects come up during conversations with friends, I have found myself saying (with pride): "My friend and colleague, George Fishman, is the preeminent expert on simulation and Monte Carlo, and I encourage you to look at his books, *Monte Carlo: Concepts, Algorithms, and Applications* and *Discrete-Event Simulation: Modeling, Programming, and Analysis*."

During our mutual retirement, George and I have made a habit of going out for coffee about once a week. We discuss music, politics, religion, economics, and, of course, the joys of grandchildren. The friendship that has developed between us is, and will continue to be, one that I cherish.

During the course of his career, George has made major contributions in laying the foundations for the field of computer simulation and in advancing that field. From 1972 to 1974, he served as the chair of the College on Simulation and Gaming of The Institute of Management Sciences (TIMS), an organization subsequently named TIMS/College on Simulation and later renamed the Simulation Society of the Institute for Operations Research and the Management Sciences (INFORMS). From 1978 to 1980, George served as the representative of TIMS/College on Simulation and Gaming on the Board of Directors of the Winter Simulation Conference. Moreover, from 1978 to 1987, he served as the founding editor of the Simulation Department of the journal *Management Science*. From the perspective provided by the passage of almost three decades, it is now clear that George's service as founding editor of the Simulation Department of *Management Science* was critical to the establishment and initial advancement of the simulation literature as we know it today. It can be argued that this development coupled with George's own research contributions to the simulation literature were two of the key factors in the survival of the field of simulation as a separate, recognizable subject with its own body of relevant theory, methodology, and applications.

From 1989 to 1992, George served on the Editorial Advisory Board of *ACM Transactions on Modeling and Computer Simulation*, a flagship publication of the Association for Computing Machinery. Further, over the years, George served on numerous committees of professional societies and international conferences.

George has received numerous forms of professional recognition for his remarkable contributions to simulation and the larger disciplines of operations research and the management sciences over the past five decades. In 1990 he received the Distinguished Service Award from TIMS/College on Simulation. For his book *Monte Carlo: Concepts, Algorithms, and Applications* (Fishman 1996), he received two awards—the Frederick W. Lanchester Prize from INFORMS in 1996, and the Outstanding Simulation Publication Award from the INFORMS Simulation Society in 1997. In 2003 George was elected a Fellow of INFORMS. In 2004 George received the INFORMS Simulation Society's Lifetime Professional Achievement Award (LPAA), the highest honor given by that organization. This award recognizes major contributions to the field of computer simulation that are sustained over a professional career, with the critical consideration being the total impact of those contributions on the field.

In a supporting letter for the LPAA nomination, Russell Cheng (University of Southampton) wrote:

> Discrete-event simulation has seen substantial growth both in terms of its theoretical base and in terms of its huge range of application. Professor Fishman has been at the forefront of these developments. On a worldwide basis, I would rank him in the top few in terms of both range and the scholastic quality of his contributions, many of which have been seminal.

John Charnes (Senior Vice President, Bank of America) concurred:

> When I look at Prof. Fishman's vita, I am amazed that he has done so much. With contributions to research in so many different fields, mentorship of so many students, and service to the profession in so many ways, he is certainly qualified to be awarded the LPAA.

2 Contributions to Computer Simulation

George is a scholar and researcher of the first rank, a stimulating colleague, a thoughtful mentor, and a steadfast friend to those individuals fortunate enough to have worked with him. Over the course of a career spanning five decades, George has made fundamental contributions to both the theory and practice of simulation, dissemination of knowledge, service to the profession, and the advancement of the status of the field. In the sections that follow, we give a synopsis of George's impact on the field of computer simulation.

2.1 Contributions to Research

George has made numerous groundbreaking contributions to the following methods for simulation output analysis:

- the spectral method;
- the autoregressive method;
- the regenerative method; and
- the batch-means method.

Moreover, George has made seminal contributions to the following areas within the field of simulation:

- efficiency-improvement (variance-reduction) techniques;
- estimation of network performability measures;
- pseudorandom number generation; and
- random variate generation.

Although George's first archival journal article (Fishman 1964) and his first book (Fishman 1969) were in the field of economics, much of his subsequent work focused on simulation analysis methodology and the design of efficient simulation experiments. In the rest of this section, we survey George's contributions in the areas listed above.

2.1.1 Spectral Method

George's first major contribution to simulation output analysis was the development with Phil Kiviat of the spectral method (Fishman and Kiviat 1967). The name of this method is based on the following property of a stationary stochastic process, presumably generated by a simulation in steady-state operation: if we compute the power spectrum of the process (that is, the Fourier cosine transform of the associated autocovariance function), then the power spectrum evaluated at frequency zero is equal to the variance parameter of the process (that is, the sum of the autocovariances at all lags). Moreover, the variance of the sample mean is asymptotically equal to the variance parameter divided by the sample size as the sample size increases. Hence, for large sample sizes, the estimation of the variance parameter and the construction of a valid confidence interval for the steady-state mean reduces to the estimation of the power spectrum of the process at zero frequency.

As formulated by Fishman and Kiviat (1967), the spectral method for simulation analysis estimates the variance parameter (and hence the variance of the sample mean) by a weighted average of estimators of the autocovariance function, with appropriately chosen weights. The methodology was revisited in the early 1980s by Heidelberger and Welch (1983), and its relationship to the methods of nonoverlapping and overlapping batch means is discussed by Welch (1987). This variance-estimation approach has regained attention recently with the development of wavelet-based methods for estimation of the power spectrum (Lada and Wilson 2006).

2.1.2 Autoregressive Method

In Fishman (1971) and Fishman (1973), George pioneered the autoregressive method for simulation output analysis. This technique is based on the following properties:

- A mixed autoregressive–moving average (ARMA) process of order (p, q) (where $0 \le p, q < \infty$) that is stationary and invertible often provides an adequate model for many time series encountered in practice—including many simulation-generated times series.
- A stationary invertible ARMA(p, q) process can be approximated in quadratic mean to any prespecified accuracy by a stationary autoregressive (AR) process of sufficiently large order p' (that is, an AR(p') model).

George formulated a comprehensive method for determining an appropriate order of the pure autoregressive model as an approximation to a given simulation-generated process and for estimating the coefficients of that model. A simplified variant of the autoregressive method has resurfaced recently as a key ingredient of automated batch-means methods for simulation analysis (Steiger et al. 2005, Lada et al. 2008, Tafazzoli et al. 2008).

2.1.3 Regenerative Method

George's next fundamental contribution to simulation analysis was the estimation of steady-state means and quantiles based on the regenerative method (Fishman 1974a, Fishman and Moore 1979, Seila 1976). If a stationary stochastic process has the regenerative property, then the process has regeneration points defined by transitions out of a distinguished state—for example in a stable single-server queueing system, a regeneration point occurs each time an arriving customer finds the system empty and thus triggers a transition out of the empty-and-idle state. At each regeneration point, the probabilistic mechanism governing the evolution of the process is restarted independently of the previous history of the process. The regenerative property enables us to obtain independent and identically distributed blocks of data on which to apply relevant central limit theorems. By coincidence, Crane and Iglehart (1974a, b) published similar results on the regenerative method for simulation analysis simultaneously with Fishman (1974a); in fact, all three papers appeared in the January 1974 issue of *Operations Research.* George relates the following story:

> A copy of my technical report on the regenerative method was sent to Gerry Lieberman, Chair of the Operations Research Department at Stanford University. Interestingly, a copy of a technical report from Stanford University, authored by Don Iglehart and Michael Crane and describing a similar methodology arrived at Yale University one week later. Apparently, the Stanford group and I had been working on the same problem, unaware of the other's work.

Since the appearance of these seminal papers on the regenerative method, hundreds of papers have been published on this approach to the analysis of simulation output.

2.1.4 Batch Means Method

George's pathbreaking contributions to the batch-means method for simulation analysis were motivated by the difficulties that arise in practical applications of the regenerative method. In general, it can be difficult to identify regeneration points in a simulation-generated output process. Moreover, even when such regeneration points can be identified, their recurrence frequency may be so small that an excessive simulation run length is required to accumulate a sufficiently large number of complete regenerative cycles so as to estimate long-run average performance measures with acceptable precision.

Although the method of batch means was known to the simulation community since the early 1960s (Conway 1963), George's groundbreaking work on this method in the 1970s (Fishman 1978a, b) provides the first comprehensive batch-means procedure for computing valid confidence intervals for the steady-state mean. In particular, the discussion on pp. 245–246 of Fishman (1978a) explains how a perusal of the plot of the batch means can give good insights into the effect of the simulation's initial condition, as well as the approach to independence and normality of the batch means with increasing batch size.

In the early 1990s, George revisited the method of batch means with the development of ABATCH and LBATCH, two implementations of the batch-means method that yield not only a consistent estimator of the variance parameter but also asymptotically valid confidence intervals for the steady-state mean as the batch size and the number of batches both tend to infinity at suitable rates (Fishman and Yarberry 1997). ABATCH and LBATCH are the only batch-means algorithms that require $O(n)$ time and $O(\log_2 n)$ space for the total sample size n accumulated with each additional iteration of the overall procedure. Although linear time complexities are known for algorithms based on fixed batch and sample sizes, the dynamic setting of the ABATCH and LBATCH algorithms offers an important additional advantage not present in static approaches: as the analysis evolves with the availability of additional data, ABATCH and LBATCH allow the user to assess visually and quantitatively how the batch-means estimate of the variance parameter converges to the desired limiting value, in linear computation time and sublinear space (that is, computer memory). Such direct assessment enables the user to gauge the quality of the variance estimate and the confidence interval for the mean.

Peter Glynn (Stanford University) applauds George's contributions in this area, while also crediting him as a codeveloper of the regenerative method.

> [George] played a major role in providing a rigorous analysis of the method of batch means, and in developing practical implementations of the method. Given the important role of this output analysis procedure in the steady-state simulation setting, this stands as a significant accomplishment. But this is only one of several fundamental contributions to the output analysis problem. George also introduced the regenerative method (along with Don Iglehart), and developed means of producing confidence bounds that are non-asymptotic (and are thereby guaranteed to hold for any fixed sample size). Each of these contributions has stimulated an extensive amount of follow-on research.

2.1.5 Efficiency-Improvement (Variance-Reduction) Techniques

Throughout his professional career, George has had a particular interest in methods to improve the efficiency of simulation experiments. In Fishman (1968), he addressed the relationship between the computing budget and the accuracy and precision of the resulting simulation-based estimators. In Fishman (1974b), George laid the foundation for the use of common random numbers and antithetic variates to improve the efficiency of estimation of linear simulation metamodels, thereby anticipating the correlation-induction strategy of Schruben and Margolin (1978) and all the follow-up work on correlation-induction strategies for simulation metamodel estimation. In his groundbreaking papers on efficiency-improvement techniques for the simulation of Markov chains (Fishman 1983a, b; Fishman and Huang 1983), George anticipated the recent explosive growth in the use of Markov Chain Monte Carlo (MCMC) (Gilks et al. 1996), especially in the implementation of Bayesian statistical methods. George's work on efficiency-improvement techniques in the 1970s and early 1980s were precursors to his influential contributions on network reliability estimation and his work in the 1990s. Fishman and Rubin (1992a, b) and Fishman et al. (1992) exploit the availability of bounds on the underlying distribution (prior information) to obtain best- and worst-case bounds on the variance and coefficient of variation of the corresponding Monte Carlo–based estimators. Fishman and Kulkarni (1992) describe necessary and sufficient conditions for MCMC sampling to perform more efficiently than Monte Carlo sampling based on independent trials.

2.1.6 Network Reliability

In the early 1980s, George focused much of his research on the development of Monte Carlo sampling plans for estimating performability measures of networks whose components have random characteristics (e.g., lengths, durations, and capacities). This development was funded by a research grant from the Air Force Office of Scientific Research and spanned a period of about ten years. The sampling plans developed by George and his collaborators typically use bounds on the measures under study to construct conditional distributions defined on a subset of the system's state space. The resulting estimators are unbiased, have substantially smaller variance than estimators based on standard Monte Carlo and equal computing effort (encompassing sample size and sampling time), and have bounded relative error as the sample size grows; see, for example, Fishman (1986a, b; 1989b, c). Although bounds existed since the early 1970s (Frank and Frisch 1970, Van Slyke and Frank 1972), they did not permit the construction of effective sampling distributions.

Fishman (1986a) and Fishman (1986b) are landmark papers on the problem of computing network reliability with binary state components. In Fishman (1986b), George exploits bounds based on disjoint minimal path sets and minimal cut sets. In Fishman (1986a), he takes advantage of bounds on the coefficients of the polynomial reliability function when all components have equal reliabilities.

To estimate the distribution of two-terminal maximum flow in networks with discrete arc capacities, George and his collaborators achieve striking increases in estimator efficiency by exploiting bounds based on minimal paths and cuts (Fishman 1987a, b; Fishman 1989a, c; Alexopoulos and Fishman 1991, 1993). An alternative approach is proposed by Fishman and Shaw (1989) and Alexopoulos and Fishman (1992) in which the bounds result from iterative partitions of the system state space, and the corresponding estimators are based on a combination of importance and stratified sampling. The work of George and his coauthors in this area is referenced prominently in the handbook chapter by Ball et al. (1995).

Concerning George's work on Monte Carlo–based analysis of network performability, Peter Glynn offers the following perspective.

> He has made fundamental contributions to the development of special-purpose methods for efficiently computing various performance measures in the stochastic networks context. This class of networks arises naturally in many applied settings, and is of great practical interest. The tools that George has developed in this setting make possible computations that would be difficult (or perhaps even impossible) if attacked using naïve algorithms.

2.1.7 Generation of Pseudorandom Numbers and Variates

George has authored three widely cited studies of pseudorandom number generators (Fishman and Moore 1982, 1986; Fishman 1990) and six papers on random variate generation. In the latter area, the paper by Fishman and Moore (1984) is worthy of special mention. Although the alias method of Walker (1977) is the most frequently cited approach for generating samples from a discrete distribution in bounded time, it fails to provide a monotonic functional dependence of each generated sample on the corresponding random number, a fundamental property required by the methods of common random numbers and antithetic variates. The cutpoint method of Fishman and Moore overcomes this limitation of the alias method, is easier to understand, and shows how to ensure a fixed bound on variate-generation time regardless of the number of points in the distribution.

Pierre L'Ecuyer (Université de Montréal) talks about the influence that George's random number generation work had on his career.

> My first and closest encounter with his work was his papers with Louis R. Moore on search and evaluation of linear congruential generators. These papers had a very strong influence on my own work on uniform random number generators, especially at the earliest stages. In particular, the figure of merit I adopted to select the famous combined [linear congruential generator] of my 1988 *CACM* paper was taken from these papers. In their 1982 *JASA* paper, they introduced a normalized spectral test measure which has become the standard for the theoretical evaluation of linear congruential and multiple recursive random number generators.

In the above quotation, L'Ecuyer (1988) is the article referred to as the "1988 *CACM* paper"; and Fishman and Moore (1982) is the article referred to as the "1982 *JASA* paper."

2.2 Dissemination of Knowledge and Advancement of the Field

Of course, many of us in the simulation community have grown up on George's simulation texts. George authored one monograph on the application of spectral methods in econometrics, three books on discrete-event simulation and two books on the Monte Carlo methodology.

His first simulation text, *Concepts and Methods in Discrete Event Digital Simulation* (Fishman 1973), served as the state-of-the-art reference for many researchers. Barry Nelson (Northwestern University) relates the following story that summarizes the value of this text.

For me, personally, George's work joins the work of Jim Wilson as having the most impact on the direction of my own career. To illustrate this point, I want to single out his 1973 book *Concepts and Methods in Discrete Event Digital Simulation*. I first encountered this book in Bruce Schmeiser's office when I was a graduate student at Purdue in 1981. Bruce used this book as a reference for his IE680 class, and it was this material (along with Bruce) that persuaded me to stay for my Ph.D. rather than leave Purdue after my M.S., as originally planned. The book was already out of print, but I made it a priority to find a copy, and I eventually obtained one from a used book store for $10 (perhaps the highest value per dollar of any purchase I have ever made!). In addition to being one of the first comprehensive books on simulation, covering modeling, programming, and analysis, *Concepts and Methods in Discrete Event Digital Simulation* was the first formal mathematical treatment of "analysis methodology" that I encountered. For many years after I left Purdue, it was one of the first books I pulled from my shelf when I tackled a new research problem.

His second simulation text, *Principles of Discrete Event Simulation*, appeared in 1978. This text incorporated the methodological advancements in simulation output analysis during the 1970s and, most importantly, contained SIMSCRIPT II.5 codes for computing point and confidence interval estimates based on the regenerative and batch-means methods.

In 2001, George published *Discrete-Event Simulation: Modeling, Programming, and Analysis*. Of particular import are the excellent chapter on efficient execution of simulation programs (appropriately entitled "Search, Space, and Time") and the chapter on output data analysis. In addition to the description and implementation of sequential methods based on batch means, the latter chapter contains an enlightening discussion regarding the potential deleterious effects of the initialization bias inherent in steady-state simulation experiments on the validity of confidence intervals obtained by the method of replication/deletion. This discussion directly inspired the paper by Alexopoulos and Goldsman (2004), recipient of the INFORMS Simulation Society's Outstanding Simulation Publication Award in 2007.

The monograph *Monte Carlo: Concepts, Algorithms, and Applications* (Fishman 1996), winner of the Frederick W. Lanchester Prize from INFORMS in 1996, as well as the INFORMS Simulation Society's Outstanding Simulation Publication Award in 1997, is perhaps the tour de force of George's career.

Stunningly complete and comprehensive, the text is a must-read for anyone in the field. In fact, the book has found great use in a number of wide-ranging fields: operations research and industrial engineering, mathematics, probability and statistics, computer science, financial engineering, and physics, just to name a few. The book is unparalleled in scope and content; yet, remarkably, it is a self-contained piece that requires little previous exposure to the field. It begins with a number of methods for estimating volumes and counts, including classical (but difficult) problems involving network reliability, multidimensional integration, sensitivity analysis, bounds for simultaneous confidence intervals, estimation of the expected value of the ratio of random variables, and sequential estimation. The book then presents an encyclopedic discussion on random variate generation techniques. Not only does the list encompass the "usual" random variables; it also includes problems involving the generation of points constrained by interesting geometric shapes including convex polytopes—thus making the book tremendously useful for researchers in mathematical optimization. The text goes on to discuss methods such as correlated sampling, control variates, importance sampling, and stratified sampling to increase sampling efficiency. These topics are in the spirit of Hammersley and Handscomb (1964), albeit at a more-rigorous and more-modern level. A particularly significant chapter is that on random tours, encompassing random walks (and generalizations thereof) on a variety of domains. The book devotes a chapter to simulation output analysis—how should one report the results of an experiment in a statistically rigorous way? The text is in fact the first to treat sequential, commercially useful output analysis methods (LBATCH and ABATCH). The tome concludes with a state-of-the-art treatment on pseudorandom number generation, in which we see how to generate the underlying uniform random variates that drive the stochastic simulation. The overall result is that the book is a beautiful compendium covering everything that one needs to know about the Monte Carlo method; the Lanchester Prize certainly underscores this obvious fact.

The primary target of his 1996 book on Monte Carlo methods is, of course, researchers. This realization led George to author the primer *A First Course on Monte Carlo* (Fishman 2005), for purposes of exposing the methodology to a wider, less-mathematically sophisticated audience. The main contributions of this text are the simplicity of the exposition, the detailed algorithmic description of the various techniques, the plethora of real-world examples from various application areas, and the inclusion of "hands-on" exercises that enable the reader to try the techniques and identify the most-effective ones for the underlying problems.

2.3 Development of Software

George and Louis Moore developed comprehensive computer programs for evaluating the performance of random-number generators using statistical and geometric tests. In addition, George and Stephen Yarberry developed and implemented the only sequential batch means algorithms (LBATCH and ABATCH) that run in linear time and logarithmic space. The algorithms in the LABATCH.2 package have been used by numerous researchers in the operations research and computer science communities. The LABATCH.2 package is available online via

`www.or.unc.edu/~gfish/labatch.2.html` .

3 George's Academic Family Tree

George has a distinguished academic pedigree. His doctoral research advisor was Dr. Robert Jennrich, who is currently a Professor in the Department of Mathematics at UCLA. In reverse chronological order, one path of George's academic ancestry includes Paul Hoel, Dunham Jackson, Edmund Landau, Georg Frobenius, Lazarus Fuchs, Ernst Kummer, Karl Weierstrass, Friedrich Bessel, Christoph Gudermann, and Carl Gauss. A second path can be traced directly to Nicolaus Copernicus and a third path originates from Georgios Gemistos Plethon, a Greek neoplatonist philosopher from Constantinople.

George served on many doctoral dissertation committees both as an advisor and reader. He advised the following students:

- Andrew F. Seila, dissertation: "Quantile estimation methods in discrete event simulations of regenerative systems," Operations Research, UNC (1976).
- Veena G. Adlakha, dissertation: "Starting and stopping rules for data collection in queueing simulations," Operations Research, UNC (1979).
- Louis R. Moore, dissertation: "Quantile estimation in regenerative processes," Statistics, UNC (1979).
- Bao-Sheng Huang, dissertation: "Antithetic sampling method: A variance reduction technique in computer simulation," Operations Research, UNC (1980).
- Kenneth J. Risko, dissertation: "Binomial population selection procedures for fixed unequal sampling costs," Statistics, UNC (1982).
- Tien-Yi Shaw, dissertation: "Monte Carlo methods for reliability analysis of stochastic flow networks," Operations Research, UNC (1988).
- Christos Alexopoulos, dissertation: "Maximum flows and critical cutsets in stochastic networks with discrete arc capacities," Operations Research, UNC (1988).
- L. Stephen Yarberry, dissertation: "Incorporating a dynamic batch size selection mechanism in a fixed-size batch means procedure," Operations Research, UNC (1993).
- M. Cristina Arguelles, dissertation: "Exploiting special structure to enhance efficiency of manufacturing simulation," Operations Research, UNC (1997).

Currently, Andy Seila is a professor emeritus in the Department of Management Information Systems of the Terry College of Business at the University of Georgia. Veena Adlakha is a professor in the Management Department of the Merrick School of Business at the University of Baltimore. Lou Moore is a senior operations researcher with the RAND Corporation in Santa Monica, California. From 1980 to 2000, Bao-Sheng Huang worked for Bell Labs, where he became a distinguished member of the technical staff and technology consultant for his contributions in the areas of network modeling and simulation; and since 2004 he has served as director of systems engineering for Wide Area Network Design Laboratory (WANDL). Kenneth J. Risko is a senior manager in the regulatory and capital markets consulting practice for the financial services industry at Deloitte & Touche LLP. Danny Shaw is an operations research specialist in the Operations Research and Development Department of the SAS Institute Inc. Christos Alexopoulos is an associate professor in the H. Milton Stewart School of Industrial and Systems Engineering at the Georgia Institute of Technology and the director of the Modeling and Simulation Research and Education Center at Georgia Tech. L. Stephen Yarberry is currently the chief information officer at Practice Plus/Arkansas Health Group and chief information security officer at Baptist Health in Little Rock, Arkansas. In addition, he owns Yarberry & Associates, a management and telecommunications consulting company; and he serves as an adjunct professor at both Webster University and the University of Central Arkansas. Cristina Arguelles Tasker is a client business manager with i2 Technologies in London.

Despite his heavy administrative duties and research agenda, George was a resourceful teacher and a wonderful academic advisor. Veena Adlakha comments on George's dedication to his students.

> I have always considered it as a great honor and privilege to have had Dr. George Fishman as my Ph.D. thesis advisor. Working with him as his research assistant and as his student was a great learning process. His office was always open and he always had time for his students—whether to debug a program, improve the flow of logic, or simply answer a question. Dr. Fishman was unrelenting in his demand for hard work and perfection, but he always guided his students with patience. I recall vividly how he made me rewrite the first chapter of my thesis seven times. It was only when I would write my own research papers later that I appreciated the effect that his persistence had on improving my writing abilities.
>
> Being Dr. Fishman's student certainly had its perks. In 1979 I attended the ORSA/TIMS conference with Dr. Fishman, where I received several job offers even though I was not seeking a job at the time, no doubt simply because I was Dr. Fishman's student.
>
> Dr. Fishman was a very gracious man and became a good family friend. I wish nothing but the best for George in his well-deserved retirement.

Stephen Yarberry offers a similarly memorable view of what it means to be one of George's former students.

> While George pushed his students to excel, he never had higher expectations of us than he had for himself. He was so much more than just an advisor and a mentor—he was, and still is, my friend. I count it a honor and a privilege to have received tutelage from a man whom many refer to as the father of our field.

4 Recapitulation

George Fishman is remarkable not only for his exceptional level of scholarly productivity sustained over five decades but also for this long-standing dedication to the advancement of the professions of simulation, operations research, and the management sciences. David Rubin gives the following assessment of George's contributions from the perspective of a faculty colleague and frequent collaborator.

> While shepherding the curriculum/department, George maintained his very active scholarly career. I need not recount here the list of monographs, texts and journal articles he wrote, nor the roster of Ph.D. theses and MS expository papers he directed over the past 40+ years. He recently commented to me about how much he enjoyed doing joint research with colleagues who had skills complementary to his own. I was a beneficiary of that outlook. It was my great pleasure to be a coauthor with him on four of those journal articles. George's work in simulation led him to questions about the size of samples needed to guarantee specified precision in estimating quantities whose exact distributions were unknown. He recognized that there were optimization problems at the base of these questions, I knew something about structured optimization problems, and a fruitful collaboration ensued.
>
> For almost 35 years I have been fortunate to count George among my colleagues, mentors, and special friends. On the occasion of this Festschrift, I wish him many more happy and productive years.

Andy Seila (University of Georgia) gives what is perhaps the best high-level summary of George's professional achievements over the past five decades.

> In order to understand George's contributions to the field of simulation, you have to go back to the early 1970s and examine the status of computing in general and simulation in particular. At that time, two processes were developing that would have a profound influence on the development of simulation as a tool for systems analysis: computers were becoming powerful enough—both in raw size and speed, and in the availability of compilers and other tools—to move simulation from a niche tool to one that would be available to engineers, statisticians and scientists with a modest amount of training. Discrete event simulation modeling tools such as SIMULA, SIMSCRIPT II.5, GPSS, and GASP were widely available and relatively easy to use. At the time, the focus was on modeling methodology, and analysis of output data (as well as input data) was an ad-hoc process. George had the insight to see that, in order for simulation to become a tool that managers would use with confidence, reliable statistical methodology had to be developed. This is the area in which George made his initial contribution and continues to contribute. Thanks to his work and that of others who were influenced by him, we now have a foundation of statistical methodology for the analysis of simulation output data that enables practitioners to compute estimates of performance measures and, more importantly, to assess the usefulness of the estimates. The combination of modeling methodology and data analysis methodology, contributed by George, has provided the fuel to power the explosion in simulation applications we have seen in the past three decades.

On a more personal level, we (the authors of this paper) can trace virtually every line of research we have pursued individually or collectively over the past three decades back to George's seminal papers and books on computer simulation that are surveyed in this article. More important, however, is that many others in the international simulation community spanning several generations of academics, practitioners, and researchers acknowledge a similar debt to the intellectual heritage they received from George Fishman.

References

Adlakha, V. G., and G. S. Fishman. 1982. Starting and stopping rules for simulation using a priori information. *European Journal of Operations Research* 10:379–394.

Alexopoulos, C., and G. S. Fishman. 1991. Characterizing stochastic flow networks using the Monte Carlo method. *Networks* 21:775–798.

Alexopoulos, C., and G. S. Fishman. 1992. Capacity expansion in stochastic flow networks. *Probability in the Engineering and Informational Sciences* 6:99–118.

Alexopoulos, C., and G. S. Fishman. 1993. Sensitivity analysis in stochastic flow networks using the Monte Carlo method. *Networks* 23:605–621.

Alexopoulos, C., and D. Goldsman. 2004. To batch or not to batch? *ACM Transactions on Modeling and Computer Simulation* 14(1):76–114.

Ball, M. O., C. J. Colburn, and J. S. Provan. 1995. Network reliability. In *Handbooks of Operations Research and Management Science: Network Models*, eds, M. O. Ball, T . L. Magnanti, C. L. Monma, and G. L. Nemhauser, Chapter 11. Amsterdam: Elsevier Science Publishers B.V.

Cohen, M.-D., and G. S. Fishman. 1980. Modeling growth-time and weight-length relationships in a single year-class fishery with examples for North Carolina pink and brown shrimp. *Canadian Journal of Fisheries and Aquatic Sciences* 37:1000–1011.

Conway, R. W. 1963. Some tactical problems in digital simulation. *Management Science* 10: 47–61.

Crane, M. A., and D. L. Iglehart. 1974a. Simulating stable stochastic systems I: General multiserver queues. *Journal of the ACM* 21(1):103–113.

Crane, M. A., and D. L. Iglehart. 1974b. Simulating stable stochastic systems II: Markov chains. *Journal of the ACM* 21(1):114–123.
Fishman, G. S. 1964. Price behavior under alternative forms of price expectations. *Quarterly Journal of Economics* 78:281–298.
Fishman, G. S. 1967. Problems in the statistical analysis of simulation experiments: The comparison of means and the length of sample records. *Communications of the ACM* 10:94–99.
Fishman, G. S. 1968. The allocation of computer time in comparing simulation experiments. *Operations Research* 16:280–295.
Fishman, G. S. 1969. *Spectral Methods in Econometrics*. Cambridge, Massachusetts: Harvard University Press.
Fishman, G. S. 1971. Estimating sample size in computer simulation experiments. *Management Science* 18:21–38.
Fishman, G. S. 1973. *Concepts and Methods in Discrete Event Digital Simulation*. New York: John Wiley & Sons.
Fishman, G. S. 1974a. Estimation in multiserver queueing simulations. *Operations Research* 22:72–78.
Fishman, G. S. 1974b. Correlated simulation experiments. *Simulation* 23.177–180.
Fishman, G. S. 1978a. *Principles of Discrete Event Simulation*. New York: John Wiley & Sons.
Fishman, G. S. 1978b. Grouping observations in digital simulation. *Management Science* 24:510–521.
Fishman, G. S. 1983a. Accelerated accuracy in the simulation of Markov chains. *Operations Research* 31:466–487.
Fishman, G. S. 1983b. Accelerated convergence in the simulation of countably infinite state Markov chains. *Operations Research* 31:1074–1089.
Fishman, G. S. 1986a. A Monte Carlo sampling plan for estimating reliability parameters and related functions. *Networks* 17:169–186.
Fishman, G. S. 1986b. A Monte Carlo sampling plan for estimating network reliability. *Operations Research* 34:581–594.
Fishman, G. S. 1987a. Maximum flow and critical cutset as descriptors of multistate systems with randomly capacitated components. *Computers and Operations Research* 14:507–520.
Fishman, G. S. 1987b. The distribution of maximum flow with applications to multistate reliability systems. *Operations Research* 35:607–618.
Fishman, G. S. 1989a. Monte Carlo estimation of the maximal flow distribution with discrete stochastic arc capacity levels. *Naval Research Logistics Quarterly* 36:829–849.
Fishman, G. S. 1989b. Estimating the s-t reliability function using importance and stratified sampling. *Operations Research* 37:462–473.
Fishman, G. S. 1989c. Monte Carlo, control variates and stochastic ordering. *SIAM Journal on Scientific and Statistical Computing* 10:187–204.
Fishman, G. S. 1990. Multiplicative congruential random number generators with modulus 2^β: An exhaustive analysis for $\beta = 32$ and a partial analysis for $\beta = 48$. *Mathematics of Computation* 54(189):331–344.
Fishman, G. S. 1996. *Monte Carlo: Concepts, Algorithms, and Applications*. New York: Springer-Verlag.
Fishman, G. S. 1999. An analysis of Swendsen-Wang and related sampling methods. *Journal of the Royal Statistical Society*, Series B, 61:623–641.
Fishman, G. S. 2001. *Discrete-Event Simulation: Modeling, Programming, and Analysis*. New York: Springer-Verlag.
Fishman, G. S. 2005. *A First Course in Monte Carlo*. Belmont, California: Duxbury.
Fishman, G. S., and B. D. Huang. 1983. Antithetic variates revisited. *Communications of the ACM* 26:964–971.
Fishman, G. S., and P. J. Kiviat. 1967. Spectral analysis of time series generated by simulation models. *Management Science* 13:525–557.

Fishman, G. S., and V. G. Kulkarni. 1992. Improving Monte Carlo efficiency by increasing variance. *Management Science* 38:1432–1444.

Fishman, G. S., and L. R. Moore. 1979. Estimating the mean of a correlated binary sequence with an application to discrete event simulation. *Journal of the ACM* 26:82–94.

Fishman, G. S., and L. R. Moore. 1982. A statistical evaluation of multiplicative congruential random number generators with modulus $2^{31} - 1$. *Journal of the American Statistical Association* 77:129–136.

Fishman, G. S., and L. R. Moore. 1984. Sampling from a discrete distribution while preserving monotonicity. *The American Statistician* 38:219–223.

Fishman, G. S., and L. R. Moore. 1986. An exhaustive analysis of multiplicative congruential random number generators with modulus $2^{31} - 1$. *SIAM Journal on Scientific and Statistical Computing* 7:24–45.

Fishman, G. S., and D. S. Rubin. 1992a. Bounding the variance in Monte Carlo experiments. *Operations Research Letters* 11:243–248.

Fishman, G. S., and D. S. Rubin. 1992b. Evaluating best-case and worst-case coefficients of variation when bounds are available. *Probability in the Engineering and Informational Sciences* 6:309–322.

Fishman, G. S., and T. Shaw. 1989. Evaluating reliability of stochastic flow networks. *Probability in the Engineering and Informational Sciences* 3:493–509.

Fishman, G. S., and L. S. Yarberry. 1997. An implementation of the batch means method. *INFORMS Journal on Computing* 9:296–310.

Fishman, G. S., B. Granovsky, and D. S. Rubin. 1992. Evaluating best-case and worst-case variances when bounds are available. *SIAM Journal on Scientific and Statistical Computing* 13:1347–1361.

Frank, H., and I. T. Frisch. Analysis and design of survivable networks. *IEEE Transactions on Communications* COM-18:501–519.

Gilks, W. R., S. Richardson, and D. J. Spiegelhalter. 1996. *Markov Chain Monte Carlo in Practice*. London: Chapman & Hall.

Hammersley, J. M., and D. C. Handscomb. 1964. *Monte Carlo Methods*. London: Chapman and Hall.

Heidelberger, P., and P. D. Welch. 1983. Simulation run length control in the presence of an initial transient. *Operations Research* 31:1109–1144.

L'Ecuyer, P. 1988. Efficient and portable combined random number generators. *Communications of the ACM* 31:742–749 and 774.

Lada, E. K., and J. R. Wilson. 2006. A wavelet-based spectral procedure for steady-state simulation analysis. *European Journal of Operational Research* 174:1769–1801.

Lada, E. K., N. M. Steiger, and J. R. Wilson. 2008. SBatch: A spaced batch means procedure for steady-state simulation analysis. *Journal of Simulation* 2(3):170–185.

Schruben, L. W., and B. H. Margolin. 1978. Pseudorandom number assignment in statistically designed simulation and distribution sampling experiments. *Journal of the American Statistical Association* 73:504–525.

Seila, A. F. 1976. Quantile estimation methods in discrete event simulations of regenerative systems. Ph.D. diss., Curriculum in Operations Research and Systems Analysis, University of North Carolina–Chapel Hill.

Steiger, N. M., E. K. Lada, J. R. Wilson, J. A. Joines, C. Alexopoulos, and D. Goldsman. 2005. ASAP3: A batch means procedure for steady-state simulation analysis. *ACM Transactions on Modeling and Computer Simulation* 15(1):39–73.

Tafazzoli, A., J. R. Wilson, E. K. Lada, and N. M. Steiger. 2008. Skart: A skewness- and autoregression-adjusted batch-means procedure for simulation analysis. In *Proceedings of the 2008 Winter Simulation Conference*, eds. S. J. Mason, R. R. Hill, L. Mönch, O. Rose, T. Jefferson and J. W. Fowler, pp. 388–395. Piscataway, New Jersey: IEEE.

Van Slyke, R. M., and H. Frank. 1972. Network reliability analysis: Part I. *Networks* 1:279–290.

Walker, A. J. 1977. An efficient method for generating discrete random variables with general distributions. *ACM Transactions on Mathematical Software* 3(3):253–257.

Welch, P. D. 1987. On the relationship between batch means, overlapping batch means and spectral estimation. In *Proceedings of the 1987 Winter Simulation Conference*, eds. A. Thesen, H. Grant and W. D. Kelton, pp. 320–323. Piscataway, New Jersey: IEEE.

A Conversation with George Fishman

Christos Alexopoulos, David Goldsman, and James R. Wilson

Abstract The following are excerpts from an extended conversation between the authors and George Fishman that was recorded over the period October 25–26, 2007.

1 Education and Early Career

This section concentrates on George's education and his experiences in industry prior to his academic career. The speakers are indicated by the initials of their first names.

C: Good morning, George. We are very pleased to see you. This is part one of your interview, and we'd like to start with your educational background.

G: In the early 1950s, there was considerable interest in science and technology, and high school students were not immune from that enthusiasm. Those who did well in technical subjects were often encouraged to go into sciences or engineering. Because I was living in Chelsea, Massachusetts, MIT was the closest university that had a good reputation in both disciplines and I decided that I would like to go there. Of course, it became clear that if I was going to have an interview, I had to know what field I wanted to enter. For reasons that still elude me today, I chose aeronautical engineering. The admissions interview committee included an admissions officer and a single faculty member. We had a long discussion about my interests, my grades, my extracurricular activities, etc., and then they asked me what I saw myself going into. I said "aeronautical engineering." Immediately the faculty member's eyes lit up. It turned out that he was in that department. At that moment I was sunk because I knew nothing about aeronautical engineering except that it had to do with airplanes and avionics. Although I thought I had flunked the

C. Alexopoulos (✉)
H. Milton Stewart School of Industrial and Systems Engineering, Georgia Institute of Technology, Atlanta, GA 30332, USA
e-mail: christos@isye.gatech.edu

C. Alexopoulos et al. (eds.), *Advancing the Frontiers of Simulation: A Festschrift in Honor of George Samuel Fishman*, International Series in Operations Research & Management Science 133, DOI 10.1007/b110059_2, © Springer Science+Business Media, LLC 2009

interview, I was admitted shortly thereafter. No more than four or five months after matriculating, I decided that engineering was not my true interest.

Economics was something that I always liked and at that point in time MIT had the leading Department of Economics in the United States. Paul Samuelson was the reigning authority with a widely adopted introductory textbook in economics. I enjoyed the economics courses that I took but, in order to graduate from MIT at that time, one was required to take a minimum of about 60% of one's courses in either science or engineering. Therefore, I also took a wide spectrum of science and engineering courses. These included quantum physics with a well-known physicist, John Slater. Although I've had zero use for this course during the last fifty years, it was more enjoyable than one might think to learn about Schrödinger's equation everyday for six months. [Laughter.]

My principal engineering courses were in electrical engineering, and that's where I learned about networks. Instructors in both economics and electrical engineering encouraged us to take probability theory and statistics. I received my degree from MIT in Course XIV, and my diploma reads economics, political science, and electrical engineering. [Laughter.] Technically, it's all three of those, but I focused on economics.

D: They are really all in the same physical location?

G: No, most lower-level economics courses were taught on the main MIT campus. More advanced ones were offered at the Sloan School, some distance away. I would leave a main-campus class immediately after the bell rang and in ten minutes would have to get to the class at the Sloan School. I usually sat in the back of the class because the teachers didn't like my heavy breathing from rushing from the main campus. Because I finished my in-class course work in December of 1959, I had six months before I could begin graduate school. Therefore, I began working for several MIT economists. These included Robert Solow, who eventually won the Nobel Prize in economics. Another, Maurice Adelman, was an authority on the automobile and oil industries.

When I asked for additional work hours, they referred me to the Political Science Department, where I went to work for Professor Ithiel de Sola Pool. Although this sounded to me like a very nonquantitative area, it turned out that Professor Pool was a protagonist for quantitative methods in political science. His main interest was in content analysis, which had to do with the propagation of information as it passed from one person to another—how fast it spreads in different kinds of societies, things of that sort.

I'm trying to get to the point of what all this had to do with simulation. It was the presidential election year, 1960, and Professor Pool was intimately involved with the Kennedy campaign. He and his collaborators, including Robert Abelson who was a psychologist at Yale, were doing studies of peoples' responses to a variety of political and economic issues. It was the first election in which that was done, and much of the work involved random sampling on a computer by very primitive methods.

When I began this work, I found that the computer programs in use were written in SAP, then the assembly language used for IBM computers. It consisted of three-letter codes, and the program, if stretched out linearly, would probably have covered the entire floor in the computing center. Moreover, it was not executing correctly and no one had been able to debug the program. At my skill level, I knew that I would not be able to as well. Fortunately, during my last semester, I had taken a course in computer programming languages. Although it focused on SAP, the instructor had mentioned that a new language called FORTRAN was considerably easier to use. I began studying the few available FORTRAN manuals and eventually suggested we rewrite the SAP programs in FORTRAN. A long debate ensued as to whether we should do this and whether or not FORTRAN was a good choice. Once we made the switch, the new program executed with a minimum of difficulty, so my opinions became more credible to others on the project.

When I announced to my employers that I had been accepted at Stanford in economics, Professors Pool and Solow told me that they were good friends of Kenneth Arrow, a leading mathematical economist at Stanford. He later also became a Nobel Prize winner. I had not thought to ask them beforehand to write recommendations for me. Once they learned of my decision to go to Stanford, they wrote letters of recommendation to Professor Arrow. When I got to there, he offered me an assistantship. Because he was (and still is) a theoretician, the amount of computation he did was minimal. He assigned me to Professor Marc Nerlove, a twenty-six-year-old econometrician who was already a full professor. He is now at Maryland. I also worked for Irma Adelman, a professor of economic development at Berkeley who was visiting Stanford that year. She had published one of the first studies on economic development using simulation.

Although I was not part of Professor Adelman's simulation work, I did learn from working with her how to go about testing models. Professor Nerlove had developed an interest in applying frequency-domain analysis to economic time series. Previously, these series were analyzed almost exclusively in the time domain. Working with him, I began to cultivate an interest in spectrum analysis.

D: Had you had any spectral courses back at MIT?

G: You can never tell which college courses are going to be helpful. Having taken electrical engineering courses, I was familiar with frequency-domain analysis. Therefore, I found it relatively easy to integrate myself into Professor Nerlove's research. We had a good working relationship in which I felt that I had something more than merely programming skill to offer. Conversely, I learned a lot from him on how to conduct a quantitative analysis, never overlooking contradictions and always giving explanations that would hold up under scrutiny. It was a good relationship and I have always been grateful to him for his guidance.

At the end of my second year at Stanford, I was offered a summer internship at the RAND Corporation, based on recommendations from Professors Arrow and Nerlove. Both had been RAND consultants. That August (1962), I was offered a permanent position at RAND which I accepted. My completed credit hours at Stanford earned me a master's degree in economics that Fall. At RAND, I joined the Logistics Department. To a great extent, logistics involves microeconomics and that fit with my education to date.

Logistics is a major component of all military organizations. I didn't know that before I went to RAND. As Jim has been in the service, he can appreciate this fact. The Air Force, RAND's principal client in 1962, had challenging problems in reliability, maintainability, inventory management, and facilities location. These intrigued me.

D: Can I make you do an aside for a second? I should know this, but how exactly is RAND related to government organizations?

G: RAND came into existence as a consequence of the actions of farsighted people in the scientific community and the U.S. Air Force. During World War II, groups of scientists had been set up in Washington, D.C., to work on defense related problems. The leader of the entire scientific effort was Vannevar Bush, whom you may never have heard of.

C: No, he's well known. I have read about him in history books.

G: Bush had become head of the Carnegie Institution in Washington, D.C. President Roosevelt made him his science advisor. He was the liaison between the scientific community and Roosevelt, and basically mobilized a considerable amount of the scientific war effort. Bush recruited people like Phil Morse from MIT to work on a whole host of problems. The Air Force recognized the value of this work; in particular, General "Hap" Arnold. He was one of a handful of five-star generals during World War II.

C: I think I have read about Vannevar Bush in relation to George Dantzig.

G: That's possible. Dantzig was at a considerably more junior level. He worked for Marshall Wood, who led a group doing analysis for the Army Air Force. Wood was one of the people who were instrumental in the armed services becoming interested in operations research. He recognized the value of OR techniques and assigned Dantzig a variety of problems related to aircraft scheduling, airlift scheduling, transportation, etc. That's where many of the problems on which Dantzig focused originated. But getting back to your original question, two important events happened at the end of the war. Bush recommended that an organization be set up with sufficient funds to sponsor a wide range of basic research. That organization came to be known as the National Science Foundation.

D: That's where NSF came from?

G: Right, it came from the recommendation that Bush made to President Truman. Bush went on to become the Chairman of the Board of MIT. I have to tell you an anecdote. Bush went to the same high school as I, where we had a high school play focused on famous alumni. Guess who I got to play? Vannevar Bush! [Laughter.]

To return to the discussion, General Arnold recognized that the Air Force would benefit from technical assistance in many areas. He arranged for the Air Force to give a contract to Douglas Aircraft to set up a research group. It was called the Research and Development Group, which we now call RAND. Starting about 1947 or 1948, the group occupied a building that Douglas Aircraft owned in Santa Monica. Several years later, RAND moved into newly constructed buildings close to its current location on Main Street in Santa Monica. At approximately that time, H. Rowan Gaither, a name that you won't recognize, served as Chairman of the Ford Foundation and as Chairman of the Board of RAND. I may be off in the date but he certainly was a presence in RAND's early development. To answer your original question, RAND was a nonprofit corporation with buildings put up with Air Force and the Ford Foundation financial assistance.

It soon became apparent that RAND's aims and objectives, especially with regard to long-term research, differed from those of Douglas Aircraft, a profit-making company, that was considerably more task-oriented. Here's what we have to do, and here's the data. RAND was not designed that way. To his credit, General Arnold recognized that you had to put people into an environment in which they could think more comprehensively. That was the RAND atmosphere that I found when I got there in 1962. For me, that was a very good experience. As I said, I was in the Logistics Department and worked on a variety of problems. In retrospect, none of my contributions were significant for solving Air Force problems. I was learning how to be a researcher and paid considerable attention to how more senior members of the research staff went about their work.

Towards the end of 1963, Murray Geisler became the head of the Logistics Department. A statistician, he was active in management science and had written several papers on simulation. RAND used simulation to study a variety of Air Force logistics problems. I never became an integral part of that effort. As chairman, Geisler had many responsibilities one of which was to assign referees to Logistics Department documents that were going to appear as external RAND publications. Each was internally reviewed by two people. I was assigned a paper having to do with methodology for running simulations, which focused on the batch means method.

I wrote a brash report that focused on all the issues that the authors had overlooked. Very brash. [Laughter.] I wouldn't write a report like that today. Geisler told me that he liked the report, which was very reassuring. He also encouraged me, saying that simulation was an emerging area and suggested that I should devote time and energy to it. Then he said "To get you started, here is an internal RAND document by Ken Arrow that's never been published."

That document gave me a considerable understanding of the area. Conceptually, Arrow understood exactly what the issues were. Over the years, I've come to realize how farsighted he was in terms of what the methodological challenges for simulation were. That's how my serious research interest in that area began. At the same time, Phil Kiviat came to work at RAND. After getting his master's degree in operations research from Cornell, he had spent two years at U.S. Steel where he developed the GASP discrete-event simulator, which was FORTRAN-based.

At the time, Cornell was, without question, the university where the basic concepts of discrete-event simulation were put into the classroom on a formal basis. Dick Conway and Bill Maxwell were responsible for that. They saw simulation as a legitimate area of inquiry, an opinion not widely shared then by others in the academic community.

D: This would be around 1963?

G: Yes, that's my recollection. When I would sit down with Phil, I didn't get the idea of a "piecemeal" field that was being put together by people with limited technical skills. It was a formal field, in which language was the major formalism.

Although his training was as a mechanical engineer, he had gravitated to the area of simulation languages. He encouraged me repeatedly, telling me there was a great need for people in the area of statistical methodology. So that was the area in which I chose to work. Moreover, the paper that I had refereed motivated me to try another method for estimating the variance of the sample mean. I took what I knew about spectrum analysis and began doing simulation experiments. Phil supplied many ideas and we worked well together. That collaboration led to our 1967 *Management Science* paper. Although that paper shifted attention from *ad hoc* evaluation to more of a methodological approach, it required considerable computing time relative to the time spent simulating.

When I first came to RAND, the computation of the spectrum at any point was a quadratic sum calculation done by computing the autocovariances and then taking their Fourier transform. That was an order n^2 operation. Since it took much longer to analyze the data than to generate them, the method had limited appeal. But in 1965, the Cooley-Tukey algorithm for the fast Fourier transform came along. It transformed not only simulation, but also many other areas. You could then perform those computations very rapidly. As a result, the technique developed a broader appeal.

Actually, Jim, I'm going to have to tell you a story about your colleague, Salah Elmaghraby. The first public presentation of the spectrum analysis paper that I gave was in Vienna, at a meeting of the International Institute of Management Sciences. The talk was limited to about fifteen minutes, restricting what I could say about the method. Salah was in the audience and, after I finished, he said to me: "That's a wonderful paper. What is it all about? You have to explain these details." [Laughter.]

D: He was around when Maxwell, Conway, and all these guys were starting to think about these concepts at Cornell.

G: Oh yes. He was one of the first Ph.D. students in the Cornell OR Program. Previously, Salah had worked for Western Electric and then went back to graduate school at Cornell.

After publication of our paper, I concluded that, computing cost aside, the setup cost to learn about spectrum analysis was too large to make the method widely attractive for estimating the variance of the sample mean in simulation experiments. I knew that an autoregressive scheme had an easily computed rational spectrum and thought this alternative approach might offer a convenient approximation for the spectrum of a queueing process. Although these processes generally do not have linear autoregressive representations, their spectra can be approximated to some degree of accuracy by rational spectra. Also, this approach got the analysis out of the unfamiliar frequency domain.

I wrote my paper on the autoregressive method in 1968 and through Phil Kiviat was invited to present it at the Second Winter Simulation Conference in New York that December. Because I was unaware that this conference was a succession of single-session events, that is, no parallel sessions, I was surprised to find an audience of three to four hundred people waiting for me to speak. This was intimidating to someone more accustomed to fewer than ten or fifteen attendees at a presentation in one of RAND's conference rooms. My talk was well received. During the presentation, I had felt that I was not making sense to the audience. But the response afterward indicated otherwise. Most of all, they were happy to see someone talking about statistical methodology. That's where and when I finally accepted Murray Geisler's assertion that statistical methodology was an important component of simulation.

In the late 1960s, Phil Kiviat and I decided to write a series of papers on the methodological aspects of discrete-event simulation. We called the series "Discrete-Event Simulation." Each would be on a different topic. I wrote one on statistics, he wrote one on languages and another on modeling alternatives, which I regard as one of the best papers in the simulation literature. That paper formalized the difference between the event-scheduling approach, the process-interaction approach, and the activity-scanning approach. I adopted many of those concepts in my 1973 book on simulation.

Most of Phil's papers never became academic publications. Some of mine did. As to the balance between languages and statistics, there's no question that when I came into the simulation field, the emphasis was on language. Moreover, the distinction between model and language was not clear. People had unusually creative ideas about languages and modeling, but the focus was on the issues of how to make event lists function efficiently.

SIMSCRIPT was an attempt to make simulation modeling more accessible by being more conversation-like than FORTRAN. By making it more forgiving during compilation, it made it easier to code, in principle. It might have prevailed as the dominant simulation language at that time, except for the fact that IBM had come out with GPSS. GPSS had two advantages. It was an IBM software product and it offered a more attractive environment for modeling. The user sat at a remote terminal—that's what we called them at the time, basically a teletype machine—and merely interacted with the code. Because of its structure, successive interactions occurred without a full-blown recompilation. No recompilation meant faster interactive responses. At that time, compiler-based simulation languages took a long time to compile, an unappealing property for people who had thousands and thousands of statements in their simulation code. GPSS internalized much of this modeling effort by using "off-the-shelf blocks" and interpretively executing the simulation program made up of these blocks.

Although eliminating compilation gave GPSS an edge, it had other limitations. I was present at many discussions about its slow execution and its lack of flexibility. Most notable were its inefficiencies in processing the current and future events chains that contained transactions that were waiting to execute.

Harry Markowitz and Phil Kiviat were the principal developers of SIMSCRIPT II at RAND. Harry had left RAND shortly after I arrived and gone into business with Herb Karr. However, he remained a consultant. He and Phil would get together several times a month. My first exposure to Harry actually was at Stanford where I took a course that Arrow offered on portfolio analysis in 1961.

D: That's what Markowitz was famous for though, right?

G: Right, the principal topic was Markowitz's book on portfolio analysis. Arrow cast the topic in a more formal setting focused on utility functions and nonlinear optimization. But Markowitz's book was definitely the essential feature. When I got to RAND, I was puzzled by what Harry was doing working on simulation. It turned out that he had very broad interests, including discrete-event simulation. His Nobel Prize award testifies to his accomplishments.

By the end of the 1960s, I realized that RAND was a plateau type of environment. Many staff were at my level, but few were senior people. I equivocated in my own mind as to whether I would enjoy being a senior person. I wasn't sure I was suited for it. The only option for me was to find something else to do. Several times, I had been encouraged by RAND colleagues to teach. I actually did teach introductory statistics at UCLA in 1965. In 1967, I finally acknowledged to myself that the only way I could teach at a university was to get a Ph.D. So here is the answer to your question about why I studied biostatistics. There were two major universities in the area that offered Ph.D. programs. UCLA which was twenty minutes away, and the University of Southern California which was thirty-five minutes away. Several RAND staff in the Logistics and Mathematics Departments had recently earned their Ph.D.'s through the Biostatistics Department at UCLA. You may know Stan Azen, who has served as the editor of the *Journal of Graphics in Statistics*, and Craig Sherbrooke, one of the major contributors to multiechelon inventory management. Both had gone through the biostatistics program. I saw that they were able to balance the demands of full-time work at RAND and the graduate program. So I enrolled in 1967 and completed my degree requirements in March 1970.

By then the RAND environment had changed dramatically. It was no longer the research organization that I had joined in 1962, partially because of the change in funding arrangements. RAND had done a good job of educating the Air Force in using analytical techniques and so it now was capable of doing analysis for itself. Therefore, its level of dependence on RAND had become less. RAND sought other sources of funding. It solicited support from nondefense government agencies whose interests were more task oriented. That implied shorter research time horizons and less time to indulge one's interest in more conceptual research.

I decided to look for an academic appointment in the Spring of 1970. I was invited to visit Northwestern University for an interview in the Department of Industrial Engineering. In late June, the department offered me a position. The chairman sent me a handwritten offer for a tenure-track position. He apologized for the informality, but student campus protests had effectively shut down the university. Shortly before that, I had received a call from Harvey Wagner at Yale University, whom I had known at Stanford. He was familiar with my published research, and we knew each other casually. He said "Would you like to be considered for a position here?" So early in July of that year, I visited New Haven, Connecticut, and gave a presentation to the faculty of the Department of Administrative Sciences. The department comprised the disciplines of organizational behavior and operations research. Shortly thereafter, Bob Fetter, the department chairman, called me from a public pay phone on a highway in New Mexico to offer me a position as an associate professor. I had made it clear to both Northwestern and Yale that I did not want to begin as an assistant professor. One of the benefits of RAND was that I was able to write a 1969 book on spectral methods in econometrics, which was basically an outgrowth of the work that I had done under Marc Nerlove. So between that and my journal publications on simulation, both places were willing to offer me an associate professorship.

The informality of both offers made me wonder as to how these universities worked. That was compounded by my next experience. I can still recall that when I got to Yale, I asked Bob Fetter "Okay, what am I to teach?" He looked at me with a smile and said "What would you like to teach?"

2 Academic Career

In this section, George discusses his first academic appointment at Yale University and his move to the University of North Carolina.

C: We are now moving to the second part of the interview which concerns your arrival at Yale and eventually the move to UNC. So you have the honors.

G: I wasn't ready for Bob Fetter's offer to teach whatever I wanted, nor was I ready when I first got to New Haven for the fact that I was going to have to prepare teaching materials. I assumed that Bob would pick the subject and that there would be a standard textbook for it. He wasn't suggesting anything—all courses were covered—and he said "Teach whatever you wish." I responded with "What if I offer a course in spectral methods in economics?" He said "That's fine, but remember we hired you principally for simulation." I said that I would do that in the spring semester, and that was agreeable. That Fall, I lectured three times a week to three or four students on spectral methods.

J: So this was the Fall of 1970?

G: In the spring semester, I taught the simulation course. During the fall semester, I had looked at the potential textbooks, but none struck me as acceptable in the methodological sense. The closest one was Tocher's book. But that was so tightly written, that it would be very hard for students to get the meaning of what was really a very rich book. So that Fall I began writing notes. When I look back, it's hard to believe that during the four fall months I prepared 400 pages of typed notes for the spring semester. Needless to say, during my first semester of teaching simulation, I was merely ahead of the students by a few hours, so to speak. They were kind, helping me with typos, etc. Those notes formed the basis for my 1973 book. It focused on modeling, programming languages, and statistics, ideas that I had adopted from my association with Phil Kiviat. Because of the separation of modeling and languages, language did not dictate modeling, but modeling dictated what features a simulation language needed.

I should say that one of my responsibilities at Yale—the way that I was hired—was as the associate director of a health services research project, a joint effort of the School of Public Health and the Administrative Sciences Department. So I had some administrative duties as well, and taught a seminar in health services research using quantitative methods.

D: Were you involved with health systems back then?

G: Yes. I taught a seminar in it. Lee Schruben was one of my students. In fact, if I remember correctly, when we recruited Lee as a student, we offered him a fellowship in the Health Services Program. Wayne Winston of Indiana University was also a member of that program.

Several things occurred during those years. The OR simulation research community became much more active. The Winter Simulation Conference was instrumental in that. It created a focus. In particular, it established an accommodation between industry and universities, which in many ways continues to this day, perhaps in different proportions. It was well understood that both domains would be respected and be part of the annual conference. The first several conferences were unusually successful. Its sponsorship progressively got broader. For a number of years, the then National Bureau of Standards was a cosponsor. The name of the fellow there slips my mind. He was very instrumental in helping with it.

J: Paul Roth?

D: He's listed on the program even to this day.

G: Yes, he arranged for the facilities of the Bureau to be made available for the conference. You've all been involved with the conference so you know it needs some kind of continuity from year to year. Well, that was one of the difficulties when it first began, first through Julian Reitman's efforts and then with the help of Arnie Ockene of IBM. In the early 1970s, the conference was attracting more academics from diverse fields such as civil engineering. Joe Sussman at MIT, whose research was in transportation, did a lot of simulation in civil engineering.

There were also more people getting interested in the statistics of simulation. I neglected to mention that during my RAND years Alan Pritsker was also instrumental in making me see the value in simulation statistical methodology as a field of research. Alan was a consultant to RAND. I often worked in my RAND office in the evenings and Alan was always there. We had many conversations and I could see that his ideas were similar to Phil's. It was not difficult to see that Alan, like Phil, had a well-thought-out view of modeling, language, and statistics as applied to discrete-event simulation. While their views may have differed in emphasis, their conceptualizations made it considerably easier for me to see where my interests could fit.

There were also people working on variate generation and random number generation. In the early 1970s, Ahrens and Dieter published several papers that described variate generating algorithms with bounded computing cost, independent of distributional parameter values. Those were intellectual ideas which I don't think are fully appreciated today, because they're lost in the mix over time.

My first sponsored research proposal was to do simulation analysis graphically on a monitor. [Laughter.] It was not funded. Although well regarded by reviewers, NSF rejected it. Afterwards, one of the reviewers told me that he recommended rejection because my proposal wasn't feasible. [Laughter.] At that time, he was probably right. Tektronix offered the most advanced graphical capability. One could take a body of data and put it on a screen, but if you tried to add or subtract something from those data, you had to rewrite the entire screen. No addressing of individual pixels or anything like that. No animation. It was in primitive form. I had not realized how slow the entire interactive process was until Tektronix gave a demonstration at Yale. By the end of the decade, graphics had become a standard part of discrete-event simulation. Graphics devices were being improved and new language constructs were making it easier. The introduction of PCs in the late 1970s made the interaction more productive per hour of effort. So I might have gone in an entirely different direction, had those developments occurred a decade earlier.

My interaction with Matt Sobel came about in an interesting way. During my first Yale year, he had been away at CORE in Brussels. In the Fall of 1971, Bob Fetter mentioned to the faculty that the department had received a request from the New Haven Housing Authority for help in managing their inventories and maintenance. Matt and I were the only faculty who expressed interest in the project. He and I spent three years consulting for the Authority. The first day we arrived on site, the director said to us "Where did you park your car?" We said that we parked in the street, and he said "Go out and get it and bring it inside the fence, inside the barbed wire." [Laughter.] The Authority was in a tough section of town. It had wonderful problems for anyone interested in OR. To put it concisely, they had the funds to buy supplies, but needed guidance on how to set reorder levels.

For example, they had a ten-year supply of Moen cartridges. [Laughter.] Moen cartridges were not in common use in those days. However, the Authority had learned that when they used regular washer faucets, leaks were not reported punctually and, as a result, it was paying for a lot of needlessly leaking water. By using Moen cartridge faucets, they could eliminate the leakage. However, that policy led to a substantial inventory of Moen faucets.

C: Did that involve simulation at all?

G: No. This experience broadened my understanding of what OR could do for people. At about the same time, I was asked to consult for the RAND Institute and the Ford Foundation in New York City. The RAND work never panned out, because we were unable to identify a specific problem calling for my expertise. Al Madansky, a statistician whom I had known at RAND, was responsible for my involvement with the Ford Foundation. Al had become the chair of the Computer Science Department at CUNY. He was also consulting for the Foundation. He felt that my interest in simulation would be helpful on one of its studies of the performing arts. Every several weeks, a group of consultants would meet at about 4 p.m. at the Foundation's headquarters in a beautiful building close to Second Avenue in New York City. The 4 p.m. time allowed those who were academics to travel, after class, into the city from their respective universities. I recall that the consultants included faculty from Yale, Princeton, CUNY, and possibly Columbia. I never saw how simulation could be a major contributor. The principal focus was on the analysis of data. Eventually, I helped edit one of the studies. Seeing how the Foundation operated was an eye-opener. Although it did not lead to much fundamental research, it did provide good conversation and many good dinners at the Foundation's expense. Dick Shelton, the Ford study leader, arranged these memorable occasions after our working meetings. On one occasion, I recall that the arrival of the bill prompted the invention of two additional attendees to justify its size. [Laughter.]

In conversations with Matt Sobel, whose interest was in queueing control, I began to think about how queueing properties affected statistical behavior in discrete-event simulation. I realized that the time-dependence within queueing sample paths was unconditional, but that there were conditions under which successive segments of the time path were independent. For example, entry into the empty-and-idle state. I also raised this issue with Madansky who pointed out to me that the empty-and-idle state is a special case of the more-general concept of renewal processes. These interactions motivated me to read about renewal processes and led to my first paper on the topic of independent sample-path segments.

I had also received a grant from the Office of Naval Research (ONR). One of the grant's stipulations was that all those who were supported by ONR in a common research area were to interact with each other. Interaction meant that you sent each other technical reports. One of these grants was at Stanford, and the only person on that grant whose name I recognized was Gerry Lieberman. I sent him a copy of my paper on what's come to be known as the regenerative method, and he promptly sent me back a paper by Don Iglehart and Michael Crane on the same topic. The topical match between them was a big surprise—two groups had come up with the same idea at the same time. Although each paper had a different twist, there was no question about the commonality of the idea. Both papers were published roughly at the same time. In the Fall of 1973, Don Iglehart and I were invited to present our papers at a TIMS meeting in San Diego at an OR-sponsored session.

By then—well, much earlier than that—I had decided that I wanted to leave Yale. I had come into a department that was truly a "warehouse" for two disparate disciplines. Early in the days of operations research, there had been this concept of having it interact with the psychology community, particularly organizational behavior. The concept of man-machine simulation was big—it was a major topic in the Logistics Department when I arrived at RAND. The Administrative Sciences Department at Yale had been established as a home for organizational behavior and OR. But the two disciplines had fundamentally different views of what constituted research. This led to a tense atmosphere that I did not enjoy. Although I suspected that it could eventually be at the expense of junior faculty members, in retrospect, I don't know of any junior faculty whose progress at Yale actually suffered because of the conflict.

In 1973, the Administrative Sciences Department was incorporated into Yale's newly established School of Organization and Management. That arrangement led to other conflicts. I decided to look for a new position. After the San Diego meeting, I gave talks at several universities. Maryland's business school and UNC's newly established Curriculum in Operations Research and Systems Analysis expressed interest.

C: But the Curriculum was still housed within the Statistics Department.

G: No. The previous year, Jim Gaskin, the Dean of the College of Arts and Sciences, had established the Curriculum as a separate academic entity.

C: And they were located in the Phillips Annex?

G: Yes. Originally, Jerry Gould was the chairman, but by then he had departed for the University of Chicago, and Jack Evans was the chairman. My interview was at the height of the 1973–74 oil crisis and gasoline was hard to come by in Chapel Hill. So I agreed to take a bus from the Raleigh-Durham Airport to the Holiday Inn in Chapel Hill. That gave me the equivalent of a Cook's tour of the Triangle,

Durham, and Chapel Hill. Jack Evans met me at the motel and said "Before dinner, we'll take you for a drive around so you can see the town." And I said "Oh, will we have time?" I had no concept how small the town was!" [Laughter.]

C: What year was that?

G: 1974. Shortly thereafter, UNC offered me a position as a tenured full professor. Although Maryland was still mulling over what they wanted to do, I had already decided that Chapel Hill was a better place for my family and me. We moved here in July, 1974—I, my children, Becky and Matt, and my wife, Sue. I quickly learned that circumstances were not as I had originally pictured them. The Curriculum was a separate freestanding unit, and it did have two tenure-track positions. But it was still not a department, and therefore, whenever we would go to college-level meetings, there were people who would say "It's a curriculum. What are you doing here?" or "Operations Research? I thought that was part of Statistics!" The OR program did not have much status on campus. Moreover, there were people who saw no reason to continue the program. Although these attitudes made me uncomfortable for a number of years, events in the late 1970s gave reason for cautious optimism.

Even though Dean Gaskin kept his commitments to the Curriculum, he wasn't providing the additional resources that the program needed to grow. A newly appointed dean, Sam Williamson, a military historian, was considerably more of an activist. Through his affiliation with the military, he knew what OR was. He recognized the peculiar situation of the Curriculum and raised the question "What should the future of this program be? Should it be eliminated or continued?" At roughly the same time, Phil Manire was appointed as the dean of the Graduate School.

C: What year was that?

G: Probably 1978 or 1979. Dean Manire was a microbiologist who had been a guiding force behind the development of the Microbiology Department at UNC. I attribute the survival of the Curriculum to him more than to anyone else. He appointed a committee to advise him on the future of the Curriculum. It consisted of faculty from the Mathematics, Computer Science, and Statistics Departments and the Business School. Jack Evans, who had left the OR program to become the assistant to the chancellor, encouraged Manire to keep the program. Also, the chairman of the Mathematics Department, Bill Smith, felt that this was a program worth supporting. John Tolle now had a joint appointment in Mathematics and in the Curriculum and Bill was familiar with John's interests and work.

Manire recommended to Williamson that the Curriculum be continued. Williamson did that and more, an action that led me to understand what a good administrator does. Not only did he allow the program to continue, he provided additional resources for it to reach its potential. In 1980, I was asked to be chairman. Before accepting I met with Williamson and asked for additional resources, and these were also granted. The Curriculum had already gotten new space (the Smith Building), new equipment, and additional positions. That was a very successful period and the next dean of Arts and Sciences, Gillian Cell, continued that support. Upon meeting her for the first time, I learned that she was a historian whose area of specialization was Labrador. I kept saying to myself, how was a person who specializes in Labrador going to know what operations research is? [Laughter.] Well, it turned out that she was also a conscientious historian. Prior to our meeting, she had read the College's file on our history.

I should also mention one other thing about Phil Manire which forever endears him to me, besides his fair-mindedness. In discussion with him, I expressed my frustration at the slow progress in recognition that the Curriculum was making on campus. I asked him how that was done. He said "By advertising. You have to keep on going around and introducing OR, showing what it can do for people on campus and stressing its academic accomplishments. No money ever flowed to a department on this or on any other campus where the chairman did not push the department." Manire exemplified that policy. When he first came here, Microbiology was also a curriculum. He turned it into a substantially first-rate department. Most UNC AIDS research is done there.

I took him seriously, and tried to take that stance whenever I was in a meeting with other groups; not to look like the junior partner but to speak with confidence. Jim, you probably found this in your own experience that when you talk like a chairman, others treat you like a chairman. If you act like a supplicant, that's how you're going to be treated. If you talk like you deserve to be there, you find faster acceptance. [Laughter.]

D: Could I ask to step back for a second, because I might have missed something here. When you got there, who was the Curriculum answering to? They were not part of another department were they? So it was treated as kind of a minidepartment then, right?

G: Right.

C: But some statisticians were also part of the OR Curriculum.

G: Well, on this campus the concept of a curriculum was as an interdisciplinary group. Other departments were encouraged to contribute faculty time. In the early 1970s, the Business School contributed courses taught by Jack Evans, Roger Blau, and Dave Rubin. At that time, its dean, Morris Lee, believed that OR was a worthwhile discipline.

Computer Science allowed its faculty to participate, but not to teach our courses. Fred Brooks, its chairman, was supportive. Don Stanat, whose area was languages, and Vic Wallace, whose interest was decomposable Markov chains for network analysis, were part of the Curriculum and had research interests that overlapped with OR. Several OR faculty including me sat on dissertation committees in that department.

We also had Jon Tolle joint with the Mathematics Department, Walter Smith from Statistics, and Dick Shachtman from Biostatistics. As I've said, the Curriculum received its first two tenure-track positions in 1973. When several other departments found out about this, they were puzzled because this wasn't the conventional UNC definition of a curriculum. In conversation, some told me that we should not have received tenure-track positions. Never mind that they were talking to one of the tenure-track faculty. [Laughter.] That issue eventually became less of a topic of conversation.

To get back to the simulation side, issues that I and others had raised began to attract more attention. Better random number generators were materializing as were better variate generation methods. More academics were expressing interest in simulation statistical analysis. Jim was one. Lee Schruben was another. Lee's appointment to Cornell in about 1976 gave me considerable satisfaction. It meant that Cornell took seriously the area of simulation statistical methodology. When Don Iglehart, with a substantial reputation in applied probability, expressed interest, that was additional verification that Murray Geisler, Phil Kiviat, and Alan Pritsker had wisely advised me.

What I'm saying is that the analysis of discrete-event simulation has matured. However, I did have a concern. Although the statistical problems were well understood, I felt the area of output analysis was becoming one of diminishing returns. I didn't see that I had much more to contribute. In conversations with my UNC colleague, Scott Provan, I revived my interest in networks. A decade earlier, Alan Pritsker had aroused my interest in the network formulation to discrete-event simulation. If Alan could have had his way, discrete-event simulation would be taught principally through a network formulation.

J: Oh yes, I think that's true.

G: What impressed me about networks was that if you started out with this intrinsic structure, you could exploit it for variance-reducing purposes. I worked with several creative students and we produced a series of papers on variance reduction in networks, in particular, on antithetic variates. Christos worked on one of the more interesting ones, the max flow–min cut distribution problem, which to this day is still an extraordinarily tough one to analyze analytically.

C: I actually still find it the toughest one I ever worked on.

G: It is, it is. Many people had useful insights that didn't carry over to great generality. One had to cater too much to the particular network being analyzed. Nevertheless, I saw that this area was worthwhile and the nature of the problems motivated me to move more toward Monte Carlo sampling methodology. I began reading the proceedings from old conferences on simulation and Monte Carlo. These were held in the late 1940s and early 1950s, some at UCLA, at IBM in the New York area, and I think, at the National Bureau of Standards, today called the National Institute of Science and Technology (NIST). Attendees came from many areas and organizations, including RAND, IBM, and the Bureau. Von Neumann's contributions were presented by George Forsythe, then of the Bureau. Ted Harris gave a talk. To a great extent he was a theoretician. Have you ever heard of Harris recurrence? That's the same Harris. [Laughter.] When I met him some years later, he was chairman of the Mathematics Department at RAND. It was clear that his was an entirely abstract view of these problems. But in his earlier incarnation, he had a much more applied view.

During the 1960s, unflattering remarks about simulation were common. Many felt that only analysts with limited analytical skills resorted to simulation for problem solving. The remarks were usually made by individuals who had not confronted problems as complex as those under study. That attitude continued well into the 1970s. However, I was reassured by the knowledge that people like Harris, von Neumann, Arrow, and Markowitz had interests in the area and recognized the challenge. Another man who was involved with it was Herman Kahn. Although you may remember Kahn from his work on variance reduction, he achieved his greatest notoriety from a book he wrote while at RAND, *On Thermonuclear War*. When I interviewed at RAND, I wondered why people were picketing outside its entrance. Kahn's book spoke of surviving a thermonuclear war and that motivated the protest.

From reading these people's remarks on simulation and Monte Carlo methodology, I realized that they didn't think of simulation as something to be tried if all other methods failed. They saw it as a methodology that could provide flexibility. That convinced me that if you started doing this in any number of problem areas, for example in networks, you ought to be able to formalize ideas, which would have much more generality for a wide range of problems. That's what encouraged me to focus more on networks and Monte Carlo.

3 Life After Being Department Chair

This section focuses on George's research during the 1990s, following his tenure as chair of the UNC Department of Operations Research.

C: We are now moving to the 1990s, and the emphasis on computational issues.

G: As I read more about Monte Carlo, I decided that I wanted to write a book to get it into an easily understandable form and to describe what was going on in particular areas. Monte Carlo was a collection of techniques, but as a formalism, it lacked coherence. My 1996 book was an attempt to overcome that limitation. In retrospect, the book turned out to be a compendium of techniques rather than a pedagogic device. Nevertheless, the compendium gave a comprehensive picture of the area.

With the batch means method—which, as I said, was around from the beginning of simulation statistical methodology—it occurred to me, as to many other people, that it was much easier to understand than autoregression or spectrum analysis. Interestingly, there is a statistical paper by Champernowne in the 1950s in a British journal that describes a variant of the batch means method.

J: I've heard of this paper but I've forgotten much about it.

G: Some of the ideas there seem to be very much related to the time series ideas that I had seen in Maurice Bartlett's papers. I wrote a computer program for my simulation class to implement batch means that allowed a user to progressively monitor convergence of the estimate of the variance of the sample mean as the sample path increased in length. I published a paper on that approach in 1978 in *Management Science*.

At that time, Lou Moore was a doctoral student of mine. We began talking about speeding up the procedure. Lou actually wrote a program that did that. For reasons that are not clear to me, we did not pursue this speedup method and I cannot recall what happened to that computer program. It was a first-rate attempt to accelerate computation.

In the early 1990s, my student Steve Yarberry got interested in this problem and we talked about how to increase batch size while reducing computing time. We came up with the square root rule as the crucial element for doing this. Our 1997 journal article describes the technique in detail. Our LABATCH and LABATCH.2 software is based on this approach. Every so often, I hear from people with diverse backgrounds who use LABATCH and LABATCH.2. They usually contact me about technical features of the code. I don't have a good picture of the extent to which it is used, but I do know that it continues to be used.

During the summer of 1986, Russell Cheng arrived for a yearlong visit to UNC. I told him that I contemplated writing a book on Monte Carlo and he was encouraging. Work on the manuscript was slow and halting at first. By then, the OR program had become a department and the demands of chairmanship made it difficult to write as much as I wanted to.

At the end of my second term as chairman in 1990, I went on a yearlong sabbatical. Because my wife, Sue, and I concluded that the time was not opportune for taking our children out of the Chapel Hill schools for a year, I searched for a local opportunity. I approached John Geweke [now at the University of Iowa] who was the chairman of the Institute of Statistics and Decision Sciences at Duke, and I asked him if he could provide a desk. John had come over to UNC several times and we had talked about simulation. He was most cordial and kindly arranged accommodations for me at the Institute. When I arrived, I learned that John had accepted a professorship at Minnesota. [Laughter.]

I can't say enough for the faculty at Duke. They were extremely warm and welcoming. Although they were not directly interested in Monte Carlo, they used it, understood the ideas, and offered many suggestions. So I began to realize that it was now becoming part of the statistics milieu. I had many talks with the Duke faculty, especially Michael Lavine. I didn't agree with everything they said, but it all had relevance.

This exposure led me to cast what I was writing into a broader format in terms of problems and techniques that would make my book more appealing to statisticians. Although I didn't want to move too far away from OR, I included examples like the eye-hair contingency table problem in Diaconis and Sturmfels.

It took several more years for me to get to the publication stage. The reality was—you may not believe it—but there were actually more manuscript pages on several different subjects that I chose not to include in the published book. I was fearful that potential publishers would be uncomfortable with a book of more than the 700 pages that I submitted. Of all my books, it's the one that's sold best. It continues to sell a substantial number of copies in Europe, and I am mystified as to who's buying them. By now there are books for statisticians that are more focused on their interests.

C: We are moving to the later stages now.

G: By the early 1990s, interactive modeling had become an essential pedagogic device for teaching discrete-event simulation. However, I wasn't prepared to make that the focus of a simulation course. I preferred to use different languages to demonstrate their features to students, because I still thought at that point that it was very important for them to understand the limitations of individual languages. Therefore, I used a mixture of SIMSCRIPT and Arena, which by then was in a form that students could easily use. If we had had an engineering audience on the UNC campus,

I probably would have moved faster. Our students principally were mathematics majors. In the late 1990s, I decided to convert simulation class notes I had recently prepared into a book. My 2001 book, *Discrete-Event Simulation*, was the result. In reality, the material in the book was dated, partially because I used SIMSCRIPT as the prototype. Although it remained a good teaching device, SIMSCRIPT didn't have the many conveniences of other online languages. By then, students were accustomed to more immediate real-time interactions with their programs.

C: Actually, the military still uses it.

G: That's because of their association with CACI, the company that owns SIMSCRIPT. In order for the students to go out and be marketable, they have to know something about a modeling language and how to use it. I picked Arena to provide a more highly interactive experience for students. Had I continued teaching, I would have switched to AutoMod. It's hard to write a well-rounded book on simulation because the language/modeling part has again become the central focus, and anything of a statistical nature is definitely peripheral. In many cases, people don't bother with it at all. Early on, they didn't bother with it, but for different reasons. Now they don't bother with statistical procedures although they are actually accessible with merely a few keystrokes. For example, Arena and AutoMod both offer these procedures.

During my retirement which began in 2001, I decided to write a lower-level book on Monte Carlo. The book was published in 2005. I made it heavily example-oriented. It seems to interest those in a wide range of disciplines. I continue to get inquiries about issues it raises. This book, as well as my 2001 one, have one particular advantage over the others I've written. I can maintain their errata online. [Laughter.]

C: So do you believe that the 2005 book will be the most read because of the wide area it's covering?

G: I hope so. However, others also are putting out lower-level books. At the time I wrote the book I looked for an example in genomics. I picked one on protein folding and felt I had to make the example clear in terms of biology. I gave a very elaborate description of the structure of proteins and how all this works in three dimensions. Since then I have come to realize that most people don't bother reading that. They rely on a more concise approach to it, and I probably could have gotten by with a much shorter account. I think I made the description a little heavy-ended.

C: Do you have a few comments for your research associates and students?

G: I was fortunate to have good students. None at UNC shortened my life. [Laughter.] Andy Seila was my first. With him, I had the benefit of someone who was very focused on getting through. He wrote a nice dissertation that had to do with quantile estimation. It looked into an issue that had been raised by Iglehart and his colleagues.

When I first came to UNC, the Curriculum essentially had a volunteer faculty. As the only full-time professor in the program, I was the most visible faculty member. By 1978, I simultaneously had four dissertation students, Andy Seila, now retired from the University of Georgia, Veena Adlakha, now on the faculty of the University of Baltimore, Bao-Sheng Huang who went to work for Bell Labs, and Lou Moore, now at the RAND Corporation. Since each had a different thesis topic, on any one day I could easily start talking about the wrong problem with a student because I was thinking about one of the other problems. [Laughter.] Although I found simultaneously advising four theses demanding, I look back fondly on the experience.

In the 1980s and 1990s, I again had a collection of good students, including Ken Risko, now at Deloitte & Touche L.L.P., Tien-Yi Shaw, now at SAS, Christos, now on the faculty at Georgia Tech, Steve Yarberry, now at Practice Plus/Arkansas Health Group, and Cristina Arguelles Tasker, who is now at i2 Technologies in London.

4 The Future of Simulation and Operations Research

In this portion of the interview, George discusses potential future research directions for the field of simulation and offers an assessment on the status and future of operations research in academia and industry.

C: Well, that brings us to probably the last two topics. The first concerns the future of simulation. Where do we stand as a research community compared to other communities, with regard to two streams: the modeling side and the theoretical side. Any comments?

G: The modeling area of discrete-event simulation is essentially cast in concrete, principally because of the substantial investment that's been made in existing proprietary software.

J: What about Petri nets or event graphs?

G: Although Petri nets have been around for some time, they have not become a central concept. There may be room for modeling using network formats, but recall that a well-established software exists for some network problems. Some of you may remember the network program called SPICE, a creation of the 1980s. One would have to go up against well-entrenched software to motivate people to consider new concepts.

J: In fact, SPICE is still very heavily used by chemical engineers everywhere.

G: Right. Many of its users have no idea of its internal structure. [Laughter.] The same is true of much of the proprietary discrete-event simulation packages. I'm saying that it's very hard to gain acceptance for new modeling ideas.

With regard to statistical methodology, the picture is mixed. For example, many people thought that the regenerative approach was going to change how the statistical analysis of discrete-event simulation output would be conducted. But anyone who experimented with the method early on realized that was not to be, especially in highly congested systems where the regeneration period got very long. Although there were proposals to increase the frequency of regeneration, discrete-event simulation often requires one to maintain the fidelity of the local rules at each point on a sample path, a limitation to increasing regeneration frequency. However, other uses of Monte Carlo do not impose that requirement. The idea is to just to come up with an end result. There are many ways to do this. A paper by Brockwell and Kadane describes how to induce more frequent regenerations. They also give an example of its use. It's an interesting approach and in certain respects is different from earlier attempts. It relies on adding an extra state to the system, and to make use of that state in a particular way. Essentially you are dealing with an augmented chain.

J: An old trick.

G: I've written a set of notes on this method and other new methodologies but have no idea what I'm ever going to do with them. A good idea is not enough to have an impact on either discrete-event simulation or Monte Carlo, more generally. You have to make a concept or idea implementable to get it adopted. This attitude was justifiable in the past and even more so now. Today we have considerably more capacity than in the past for expeditiously translating ideas into usable products and testing them.

C: As a summary, do you believe that there is a good future for the statistical side of the simulation community?

G: Only if simulation methodologists broaden the problem set on which they work! For the last twenty years, we have been working to the point of diminishing returns. You may not want to hear that but unfortunately that's the truth. It's harder to get something new in these areas implemented today because the off-the-shelf products that are often available do a reasonably adequate job. Certainly in random number generation, we now have at least one random number generator that's equidistributed in 624 dimensions, a world apart from where we started years ago. That's been a big contribution.

J: What about the larger future of operations research itself, not just the field of simulation? What's your take on the future of OR as a discipline?

G: As a discipline, OR receives less visibility today than it did thirty years ago. At universities, it's been merged with other programs. I cannot explain why. To a very great extent, business schools have abandoned OR or at least incorporated it into their multifaceted quantitative methods courses. Some have eliminated their quantitative methods courses, replacing them with hands-on experience on a computer.

In engineering circles, there are specific classes of problems that rely on OR techniques. There's a healthy respect across engineering disciplines for what OR can do, provided it's oriented towards their problems. As an overall area of methodology, OR doesn't seem to have the visibility that we'd all like it to have. That's certainly true of simulation methodology. Many of the developments of the last forty years in discrete-event simulation hardly, if ever, get acknowledged, particularly in computer science and statistics.

C: Well, this brings me to the last question that I had, which returns to the status and visibility of the simulation community. When you talk to people in statistics or stochastics, they tell you that simulation is an applications area. For instance, statisticians will tell you that batch means or other output analysis methods are simply L_2 estimation. You talk to computer scientists, and they tell you that you are doing statistics. We're right in the middle, and I'm very concerned about what we need to do to shake this perception.

G: We're in the middle because different people have carved out sections of what was once our discipline. They often have good ideas and make important contributions. But there's still room for new OR-related ideas to play a role. The problem that I'm looking at now—counting using simulated annealing—is an example. The computer science approach focuses on complexity rather than the intrinsic opportunities within the problem for devising a near optimal sampling plan. More of a focus on OR may well lead to more appealing solutions.

So Jim, to answer your question about the profession, I think OR, regrettably, has not received the recognition that it deserves. Certainly it's realized its potential in some areas. The names of several OR techniques have become common "household terms" so that we and others no longer assign authorship or provide citations to many of them.

I think the professional societies have tried to do something about it, but I wonder how successful it's been. It's been pointed out to me that the way to judge the success of a profession is by how good the salaries are that its students receive when they enter professional life. By that standard, I think all is fine. [Laughter.]

D: How do we compare to other engineers?

J: Industrial engineers and OR types compare very favorably to civil engineers, for example, and electrical engineers. I don't really know about chemical engineers.

G: Industrial engineers, for many years, were at the top.

J: Well, they are, certainly, in terms of starting salaries, at least at NC State. They compare very favorably against almost all disciplines, including computer science, interestingly enough.

D: Not at Georgia Tech.

G: Computer science has basically held its own to a great extent in terms of what they can command, and resources they demand when they go to universities as a prize for coming there.

D: Although, apparently, computer science degrees have gone down a little bit in the last few years; I don't know why.

J: Not just a little bit.

C: Well, we're getting towards the end. Let me just ask a question. You've had a distinguished career. It's a fact. Going back, is there anything you would've done different?

G: I've been very fortunate in as much as a lot of good things came my way. I happened to have been in places where I could benefit in one way or another from contact with many accomplished people. My experience has been more favorable than others I know who didn't have the same good luck.

In terms of what I would've done differently, there were times here at UNC that I wish I had done things differently with regard to the OR program. Perhaps I should have encouraged a different academic emphasis for the program. I focused on becoming a highly methodological department at a point in time when there was a major shift towards PCs and hands-on work, and we didn't make that transition as rapidly as we perhaps should have in many areas. We didn't have any part of the manufacturing activity of the 1980s. We didn't have any part of the financial modeling of the 1990s. If I were to do things again, I would try to reconfigure the faculty into a form that would've allowed us to move more easily into those and other applied areas. Not having done so eventually became a limitation for us in terms of what our reputation was and our ability to attract people.

J: I've got one last question that I'd like to ask. What sort of advice would you give to people pursuing careers in operations research and simulation? Do you have a set of principles that you'd offer someone to bear in mind in pursuing a career in that area?

G: I am not capable of doing that. To my mind, my career was a series of good opportunities. It was partly—I have no illusions about it—attributable to the times. I was graduated from college right after Sputnik. Funding had grown considerably for research. Suddenly, the concept of a son doing operations research bordered on having a son that was a scientist, a doctor, or a lawyer. I'm sure your mothers would appreciate that. [Laughter.] So that made my life easy. There was a demand for what I wanted to do. Nevertheless, there were times that I had doubts. My first day of academic life began with a memo from the president of Yale saying, because of a budget deficit, he was freezing salaries. [Laughter.] It soon became apparent that budgets were universally tight across all academic institutions, including places like RAND. Therefore, I stuck with it, and the situation improved.

Today university life is not as I found it when I began. It has a much higher level of accountability. I can't speak for all disciplines, but from what I've seen in the mathematical sciences at UNC and other places, there are many more demands placed on faculty, and especially junior faculty. Regrettably, the junior faculty today have little awareness of the lower level of accountability of the past, and so they don't know what they're missing. You are all old enough to appreciate that difference. It's lamentable, but the truth is that future university life will continue to diverge from my experience.

I don't think I would necessarily encourage a young person to go into academia. Nor would I discourage them. I've had this debate with several people in the department who have encouraged students who seem perfect for academic life to go into it, although the student is not inclined to. I don't think that faculty encouragement serves students well. It overlooks the fact that students may have a considerably better perspective on what's right for them. They observe departmental and university governance which are much different today than when we first came to academia. I encourage students to keep their eyes open.

C: Let me end this conversation as the former student. George, it was an honor for us to have you speak with us. You and I have had parts of this conversation several times, but it was the first time I was able to get the whole nine yards. Due to the digital recording, it will live for eternity. So it was wonderful. Thank you so much for the hospitality.

G: Well, I'm glad you enjoyed it.

Computer Intensive Statistical Model Building

Russell Cheng

Abstract We consider resampling techniques in multiple linear regression where the objective is to identify a subset of the full set of explanatory variables that best captures the behaviour of the dependent variable, but using as few explanatory variables as possible. The total number of possible subsets or models grows exponentially with the number of explanatory variables, so a full examination of all possible models rapidly becomes intractable. The standard approach to this problem is to use a sequential selection procedure which avoids having to examine all subsets. When the number of explanatory variables is large there is a possible concern that good models might be missed. It is also important to examine whether the selected "best" model is the only good choice or whether other models might be equally satisfactory. We show how bootstrap resampling can handle both concerns in a simple way. In particular resampling enables a tractably small subset of good possible models to be selected as well as providing a method for comparing these models systematically. We describe the methodology and provide two numerical examples.

1 Introduction

This paper discusses the use of bootstrap (BS) resampling for tackling the well-known, but awkward, problem of model selection in multiple regression, when the number of possible explanatory variables is large. Our claim is that BS resampling is a simple and effective approach for this problem with distinct advantages over standard sequential methods that are often advocated and employed.

The ideas discussed in this paper were originally suggested for the exploratory study of a complex system using discrete event simulation. The basic methods were discussed by Cheng (2008) in that context. In this paper we discuss the methodology in more detail and more generally. In particular we consider the rationale of the methodology more fully and how to use it with the Mallows C_p criterion which is often suggested for handling this problem (see Krzanowski 1998 or Wu

R. Cheng (✉)
School of Mathematics, University of Southampton, Southampton, SO17 1BJ, UK
e-mail: r.c.h.cheng@soton.ac.uk

C. Alexopoulos et al. (eds.), *Advancing the Frontiers of Simulation: A Festschrift in Honor of George Samuel Fishman*, International Series in Operations Research & Management Science 133, DOI 10.1007/b110059_3,

and Hamada 2000, for example). The theoretical underpinning of bootstrapping in multiple regression is well-established; we will collect together the key results to underpin the methodology that we propose.

We suppose the dependent variable of interest is a (scalar) continuous random variable denoted by Y and that Y is linearly dependent on P explanatory variables X_j, $j = 1, 2, \ldots, P$. We are concerned with the *model selection problem* where we are interested in identifying simpler models in which some of the explanatory variables are omitted because they are actually unimportant. To avoid confusion we shall, from now on, use the term "model" to indicate that we are selecting a subset of explanatory variables, or *factors*, from the full set available, and use the term "full model" to indicate when all P explanatory variables are included. There are a total of 2^P distinct subsets of the explanatory variables, so that this is the number of models that we can choose from.

(Many authors exclude the null model $\mathbf{y} = \boldsymbol{\varepsilon}$ and so take the total number of distinct models to be $2^P - 1$ rather than 2^P. However, though the null model is very unlikely to be the best fit in applications, there seems no real reason for excluding it, and its exclusion can lead to misinterpretation of results if it happens to be the most appropriate model. We therefore do not exclude this possibility even if it is remote, and so take the total number of possible models as 2^P throughout this paper.)

Though the model selection problem is well known, the usually accepted methods of handling it are not always satisfactory. Wu and Hamada (2000) have discussed this problem at length. They considered the very well-known backward, forward and stepwise explanatory variable selection methods and also Bayesian strategies. The main problems with these methods are as follows.

The backward, forward and stepwise selection methods are all sequential, in which explanatory variables are considered one at a time for possible inclusion, or elimination. It is therefore possible, with non-orthogonally designed experiments, simply because of the order in which explanatory variables are considered, to end up with a selected model that does not include all those explanatory variables that are important.

Use of a Bayesian approach avoids this difficulty, but a prior distribution for explanatory variable coefficient values has to be chosen and there are also technical implementation issues, such as deciding on the length of "burn-in" period and deciding when sufficient sampling has been carried out to ensure that adequate convergence to the posterior distribution has taken place.

In this paper we consider the use of BS resampling methods to generate a large number of data sets each with the *same* statistical distributional properties, at least asymptotically, as the original data set. We can therefore deploy whatever method we wish for selecting the model that best fits the original data sample (in some sense, to be defined), and then gauge the adequacy of the selected model by studying how consistently it is selected as the best fit in the BS samples, and how well it fits these samples.

We shall use the C_p statistic introduced by Mallows (1973, 1995) as the selection criterion for choosing between different models, as it is readily calculated in terms of ANOVA sums of squares and has a direct interpretation in terms of the prediction error, making it easy to understand and use. Several other criteria are asymptotically equivalent (see Nishii 1984).

We also consider in this paper the problem of checking whether the selected model is a sufficiently good fit. We are especially interested in the situation where there are a large number of factors. There is the strong possibility that there will be a number of models that are a satisfactory fit to the data. We need therefore to have some means for gauging the adequacy of competing fitted models. Of the existing methods that we have already mentioned, the Bayesian approach seems most satisfactory in that a posterior distribution is obtained for the possible models, so that it will be clear whether there is one single best model choice or whether several competing models are equally or nearly as good. The Bayesian approach is not entirely satisfactory in that it does not provide immediate information on whether the models with the highest posterior probabilities are adequate or not.

In this paper we propose an alternative approach, based on bootstrapping, to gauge the adequacy of selected models, as bootstrapping provides a natural way of demonstrating when there is little to choose amongst several, possibly many, models. It is of interest to note that such methods are now beginning to be recognized as very appropriate for model selection in simulation work. A good example is given by Fishman (2006, Section 2.8).

In Section 2 we describe the linear statistical model that we will use and discuss selection criteria for choosing between models. In Section 3 we discuss the Mallows C_p statistic and model selection. In Section 4 we discuss two ways of generating BS samples. We also give two methods using BS resampling for identifying a small but targeted number of promising models out of the full set of 2^P possible models for fitting to the original data. We also show how bootstrapping can also be used to assess the quality of models that seem to be a good fit to the original sample. Two numerical examples are given in Section 5, and a summary is provided in Section 6.

2 The Linear Model

We consider the (full) linear model

$$\begin{bmatrix} Y_1 \\ Y_2 \\ \vdots \\ Y_n \end{bmatrix} = \begin{bmatrix} 1 & X_{12} & X_{13} & \dots & X_{1P} \\ 1 & X_{22} & X_{23} & \dots & X_{2P} \\ \vdots & \vdots & \vdots & \ddots & \vdots \\ 1 & X_{n2} & X_{n3} & \dots & X_{nP} \end{bmatrix} \begin{bmatrix} b_1 \\ b_2 \\ \vdots \\ b_P \end{bmatrix} + \begin{bmatrix} \varepsilon_1 \\ \varepsilon_2 \\ \vdots \\ \varepsilon_n \end{bmatrix}, \tag{1}$$

where Y_i, $i = 1, 2, \dots, n$ are the observed output values obtained from n simulation runs; X_{ij} are the explanatory variable values in each of the n runs; b_j, $j = 1, 2, \dots, P$ are the unknown coefficients corresponding to each of the P explanatory variables; and ε_i, $i = 1, 2, \dots, n$ are random errors. We have taken $X_{i1} = 1$, $i = 1, 2, \dots, n$ so that b_1 corresponds to a general mean. We thus treat the mean as a coefficient, so that, as far as the model selection and fitting process is concerned, we do not treat it differently from the other coefficients. In what follows, when we refer to a "factor" it is to be understood that this includes the general mean.

We shall assume that the ε_i, $i = 1, 2, \ldots, n$ are identically distributed with mean zero and variance

$$\mathrm{Var}(\varepsilon) = \sigma^2 . \tag{2}$$

Such random errors are often assumed to be normally distributed, but we do not assume that this is necessarily so in our formulation.

We shall, where convenient, write (1) in the alternative matrix form

$$\mathbf{Y} = \mathbf{X}\mathbf{b} + \boldsymbol{\varepsilon} . \tag{3}$$

Equation (1) is the full model in which all explanatory variables are included. We shall define a *model* as

$$m = \{j_1, j_2, \ldots, j_p\} \tag{4}$$

containing just the factor indices

$$j_1 < j_2 < \cdots < j_p, \ p \le P ,$$

if (and only if)

$$b_{j_1} \neq 0, \ \ b_{j_2} \neq 0, \ \ \ldots, \ \ b_{j_p} \neq 0, \ \text{and all other } b_j = 0 .$$

We shall write the observations corresponding to this model as

$$\begin{bmatrix} Y_1 \\ Y_2 \\ \vdots \\ Y_n \end{bmatrix} = \begin{bmatrix} X_{1j_1} & X_{1j_2} & \ldots & X_{1j_P} \\ X_{2j_1} & X_{2j_2} & \ldots & X_{2j_P} \\ \vdots & \vdots & \ddots & \vdots \\ X_{nj_1} & X_{nj_2} & \ldots & X_{nj_P} \end{bmatrix} \begin{bmatrix} b_{j_1} \\ b_{j_2} \\ \vdots \\ b_{j_P} \end{bmatrix} + \begin{bmatrix} \varepsilon_1 \\ \varepsilon_2 \\ \vdots \\ \varepsilon_n \end{bmatrix} \tag{5}$$

or in the matrix form

$$\mathbf{Y} = \mathbf{X}(m)\mathbf{b}(m) + \boldsymbol{\varepsilon}. \tag{6}$$

Where necessary we shall also write

$$p(m) = p \tag{7}$$

for the number of nonzero coefficients in the model m. Also we will denote the full model by M, so that $p(M) = P$.

When we fit the model m we shall use the least squares estimates (see Searle 1971, for example)

$$\hat{\mathbf{b}}(m) = \left[\mathbf{X}^{\mathrm{T}}(m)\mathbf{X}(m)\right]^{-1}\mathbf{X}^{\mathrm{T}}(m)\mathbf{Y} \tag{8}$$

for the unknown coefficient values, and

$$\begin{aligned}\hat{\sigma}^2(m) &= [n - p(m)]^{-1}\sum_{i=1}^{n}\left(Y_i - \hat{Y}_i\right)^2 \\ &= [n - p(m)]^{-1}\left[\mathbf{Y} - \mathbf{X}(m)\hat{\mathbf{b}}(m)\right]^{\mathrm{T}}\left[\mathbf{Y} - \mathbf{X}(m)\hat{\mathbf{b}}(m)\right]\end{aligned} \tag{9}$$

for the unbiased estimate of the variance of the ε_i.

3 Methods for Selecting the Best Model

3.1 "Min C_p" and "Unbiased Min p" Selection Methods

In this section we consider various issues involved in selecting which model we think is the best. The first is the statistic to be used for measuring how well a given model fits the sample. One of the most popular is the C_p statistic proposed by Mallows (1973) which is an estimate of the expected prediction error taking into account the variance and bias of the fitted model. It is defined as

$$C_p(m) = [n - p(m)]\hat{\sigma}^2(m)/\hat{\sigma}^2(M) + 2p(m) - n\ . \tag{10}$$

An alternative statistic is the Akaike Information Criterion (Akaike 1970), which for the linear model reduces to $\mathrm{AIC}(m) = n\log[\hat{\sigma}^2(m)] + 2p(m)$, up to a constant depending on n but not on m. Asymptotically C_p and AIC have the same distribution (see Nishii 1984). However C_p is perhaps more satisfactory for our purpose because of its ease of interpretation. Mallows (1973) shows that if the model m (with p factors) is satisfactory in the sense that it has no bias, then the expected value of C_p is close to p, that is:

$$C_p \approx p\ . \tag{11}$$

However, if not all important factors are included, the expected value of C_p will be larger than p. A simple selection method is therefore the following.

"Min C_p" Model Selection Method

(i) Consider each of the 2^P possible models of (1) and for each model m calculate $C_p(m)$.
(ii) Select as the best model that m for which $C_p(m)$ is minimum, with the expectation that this model will be satisfactory if $C_p(m) \leq p$.

This provides a simple selection method if we are able to examine all possible models.

As mentioned previously, an exhaustive search of all possible models can be avoided by using a sequential procedure, several of which are cited by Mallows (1995). Mallows points out that if the "min C_p" method is used in a sequential procedure, and if m^+ is a model containing one factor additional to those already in a model m, then the extra factor would be worth including if

$$C_{p+1}(m^+) - C_p(m) = 2 - (S_1/\hat{\sigma}^2(M)) < 0 ,$$

where S_1 is the 1-df sum of squares due to the additional factor. This criterion for inclusion is therefore equivalent to carrying out a t-test, with the factor included if

$$t^2 = S_1/\hat{\sigma}^2(M) > 2 . \tag{12}$$

In the non-orthogonal case the final selected model is dependent on the order in which factors are considered, but for an orthogonal design the sum of squares S_1 corresponding to each factor does not depend on the model fitted. Thus there is no need for a sequential procedure in this latter case. The minimum C_p is easily obtained by fitting the full model and then applying the test (12) in "blanket" fashion, i.e. simultaneously, to every factor sum of squares. The "min C_p" model then includes just those factors that satisfy (12).

The attraction of the orthogonal case is that the inclusion or exclusion of each factor is decided just from fitting the full model. We shall consider use of the same procedure in the non-orthogonal case, together with an adjustment to deal with the problem of including too many unimportant factors. We still fit the full model, and for each factor j calculate the so called t-value of its fitted coefficient $\hat{b}_j$:

$$t_j = \hat{b}_j/s_j , \tag{13}$$

where $s_j = \sqrt{d_j\hat{\sigma}^2(M)}$ is the estimated standard deviation of $\hat{b}_j$, with d_j the jth entry in the main diagonal of the dispersion matrix, i.e.

$$d_j = \left[(\mathbf{X}^{\mathrm{T}}\mathbf{X})^{-1}\right]_{jj} . \tag{14}$$

Our selected model is the one that includes only those factors j for which

$$|t_j| > a , \tag{15}$$

where a is a chosen critical level. If the true value of b_j is $b_j = 0$, then t_j has Student's t-distribution with $n - P$ degrees of freedom. If we therefore denote the complementary distribution function for the absolute value $|t_j|$ by $\bar{T}_{n-P}(\cdot)$, then the probability of success of the test (15) under the assumption that $b_j = 0$, is

$$\pi_a = \Pr\{|t_j| > a\} = \bar{T}_{n-P}(a) .$$

A common alternative way of carrying out this test is to report the so-called p-value of the estimate $\hat{b}_j$, namely $\bar{T}_{n-P}(|t_j|)$, so that the factor j is retained if

$$\bar{T}_{n-P}(|t_j|) \ < \pi_a \ . \tag{16}$$

It will be seen that (12) is the special case of (15) or (16) where $a = \sqrt{2}$, with a corresponding critical p-value in (16), when $n - P$ is large, of $\pi_a = 0.1573$. This highlights a problem with using the "min C_p" method for selecting a model when the initial number of factors under consideration is large but where the (unknown) true values of many coefficients are at or near zero, as the selection test (16) would then include nearly 16% of such negligible coefficients in the model.

The effect of varying a can be seen more fully by considering the asymptotic probability that a factor with coefficient of size $b_j = bs_j$ is selected, when we allow b to vary also. For simplicity we assume that $n - P$ is large as then s_j can be treated essentially as being a known constant, so that $\hat{b}_j \sim N\left(bs_j, s_j^2\right)$. The probability we would include the factor is then

$$\begin{aligned} \Pr\{\text{Factor } j \text{ is included in model}\} &= 1 - \Pr\{-as_j < \hat{b} < as_j\} \\ &= 1 - \Pr\{-a - b < (\hat{b} - bs_j)/s_j < a - b\} \\ &= 1 - \Phi(a - b) + \Phi(-a - b) \ , \end{aligned} \tag{17}$$

where $\Phi(\cdot)$ is the standard normal distribution function. Figure 1 shows how this probability varies as a function of b for different selected a. It will be seen that somewhat larger values than $a = \sqrt{2}$ in (12), such as $a = \sqrt{6}$ or $a = 3$ might be more appropriate in exploratory studies where we are only interested in identifying

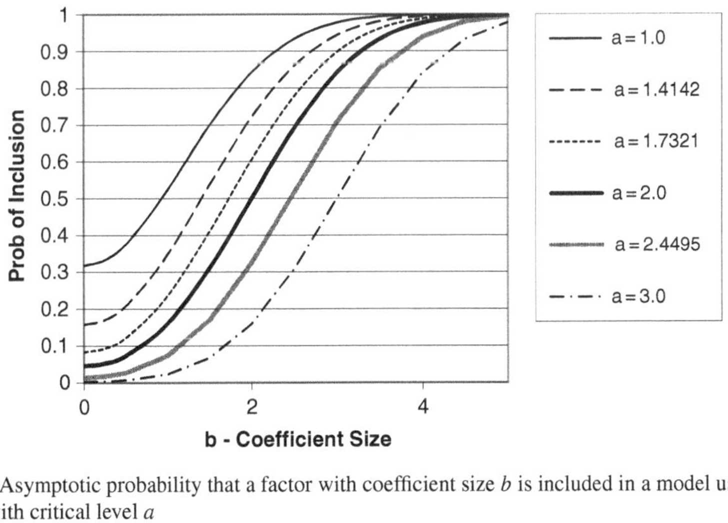

Fig. 1 Asymptotic probability that a factor with coefficient size b is included in a model using the t test with critical level a

significantly large b and would prefer the probability of retaining a zero coefficient to be much smaller than 16%.

The problem of unnecessarily including factors with zero coefficients is controlled by choice of a suitably large a. We can underline this choice by using a modified form of the "min C_p" model selection method. From (11) we know it is undesirable to select a model for which $C_p > p$. This suggests the following model selection procedure:

"Unbiased Min p" Model Selection Method

(i) Find the smallest p for which there are models m satisfying $C_p(m) \leq p$ and let

$$p_0 = \min\{p : C_p(m) \leq p\} . \tag{18}$$

(ii) Amongst all such models m, with $p(m) = p_0$, find the one for which $C_p(m)$ is minimum.

A simple way, at least in principle, of identifying this model is to plot C_p versus p for all possible models and look at the lower envelope of this scatterplot of points. For the orthogonal case where there are a large number of factors with coefficient values uniformly distributed in the neighbourhood of zero with density λ, Mallows (1995) has shown that the scatterplot has a lower boundary that is the (convex) cubic polynomial in p

$$C_p - P \approx \frac{(P-p)^3}{12\lambda^2} - 2(P-p) , \tag{19}$$

and that this boundary intersects the line $C_p = p$ at $P - p = 2\sqrt{3}\lambda$. Figure 2 depicts the scatterplot for the first example involving epoxide bonding that we will be discussing in Section 5, and this boundary and its intersection with the line $C_p = p$ are clearly distinguishable.

Our selection method (18) will clearly select a model corresponding to a point near this intersection. Specifically (18) requires finding the smallest p, p_0, for which there are points of the scatterplot below the line $C_p = p$ and then finding amongst those models with $p = p_0$, the one with minimum C_p.

In the orthogonal case, models at, or near, this intersection point will tend to include just those factors for which (15) is satisfied with $a = \sqrt{3}$, which is equivalent to using (16) to include just those factors whose estimated coefficients have p-value less than $\pi_a = 0.083$.

The condition (11) that $C_p \approx p$, obtains when the model contains no bias so that the model is completely appropriate whilst having the smallest p possible. For this reason we call (18) the "unbiased min p" method.

We delay discussion of how precisely to implement this method of model selection until we have discussed bootstrapping, as our proposed implementation will involve bootstrapping intimately.

3.2 Dimensionality Problem

As previously mentioned a critical issue that arises in model selection is the *dimensionality problem.* Because the total number of possible models, 2^P, grows exponentially with P, inspection of all models is tractable only when P is small. Thus even with just 20 explanatory variables there are already 1,048,576 models. Our approach is to identify a set of *promising models* using bootstrap resampling. The number of models in this set is easily controlled and so can be made much smaller than 2^P. But we shall show that it will almost certainly contain many good candidate models. It is thus satisfactory to select a "best" model from this subset.

We discuss bootstrapping in the next section.

4 Bootstrap Analysis

We shall use bootstrapping for two distinct purposes. Firstly, as already mentioned in the previous section, it can be used for identifying a set of promising models. However we shall also use bootstrapping to deal with the following second problem.

Once a model has been selected as being the best fit to a data set, we have the problem of determining what might be termed the *quality of the selected model.* For example, if we have used the "min C_p" method to select the model, there may be several models with values of $C_p(m)$ close to that of the best, so that we may not be sure which model really is the best. This question would be answered if we had many (independent but identically distributed) data samples and not just the one original sample, as we could determine the best model for each sample and see if the same model is best for all the samples. BS resampling enables such additional data samples to be generated.

We first outline how BS samples are generated in the next subsection, before going on to describe our two distinct uses of bootstrapping.

4.1 Bootstrap Samples

We describe first two ways of generating BS samples that asymptotically have the same form as (1). The standard way is described, for example, by Davison and Hinkley (1997). We take the modified residuals

$$r_i = (Y_i - \hat{Y}_i)/(1 - h_{ii})^{1/2}, \quad i = 1, 2, \ldots, n \tag{20}$$

obtained from the fitting the full model M to the original data, where $\hat{\mathbf{Y}} = \mathbf{X}\hat{\mathbf{b}}$ and h_{ii} is the ith main diagonal entry in the "hat" matrix

$$\mathbf{H} = \mathbf{X}(\mathbf{X}^{\mathrm{T}}\mathbf{X})^{-1}\mathbf{X}^{\mathrm{T}}.$$

We then centre these so that their average is zero:

$$e_i = r_i - \bar{r}, \quad i = 1, 2, \ldots, n. \tag{21}$$

A BS sample is then obtained by forming

$$Y_i^* = \hat{Y}_i + e_i^*, i = 1, 2, \ldots, n\ , \tag{22}$$

where the e_i^*, $i = 1, 2, \ldots, n$ are a random sample obtained by sampling with replacement from the e_i, $i = 1, 2, \ldots, n$.

A second way of resampling, *parametric bootstrapping*, is possible, if it can be assumed that the random errors ε_i, $i = 1, 2, \ldots, n$ in (1) are normally distributed and independent. The BS sample still takes the same form as (22) only now the e_i^*, $i = 1, 2, \ldots, n$ are a random sample from the fitted normal distribution, i.e.

$$e_i^* \sim N\big(0, \hat{\sigma}^2\big), \quad i = 1, 2, \ldots, n\ . \tag{23}$$

In either case we write $\hat{\mathbf{b}}^*$ and $\hat{\sigma}^{*2}$ for the estimates (8) and (9) obtained from fitting the model (1) to the BS observations (22).

The justification for bootstrapping is provided by Freedman (1981, Theorem 2.2). Assume that (1) and (2) hold and that $\mathbf{X}(n)$ is not random with

$$\frac{1}{n}\mathbf{X}^{\mathrm{T}}(n)\mathbf{X}(n) \to \mathbf{V} \text{ which is positive definite} \tag{24}$$

as $n \to \infty$. Then

$$\sqrt{n}\big\{\hat{\mathbf{b}}^*(n) - \hat{\mathbf{b}}(n)\big\} \text{ converges to } N\big(\mathbf{0}, \sigma^2\mathbf{V}^{-1}\big) \tag{25}$$

and

$$\hat{\sigma}^*(n) \text{ converges to a point mass at } \sigma\ . \tag{26}$$

The above result assumes that P is fixed as $n \to \infty$. We shall tacitly assume this in what follows. However a more refined treatment would allow P to become large as $n \to \infty$. Shibata (1981) has discussed the selection of factors for this regression problem. We shall not discuss this case explicitly here.

4.2 Bootstrap Generation of a Set of Promising Models

The "unbiased min p" method of selecting a best model does not require consideration of all 2^P models but only those near p_0, as defined in (18). Our first use of bootstrapping is therefore to generate a set of *promising models*. The number of models in this set does not need to be anywhere near 2^P, but it does need to be large enough to enable the lower boundary (19) to be clearly identified, at least near its intersection with $C_p = p$. Ideally it needs to contain all the models with scatterplot points near this intersection point. With these considerations in mind, our first proposed methods is:

"One Model per Sample" Generation of Promising Models by Bootstrapping

Step (1) Fit the full model to the original data and use this fitted full model to generate B BS samples each of the form (22).

Step (2) Set a critical t-value, a, (in view of our discussion in Section 3.1, we used $a = \sqrt{3}$) and construct one *promising* model for each BS sample as follows.

(i) Fit the full model, M, to the sample and calculate the t-value, t_j, as defined in (13), of each of the fitted coefficients, $\hat{b}_j$, $j = 1, 2, \ldots, P$.

(ii) Include in the promising model just those factors with t_j satisfying

$$|t_j| \geq a\ ;$$

see (15) above. Not all the promising models obtained in the above process will be distinct (in the sense of each model containing a subset of factors that is different from those of all other selected models). Denote the set of distinct models by S.

The above method produces at most B promising models, but can be far fewer, if the same model is repeatedly obtained from different BS Samples. If it were felt that the number of models needs to be increased, especially as we would want to include most if not all models satisfying $p = p_0$ and $C_p \approx p$, then the following variant of the "One model per sample" adds models in a straightforward way.

"Many Models per Sample" Generation of Promising Models by Bootstrapping

Step (1) Fit the full model to the original data and use this fitted full model to generate B BS samples, each of the form (22).

Step (2) For each BS sample:

(i) Fit the full model, M, to the sample and determine, as defined in (13), the t-value, t_j, of each of the fitted coefficients, $\hat{b}_j$, $j = 1, 2, \ldots, P$.

(ii) Order the coefficients by their $|t_j|$ values:

$$|t_{j_1}| \geq |t_{j_2}| \geq \cdots \geq |t_{j_P}|, \tag{27}$$

so that $\hat{b}_{j_1}$ is the most significant.

(iii) Set a critical t-value, a (we used $a = \sqrt{3}$ as before), and include all the following models in the promising set S:

$$\begin{aligned} m_1 &= \{j_1\} \\ m_2 &= \{j_1, j_2\} \\ &\vdots \\ m_k &= \{j_1, j_2, \ldots, j_k\}, \end{aligned} \tag{28}$$

where the last factor j_k satisfies

$$|t_{j_k}| \geq a > |t_{j_{k+1}}|. \tag{29}$$

Thus the model m_i is the one where the i most significant factors have been retained, with a cutoff that only factors with t-level greater than a are allowed in a model. So the last model, m_k in (28) is the one that includes just those coefficients with $|t|$-value a or greater, this being the sole model selected in Step (2) of the "One model per sample" method.

4.3 Bootstrap Quality Assessment of Selected Best Model

Once a set of promising models has been obtained, we can use the "unbiased min p" method to select the "best" model. That is we fit each promising model to the original data set, calculating C_p for each model; then we identify p_0 as in (18), and select as the best model the one with the smallest C_p subject to $p \leq p_0$ (checking that it satisfies the condition $C_p \leq p$).

We can now use bootstrapping to study the quality of the selected model. This is most easily done by adding the following steps to either of the bootstrap methods proposed in the previous section for generating a set of promising models.

Bootstrap Assessment of Selected Best Model:

Step (3) For each of the B BS samples, fit the set S of promising models, subject to the restriction that only models where $p \leq p_0$ are considered (we shall denote this restricted set of promising models by S_0) and calculate the C_p value for each model, selecting as the best model for this sample, that which minimizes C_p.

Step (4) Display the models of S_0, ranked in order of the frequency with which they are selected as being the best model in the B BS samples, displaying these frequencies as well.

Step (5) Display the empirical distribution functions of the C_p values of a selected number of those models in S_0 most frequently selected as being the best.

Let $\alpha(m)$ be the probability that model m will be selected as the best model in the sense of minimizing C_p amongst all models with $C_p \leq p_0$. Step (3) estimates these probabilities by fitting all the models in the restricted set S_0 of promising models to each of the BS samples and then selecting the best model (for the given BS sample) from this set. Note that, out of the full set of 2^p models, those that are not a good fit will have very little probability of being included in the set S, because of the way S is constructed. Hence they would not be considered for possible inclusion in S_0. Nevertheless every model has a *positive* probability of being included in S. Thus asymptotically, as $B \to \infty$, the restricted set S_0 of promising models considered in Step (3) above must tend to the full set of all models with $p \leq p_0$. This holds for either method of generating the set of promising models described in Section 4.2. Thus, as $B \to \infty$, Step (3) will converge to the exact situation where every model satisfying $p \leq p_0$ is considered for possible selection as the best. Hence, for each model m, $\alpha(m)$ can reasonably be estimated from the frequency with which m is selected as being the best model in Step (3).

The version of Step (3) given above concentrates on models with $p \leq p_0$, where p_0 is calculated from the original sample, in order to check how well this value p_0 performs. Different variants of Step (3) are possible. An alternative would be not to impose the condition $p \leq p_0$ at all, but instead simply to apply the "unbiased min p" selection procedure to each BS sample separately, using the full set S of promising models with each BS sample.

In Step (4) we simply display those models that have been most frequently selected as being the best fit.

The point of Step (5) is to assess the behaviour of the C_p values of those models that have been most frequently selected as being the best fit. For such a model to be satisfactory one would expect the distribution of its C_p value, over the BS samples, to be concentrated mainly in the region where $C_p \leq p$.

5 Numerical Examples

We give two examples. Both involve readily accessible real data samples. The first is a data set where the design matrix is orthogonal. As already remarked immediately after (12), the obvious strategy in this case of applying a test such as (12) simultaneously to each estimated coefficient gives an unambiguous selection strategy and cannot really be bettered. The analysis is thus straightforward in this case. However we include the example simply to demonstrate the way the resulting bootstrap analysis works. The second was discussed by Cheng (2008). Here we discuss the selection method more fully.

5.1 Epoxide Bond Example

The first example is data given by Williams (1968) and reproduced in Wu and Hamada (2000, Table 8.6). This measured the adhesion of an epoxide bonding system in an orthogonally designed experiment with, including a general mean, 25 factors, and 28 observations.

Our analysis is in two stages as set out in Sections 4.2 and 4.3.

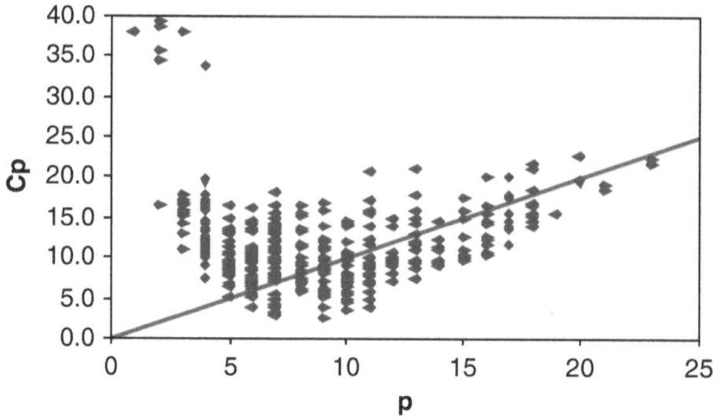

Fig. 2 C_p versus p plot of 320 promising models found for the Epoxide Bond data using the "One model per sample" BS method. The C_p values are those obtained when the promising models are fitted to the original sample

The first stage generates a set of promising models. We used the "One model per sample" BS method of Section 4.2 with B =500, $\pi_a = 0.083$, and with no limit placed on the maximum number of factors that can be included in a fitted model.

Step (1) of the analysis produced an initial subset of 320 promising models. The C_p values of these models when they were fitted to the *original* data are plotted against p in Fig. 2. Applying (18) gives $p_0 = 6$.

In the second stage we could have just used the previously generated promising models, but removing those with $p > p_0$. Instead we increased the number of promising models by using the "Many models per sample" BS method with $B = 500$, $\pi_a = 0.083$ but with the maximum number of factors permitted in a model limited to $p_0 = 6$. This yielded a set, S_0, of 488 promising models. Then, as described in Section 4.3, the models of S_0 were fitted to each BS sample, and the model with the minimum C_p was selected as being the best model for that BS sample. This yielded 235 different best models. The plot of the C_p values obtained by fitting each of these models to the *original* data set is given in Fig. 3.

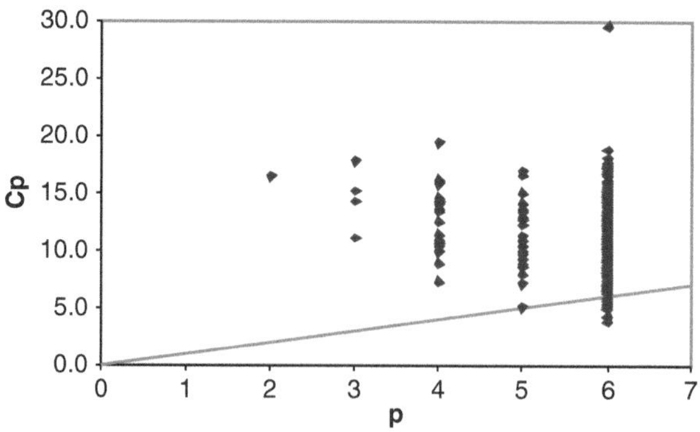

Fig. 3 C_p versus p for the 235 "best" models found by the Bootstrap Quality Assessment Method described in Section 4.3. The C_p values are those when the models are fitted to the original data

The frequency with which each of the models of S_0 was selected as being the best varies with model. The 25 models selected most frequently are displayed in Table 1. The model selected most frequently was

$$X_0, \ X_4, \ X_{14}, \ X_{15}, \ X_{16}, \ X_{19} \ ,$$

where X_0 is the mean. This model was selected as being the best model (i.e., with the smallest C_p) in 14 of the BS samples.

The model with the lowest C_p value in Fig. 3 was

$$X_0, \ X_4, \ X_{15}, \ X_{16}, \ X_{19}, \ X_{21},$$

which was the 4th best in terms of the number of times it was selected as best in the BS samples. There is little to choose between the top few models. Taken together it is fairly clear that factors X_0, X_{15}, X_{19} are the most important followed by X_4, X_{16}, and X_{21}.

5.2 Bank Data Example

The second example is taken from Makridakis et al. (1998, Table 6-8). The data is monthly. The variable of interest, Y, is the first difference, D(EOM), between the successive end of month (EOM) balances of a mutual savings bank. There are three primary X-variables: X_1 is a composite triple bond rate (AAA), X_2 is a composite (3-4) year US Government bond rate, X_3 is D(3-4), the monthly change in X_2. There were in addition 11 monthly seasonal explanatory variables (D1–D11), and three further variables, time t and its square and cube t^2, t^3, making 17 initial explanatory variables. We do not reproduce the data here as the three key variables, (EOM), (AAA) and (3-4), for 60 months, are downloadable from the Web site `www-personal.buseco.monash.edu.au/~hyndman/TSDL/`.

In our analysis we followed Makridakis et al. (1998, Table 6-8) and express Y in thousands of dollars and analysed only the first 53 months of data. We have also added a general mean X_0 as an additional factor so that we work with 18 explanatory factors. There are thus $2^{18} = 262{,}144$ distinct models to select from; a somewhat large number of models to comfortably work through.

Using a best subset analysis with an adjusted coefficient of determination, $\bar{R}^2$, for selection criterion Makridakis et al. found the best model overall was

$$X_0 \ X_1 \ X_2 \ X_3 \ D_2 \ D_3 \ D_4 \ D_5 \ D_6 \ D_7 \ D_8 \ D_9 \ D_{10} \ D_{11} \ t^3 \qquad (30)$$

and, using a stepwise regression, that the best model was

$$X_0 \ X_1 \ X_2 \ X_3 \ D_2 \ D_4 \ D_6 \ D_7 \ D_8 \ D_9 \ D_{10} \ D_{11} \ t^3. \qquad (31)$$

Table 1 Epoxide bond data, top 25 of final selection of 235 "best" models

	Mean	X1	X2	X3	X4	X5	X6	X7	X8	X9	X10	X11	X12	X13	X14	X15	X16	X17	X18	X19	X20	X21	X22	X23	X24	# b's	Cp	Freq
Pval	0.01	0.24	0.34	0.73	0.13	0.53	0.56	0.50	0.20	0.46	0.48	0.85	0.44	0.29	0.18	0.02	0.10	0.50	0.86	0.07	0.67	0.17	0.55	0.54	0.51			
1	1	0	0	0	1	0	0	0	0	0	0	0	0	0	1	1	1	0	0	1	0	0	0	0	0	6	4.2	14
2	1	0	0	0	1	0	0	0	1	0	0	0	0	0	0	1	1	0	0	1	0	0	0	0	0	6	4.4	12
3	1	1	0	0	1	0	0	0	0	0	0	0	0	0	0	1	1	0	0	1	0	0	0	0	0	6	5	1'
4	1	0	0	0	1	0	0	0	0	0	0	0	0	0	0	1	1	0	0	1	0	1	0	0	0	6	3.9	1'
5	1	0	0	0	1	0	0	0	0	0	0	0	0	1	0	1	1	0	0	1	0	0	0	0	0	6	5.5	10
6	1	0	1	0	1	0	0	0	0	0	0	0	0	0	1	1	0	0	0	1	0	0	0	0	0	6	8.7	8
7	1	0	0	0	1	0	0	0	1	0	0	0	0	0	0	1	0	0	0	1	0	1	0	0	0	6	6.9	8
8	1	1	0	0	0	0	0	0	1	0	0	0	0	0	0	1	1	0	0	1	0	0	0	0	0	6	6.5	8
9	1	1	0	0	1	0	0	0	0	0	0	0	0	0	1	1	0	0	0	1	0	0	0	0	0	6	7.8	7
10	1	1	0	0	0	0	0	0	0	0	0	0	0	0	1	1	0	0	0	1	0	1	0	0	0	6	8.7	7
11	1	0	0	0	1	0	0	0	0	0	0	0	0	0	0	1	1	1	0	1	0	0	0	0	0	6	6.6	6
12	1	0	0	0	0	0	0	0	1	0	0	0	0	0	0	1	1	0	0	1	0	1	0	0	0	6	5.3	6
13	1	0	0	0	0	0	0	0	0	0	0	0	0	0	1	1	1	0	0	1	0	1	0	0	0	6	5.1	6
14	1	0	1	0	1	0	0	0	0	0	0	0	0	0	0	1	1	0	0	1	0	0	0	0	0	6	5.9	5
15	1	0	0	0	1	0	0	0	0	1	0	0	0	0	1	1	0	0	0	1	0	0	0	0	0	6	9.2	5
16	1	0	1	0	0	0	0	0	1	0	0	0	0	0	0	1	1	0	0	1	0	0	0	0	0	6	7.3	5
17	1	1	0	0	1	0	0	0	1	0	0	0	0	0	0	1	0	0	0	1	0	0	0	0	0	6	8	5
18	1	0	0	0	0	0	0	0	0	1	0	0	0	0	0	1	1	0	0	1	0	1	0	0	0	6	7.4	4
19	1	0	0	0	0	0	0	1	0	0	0	0	0	0	0	1	1	0	0	1	0	1	0	0	0	6	7.5	4
20	1	0	0	0	0	0	0	0	1	0	0	0	0	0	0	1	0	0	0	1	1	0	0	0	1	6	14	4
21	1	0	0	0	0	0	1	0	0	0	0	0	0	0	0	1	1	0	0	1	0	1	0	0	0	6	7.7	4
22	1	1	0	0	0	0	0	0	0	0	0	0	0	0	0	1	0	0	0	1	0	1	0	0	0	5	9.7	4
23	1	1	0	0	1	0	0	0	0	0	0	0	0	0	0	1	1	0	0	0	0	1	0	0	0	6	9.3	4
24	1	0	0	0	0	0	0	0	0	0	0	0	0	1	0	1	1	0	0	1	0	0	0	0	1	6	9.1	4
25	1	0	0	0	0	0	1	0	0	0	0	0	0	1	1	1	0	0	0	1	0	0	0	0	0	6	12	4

However as the list in their Table 6-10 shows, there are many competing models with similar values for $\bar{R}^2$.

We have carried out the same analysis as in the epoxide bond example. We used the "One model per sample" method to generate a set of promising models, with $B = 500$, a critical p-value of $\pi_a = 0.083$, and with no limit to the number of factors that could be included in the model fitted to each BS sample. This led to the generation of a set of just 189 promising models at the end of Step (1). The C_p versus p plot of these models fitted to the original data is shown Fig. 4. From this we took the "unbiased min p" as $p_0 = 13$.

As in the epoxide bond example, to increase the number of promising models with $p \leq p_0$ that we examine, we used the "Many models per sample" method to generate a fresh set of promising models only with p restricted to being no greater than $p_0 = 13$, but still with $\pi_a = 0.083$ and $B = 500$. This gave a set 598 promising models. We then used the "Bootstrap Assessment of Selected Best Model" method of Section 4.3 to fit all the promising models to each BS sample, selecting the "min C_p" model for each sample, as the "best" model for that sample. This yielded just 74 "best" models. The C_p versus p plot of these models when fitted to the original data is shown in Fig. 5. The top 25 models are listed in Table 2 with their original C_p values as well as the number of times they were selected as the best model in the BS samples.

The most frequently selected model was

$$X_1\ X_2\ X_3\ D_2\ D_4\ D_6\ D_7\ D_8\ D_9\ D_{10}\ D_{11}\ t\ t^2\ .$$

This was selected as the best model in 75 of the 500 BS samples. This was also the model with the smallest C_p amongst all 74 "best" models when fitted to the original data.

The stepwise regression model

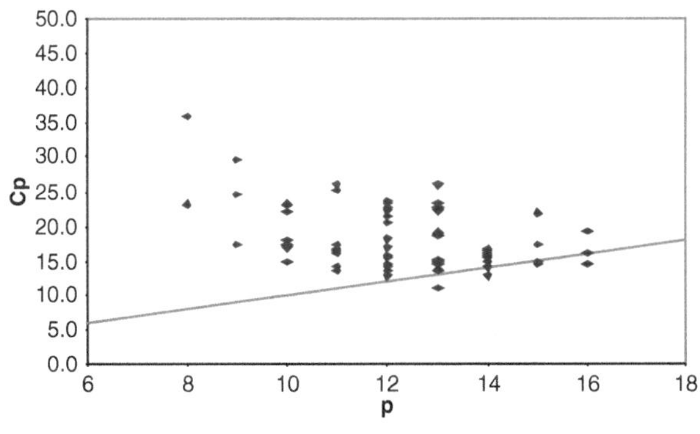

Fig. 4 C_p versus p plot of 320 promising models found for the Bank data using the "One model per sample" BS method. The C_p values are those obtained when the promising models are fitted to the original sample

Table 2 Bank data, top 25 of final selection of 235 "best" models

	Mean	AAA	3to4	D3to4	D1	D2	D3	D4	D5	D6	D7	D8	D9	D10	D11	t	t^2	t^3	# b's	Cp	Freq
Pval	0.66	0.030	0.000	0.003	0.611	0.000	0.169	0.002	0.172	0.002	0.000	0.000	0.030	0.007	0.116	0.729	0.834	0.992			
1	0	1	1	1	0	1	0	1	0	1	1	1	1	1	1	1	1	0	13	11	75
2	0	1	1	1	1	1	1	1	1	1	1	1	1	1	0	0	0	0	13	15	39
3	1	1	1	1	0	1	0	1	0	1	1	1	1	1	1	0	0	1	13	12	35
4	0	1	1	1	0	1	0	1	0	1	1	1	0	1	1	1	1	0	12	13	25
5	1	0	1	1	0	1	0	1	0	1	1	1	1	1	1	0	1	1	13	15	25
6	0	1	1	1	0	1	1	1	1	1	1	1	1	1	1	0	0	0	13	14	24
7	0	1	1	1	0	1	0	1	0	1	1	1	0	1	1	0	1	1	12	13	19
8	0	1	1	1	0	1	1	1	1	1	1	1	1	1	0	0	1	0	13	18	14
9	0	1	1	1	0	1	0	1	0	1	1	1	1	1	1	0	0	0	11	14	14
10	1	1	1	1	0	1	0	1	0	1	1	1	1	0	1	0	0	1	12	16	13
11	1	1	1	1	0	1	0	1	1	1	1	1	1	1	1	0	0	0	13	14	12
12	0	1	1	0	0	1	1	1	1	1	1	1	1	1	0	0	1	1	13	23	11
13	1	1	1	1	0	1	1	1	0	1	1	1	0	1	1	0	0	1	13	15	10
14	0	1	1	1	0	1	0	1	1	1	1	1	1	1	0	0	1	1	13	19	9
15	0	1	1	1	0	1	1	1	0	1	1	1	1	1	0	1	1	0	13	18	9
16	1	1	1	1	0	1	1	1	0	1	1	1	1	1	1	0	0	0	13	15	8
17	1	1	1	1	0	1	0	1	0	1	1	1	0	1	1	0	0	0	11	14	8
18	0	1	1	1	0	1	0	1	0	1	1	1	0	0	1	0	0	0	9	17	7
19	0	1	1	1	0	1	1	1	1	1	1	1	1	1	0	0	0	0	12	17	7
20	0	1	1	1	1	1	0	1	0	1	1	1	0	0	1	0	0	0	10	18	6
21	0	1	1	1	0	1	0	1	0	1	1	1	0	1	1	0	0	0	10	15	6
22	0	1	1	1	0	1	0	1	1	1	1	1	1	1	0	1	1	0	13	19	6
23	0	1	1	1	0	1	0	1	1	1	1	1	0	1	1	0	0	0	11	16	6
24	1	1	1	1	0	1	1	1	1	1	1	1	1	1	0	0	0	0	13	17	6
25	1	1	1	1	0	1	0	1	0	1	1	1	0	0	1	0	0	0	10	17	5

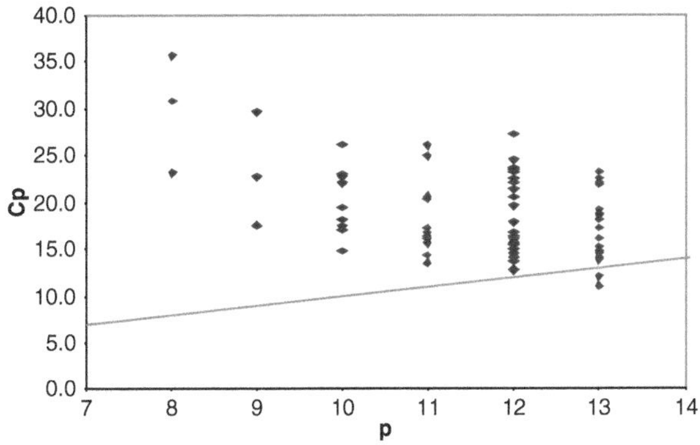

Fig. 5 C_p versus p for the 74 "best" models found by the Bootstrap Quality Assessment Method described in Section 4.3. in the Bank Data Example. The C_p values are those when the models are fitted to the original data

$$X_0\ X_1\ X_2\ X_3\ D_2\ D_4\ D_6\ D_7\ D_8\ D_9\ D_{10}\ D_{11}\ t^3$$

was the third most frequently selected, being selected 35 times.

The results suggest that it is not very important whether the mean is fitted or not. In fact, when the full model is fitted to the original sample, the p-value for the mean is 0.66, showing that the general mean is not at all close to being statistically significantly different from zero for the original data.

For all the 74 "best" models that were selected, the three main explanatory variables X_1 (AAA), X_2 (3-4), X_3 D(3-4) were clearly important, as were the seasonal variables D_2, D_4, D_6, D_7, D_8 and of the others D_9, D_{10} and D_{11} seemed marginally less important. The remaining three D_1, D_3, D_5 did not seem very important.

It seemed worth including a time variable, but it is unclear if any one of them is to be preferred given the rather random way that different time variables appear in the different models; this is similar to variations listed in Table 6-10 of Makridakis et al. (1998).

Though the details are a little different, in broad terms the BS results are very similar to the results reported by Makridakis et al.

Finally it is interesting to see how the C_p values of the top-performing models in Table 2 varied across all 500 BS samples to which they were fitted. Figure 6 shows the empirical distribution function of the sample of 500 C_p values for the top 5 models. The result shows the inherent variability of the statistic for data of this type.

6 Conclusions

We have discussed how bootstrapping can be used to analyse the selection and fitting of linear models in multiple regression. We have shown how bootstrapping can be used for two purposes.

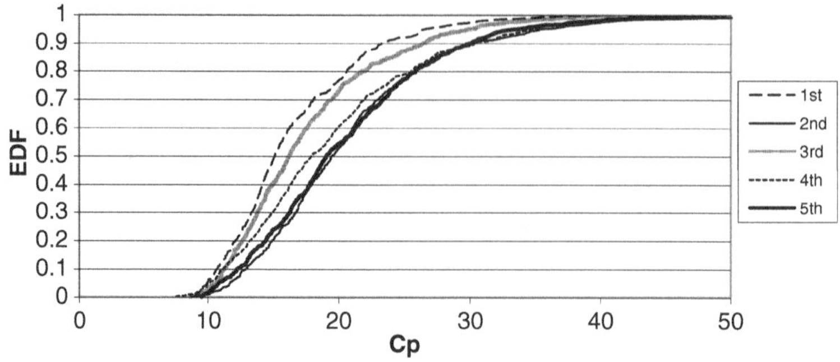

Fig. 6 Empirical Distribution Functions of the C_p value of the top 5 "best" models when fitted to 500 BS samples of the Bank Data

First it can be used to identify promising models out of the set of 2^P possible models. The "One model per sample" method yields just one model per BS sample, so that the largest number of possible models is B, the number of BS samples generated, though because of duplication, the number of distinct models (i.e., whose subsets of factors are different) is likely to be rather smaller. The "Many models per sample" method produces a maximum of BP models, though again duplication means the number of distinct models is usually significantly smaller.

The way that the set of promising models is constructed means that models with a small C_p value are likely to be identified, as is borne out in the two numerical examples. Thus bootstrapping seems attractive in enabling promising models to be tractably identified out of the full set of all possible models when the number of factors is large.

The bootstrapping also allows an assessment to be made of how stable the models estimated as being the best, or a good fit to the original data, actually are, in the sense of seeing how often that model is selected as being the best when a large number of promising models are fitted to a number of BS samples with the same form as the original data. Such information is not available using a standard best subset analysis or a stepwise regression analysis.

An Excel workbook implementing both bootstrap methods is available at `http://www.personal.soton.ac.uk/rchc/BestLinModel.htm`.

References

Akaike, H. 1970. Statistical predictor identification. *Ann. Inst. Statist. Math.* 22:203–217.

Cheng, R. C. H. 2008. Selecting the best linear simulation metamodel. In *Proceedings of the 2008 Winter Simulation Conference*, eds. S. J. Mason, R. R. Hill, L. Mönch, O. Rose, T. Jefferson, and J. W. Fowler, pp. 371–378. Piscataway, New Jersey: Institute of Electrical and Electronics Engineers. Available online via `www.informs-sim.org/wsc08papers/043.pdf` [accessed December 29, 2008].

Davison, A. C., and D. V. Hinkley. 1997. *Bootstrap Methods and Their Application.* Cambridge: Cambridge University Press.

Fishman, G. S. 2006. *A First Course in Monte Carlo.* Australia: Thomson Brooks/Cole.
Freedman, D. A. 1981. Bootstrapping regression models. *Ann. Statist.* 9:1218–1228.
Hald, A. 1952. *Statistical Theory with Engineering Applications.* New York: Wiley.
Krzanowski, W. J. 1998. *An Introduction to Statistical Modelling.* London: Arnold.
Makridakis, S., S. C. Wheelwright, and R. J. Hyndman. 1998. *Forecasting Methods and Applications*, 3rd edition New York: Wiley.
Mallows, C. L. 1973. Some comments on C_p. *Technometrics* 15:661–675.
Mallows, C. L. 1995. More comments on C_p. *Technometrics* 37:362–372.
Nishii, R. 1984. Asymptotic properties of criteria for selection of variables in multiple regression. *The Annals of Statistics* 12:758–765.
Searle, S. R. 1971. *Linear Models.* New York: Wiley.
Shibata, R. 1981. An optimal selection of regression variables. *Biometrika* 68:45–54.
Williams, K. R. 1968. Designed experiments. *Rubber Age* 100:65–71.
Wu, C. F. J., and M. Hamada. 2000. *Experiments: Planning, Analysis, and Parameter Design Optimization.* New York: Wiley.

Patchwork Distributions

Soumyadip Ghosh and Shane G. Henderson

Abstract Patchwork distributions are a class of distributions for use in simulation that can be used to model finite-dimensional random vectors with given marginal distributions and dependence properties. They are an extension of the previously developed chessboard distributions. We show how patchwork distributions can be selected to match several user-specified properties of the joint distribution. In constructing a patchwork distribution, one must solve a linear program that is potentially large. We develop results that shed light on the size of the linear program that one must solve. These results suggest that patchwork distributions should only be used to model random vectors with low dimension, say less than or equal to 5.

1 Introduction

Is there a part of stochastic simulation that George Fishman has not contributed to? If so, it is well hidden! His breadth of work, when multiplied by the length of time that he has been a major force, give rise to a very large area of contributions to the field. (Our apologies to the dimensional analysts that are trying to make sense of the last sentence.) So it is indeed an honour and a privilege to contribute to this volume in George Fishman's honour. Our contribution is in the area of input modeling. The one-dimensional case is well understood—see, for example, Fishman (2001, Chapter 10). But when we turn to higher dimensions, the situation is far less satisfactory.

The key issue is *statistical dependence* and its impact on performance measures. Indeed, much effort has been devoted to this problem in recent times. Recent applications include generating test problems for numerical algorithms (Hill and Reilly 2000), cost analysis (Lurie and Goldberg 1998), crop insurance pricing (Nelson 2004), and arrival process modeling (Avramidis et al. 2004). There are many classes of distributions that can be used to model (finite-dimensional) random vectors with given properties. For surveys, see Devroye (1986), Johnson (1987), and

S. Ghosh (✉)
IBM T. J. Watson Research Center, Yorktown Heights, NY 10598, USA
e-mail: ghoshs@us.ibm.com

C. Alexopoulos et al. (eds.), *Advancing the Frontiers of Simulation: A Festschrift in Honor of George Samuel Fishman*, International Series in Operations Research & Management Science 133, DOI 10.1007/b110059_4,

Biller and Ghosh (2004, 2006), and for a discussion of the modeling process, see Henderson (2005). In this paper we develop a new class of distributions of random vectors that we call *patchwork distributions*.

Relative to other methods, patchwork distributions have the desirable property that they afford considerable flexibility to the modeler in their ability to match properties of the distribution of the random vector. In particular, they can simultaneously match the marginal distributions of the components of the random vector, the covariance matrix, and the probability that the random vector lies within certain regions. This flexibility comes at a price: As we shall see, it can be computationally difficult to construct the desired distribution when the random vector has moderate to large dimension. Therefore, practically speaking, patchwork distributions are limited to low dimensions, say 5 or less.

Patchwork distributions are an extension of a certain class of distributions known as chessboard distributions (Ghosh and Henderson 2001, 2002), or "piecewise-uniform copulae" (Mackenzie 1994). They are constructed as follows. One assumes that the desired random vector $X = (X_1, \ldots, X_d)$ is given by $(F_1^{-1}(U_1), \ldots, F_d^{-1}(U_d))$, where F_i is the desired marginal distribution function of X_i, and U_i is a uniform random variable on $(0, 1)$. The problem then reduces to constructing the joint distribution of $(U_1, \ldots, U_d)$ on $(0, 1)^d$. (The joint distribution of $U = (U_1, \ldots, U_d)$ is known as a *copula*, and this approach of first generating U, and then constructing X from U using inversion is very common. The term "copula" was coined in Sklar 1959. See, e.g., Nelsen 1999 for background on copulas.) We break the unit cube $(0, 1)^d$ down into a grid of cells. Each cell is a hypercube with faces aligned with the axes. The conditional distribution of U given that it lies in one of the cells has a distribution with uniform marginals, which can vary from cell to cell. We call the distribution of X a *patchwork distribution*, and the distribution of the corresponding uniform random vector U a *patchwork copula*.

It is useful to allow cell-specific base copulas. For example, this allows patchwork distributions to match the extreme behaviour of X when all components are likely to move jointly. See, e.g., the discussion of tail dependence in Biller (2009).

Patchwork distributions generalize chessboard distributions, which have conditional (joint) uniform distributions, given that they lie in a fixed cell. The conditional distributions in the cells are fixed in advance heuristically, using any prior information about the joint distribution, and one then determines the vector giving the probabilities that U lies in each of the cells. This probability vector can be found by solving a certain linear program, the constraints of which reflect the desired properties of the joint distribution.

A natural property of a joint distribution that one might attempt to match is the correlation matrix, where the correlations could be Pearson product-moment, Spearman rank, or Kendall's tau correlations. We focus on Spearman rank correlations. As we will see, some rank correlation matrices are easier to match than others, in the sense that the size of the linear program that needs to be solved depends on the correlation matrix. Therefore, the computational effort required to construct a patchwork distribution is related to the correlation matrix.

The primary contributions of this paper are:

1. a characterization of the set of correlation matrices that patchwork distributions can match,
2. a theoretical analysis that relates the computational effort (size of the linear program) to a given correlation matrix, and
3. a computational study to shed further light on the results of the theoretical analysis.

The remainder of this chapter is organized as follows. In Section 2 we define patchwork distributions more carefully, and describe how one can select them to match the desired properties of X. We also describe how they can be generated. Then, in Section 3, we extend known results on the modeling power of chessboard distributions to patchwork distributions. Sections 4 and 5 contain, respectively, our theoretical and computational results on the size of the linear programs that must be solved to match a given correlation matrix. Finally, we offer some concluding remarks in Section 6.

2 Patchwork Distributions

In this section we define patchwork distributions, explain how they are constructed, and describe how to generate samples from them. For notational simplicity we mostly confine the discussion to 3-dimensional random vectors, but analogous results hold for $d \geq 2$ dimensions. Many of these results were proved for the special case of chessboard distributions in Ghosh and Henderson (2002). We give proofs of some of the extensions, even though they are usually similar to the special case of chessboard distributions, because they are helpful in understanding the structure of patchwork distributions.

We say that $X = (X_1, X_2, X_3)$ has a *patchwork distribution* if

$$X \stackrel{D}{=} (F_i^{-1}(U_i) : i = 1, 2, 3),$$

where F_i is the marginal distribution function of X_i, $i = 1, 2, 3$, and the distribution of $U = (U_1, U_2, U_3)$ is a *patchwork copula*, as described below.

Let $n \geq 1$, and divide $(0, 1]^3$ into a grid of n^3 equal-sized cubes (cells) with sides parallel to the coordinate axes. Let the cells be given by $C(j_1, j_2, j_3)$, with $j_1, j_2, j_3 = 1, \ldots, n$, so that

$$C(j_1, j_2, j_3) = \left\{(x_1, x_2, x_3) : \frac{j_i - 1}{n} < x_i \leq \frac{j_i}{n}, i = 1, 2, 3\right\}.$$

Conditional on lying in cell $C(j_1, j_2, j_3)$, U follows an appropriately scaled and translated version of a copula $\mathcal{C}(j_1, j_2, j_3)$, which can vary by cell. We call this copula the (j_1, j_2, j_3) *base* copula. To be more precise, let

$$(Z(j_1, j_2, j_3) : 1 \leq j_1, j_2, j_3 \leq n)$$

be independent random vectors where $Z(j_1, j_2, j_3)$ is distributed as $\mathcal{C}(j_1, j_2, j_3)$. Then, conditional on being in cell $C(j_1, j_2, j_3)$, the vector U is defined by

$$U_i = \frac{Z_i(j_1, j_2, j_3)}{n} + \frac{j_i - 1}{n}, \quad i = 1, 2, 3. \tag{1}$$

We allow the mass of each cell to vary. Let

$$q(j_1, j_2, j_3) = P(U \in C(j_1, j_2, j_3))$$

be the mass assigned to cell $C(j_1, j_2, j_3)$. We require that the $q(j_1, j_2, j_3)$s satisfy

$$\begin{aligned}
\sum_{j_2, j_3=1}^{n} q(j_1, j_2, j_3) &= P\left(U_1 \in \left(\frac{j_1 - 1}{n}, \frac{j_1}{n}\right]\right) = \frac{1}{n}, \quad \forall j_1 = 1, \ldots, n,\\
\sum_{j_1, j_3=1}^{n} q(j_1, j_2, j_3) &= P\left(U_2 \in \left(\frac{j_2 - 1}{n}, \frac{j_2}{n}\right]\right) = \frac{1}{n}, \quad \forall j_2 = 1, \ldots, n, \qquad (2)\\
\sum_{j_1, j_2=1}^{n} q(j_1, j_2, j_3) &= P\left(U_3 \in \left(\frac{j_3 - 1}{n}, \frac{j_3}{n}\right]\right) = \frac{1}{n}, \quad \forall j_3 = 1, \ldots, n,\\
q(j_1, j_2, j_3) &\geq 0 \quad \forall j_1, j_2, j_3 = 1, \ldots, n.
\end{aligned}$$

We call the distribution of U, as constructed above, a *patchwork copula*. Theorem 1 below proves that the distribution is indeed a copula, and therefore that X has the desired marginal distributions.

Theorem 1 *If U is constructed as above, with cell probabilities q satisfying the constraints* (2), *then U has uniform marginals. Consequently, X has the desired marginals.*

Proof Let the marginal distribution function of U_i be denoted by $G_i(\cdot)$. We show that $G_1(x) = x$ for $x \in (0, 1]$, and the proof for dimensions 2 and 3 is exactly the same. We rely on the conditional relationship (1). For any $x \in (i - 1, i]/n$, we have that

$$\begin{aligned}
G_1(x) = &\sum_{j_1 \leq i-1} \sum_{j_2, j_3=1}^{n} q(j_1, j_2, j_3) + \\
&\sum_{j_2, j_3=1}^{n} P\big((i - 1)/n < U_1 \leq x | U \in C(i, j_2, j_3)\big) q(i, j_2, j_3)\\
= &\frac{i - 1}{n} + \sum_{j_2, j_3=1}^{n} P\big(0 < Z_1(i, j_2, j_3) \leq n(x - (i - 1)/n)\big) q(i, j_2, j_3)
\end{aligned}$$

$$= \frac{i-1}{n} + n\left(x - \frac{i-1}{n}\right) \sum_{j_2, j_3=1}^{n} q(i, j_2, j_3)$$

$$= \frac{i-1}{n} + x - \frac{i-1}{n} = x$$

as required. Hence, U has uniform marginals. Since X_i is obtained from U_i via the probability integral transform, it follows that X_i has the desired marginal distribution, $i = 1, 2, 3$. □

Remark 1 Chessboard copulas, as introduced in Ghosh and Henderson (2002), are patchwork copulas where all the base copulas are derived from independent uniform random variables. They coincide with the "piecewise-uniform copulae" developed by Mackenzie (1994). Cloned distributions (Johnson and Kotz 2004) are bivariate patchwork copulas where the base copula is the same for all cells, and all cells have the same probability.

The constraints (2) are sufficient to ensure that U has uniform marginals. So long as those constraints hold, we are then free to choose the cell probabilities to match other desired properties of the joint distribution. Covariance is one such property.

We believe that for non-Gaussian marginals, it is usually more appropriate to use rank covariance than product-moment covariance as a measure of dependence. Recall that the rank covariance between two random variables X_1 and X_2 with distribution functions F_1 and F_2 respectively is given by $E[F_1(X_1)F_2(X_2)] - E[F_1(X_1)]E[F_2(X_2)]$. Our preference for rank covariance over product-moment covariance stems from the facts that rank covariance is always well defined, irrespective of whether the X_is have finite second moments or not, and that rank covariance is invariant to strictly increasing transformations of the random variables. In the case where F_i is continuous, $F_i(X_i)$ is uniformly distributed on (0, 1]. Indeed, if X_1 and X_2 are components of a patchwork random vector with continuous marginal distribution functions, then the rank covariance between X_1 and X_2 equals the product-moment covariance between U_1 and U_2, from which X was constructed. Hence, we can reduce a study of rank covariance of patchwork distributions with arbitrary continuous marginals to one of product-moment covariance of uniform random variables on (0, 1] (rank and product-moment covariances coincide for uniform marginals).

Remark 2 When some of the marginal distributions are not continuous, this convenient relationship does not hold, and one must then attempt to match the desired correlations using more-complicated methods; see Avramidis et al. (2009).

We need an expression for the product-moment covariance of two components of a patchwork copula. Let U be distributed according to the patchwork copula. Then

$$\begin{aligned}\Sigma_{12}^U = \mathrm{Cov}(U_1, U_2) &= E[U_1 U_2] - 1/4 \\ &= \sum_{j_1, j_2, j_3} q(j_1, j_2, j_3) E[U_1 U_2 | U \in C(j_1, j_2, j_3)] - 1/4 \\ &= \sum_{j_1, j_2, j_3} q(j_1, j_2, j_3) \mu_{12}(j_1, j_2, j_3) - 1/4, \end{aligned} \tag{3}$$

where

$$\mu_{12}(j_1, j_2, j_3) = E[U_1 U_2 | U \in C(j_1, j_2, j_3)].$$

The μ terms are constants that depend on the base copulas, but not on q. It follows that the covariance between U_i and U_j is a linear function of q, for each $i, j = 1, \ldots, n$.

Suppose now that we want to match the true covariance matrix Σ^U to a desired covariance matrix Σ. The diagonal terms are all equal to $1/12$, and covariance matrices are symmetric, so we can measure the error $r(\Sigma^U, \Sigma)$ as

$$r(\Sigma^U, \Sigma) = \sum_{1 \le i < j \le 3} \left| \Sigma_{ij}^U - \Sigma_{ij} \right|.$$

It immediately follows that we can attempt to match Σ using the linear program

$$\min \quad \sum_{i=1}^{2} \sum_{j=i+1}^{3} (z_{ij}^+ + z_{ij}^-) \tag{4}$$

$$\text{subject to } \Sigma_{ij}^U - \Sigma_{ij} = z_{ij}^+ - z_{ij}^-, \; i = 1{,}2 \text{ and } j = i + 1 \text{ to } 3$$

$$z_{ij}^+ \ge 0, z_{ij}^- \ge 0, \text{ together with constraints (2) and (3).}$$

Remark 3 The linear program (4) is always feasible since $q(j_1, j_2, j_3) = n^{-3}$, for all j_1, j_2, j_3, is feasible. Also, the objective function is bounded below by 0, so an optimal solution exists. If the optimal objective value is 0, then we exactly match the desired properties.

Clemen et al. (2000) discussed a number of other properties that one might elicit from users about joint distributions and therefore want to match. Several of these are easily matched when X has a patchwork distribution. Ghosh and Henderson (2001) described how to match such properties using chessboard distributions, and the methods extend to patchwork distributions. For example, probabilities of the form $P(X \in A)$ for various regions A can be expressed as linear functions of q, and therefore can be matched using linear programming as above. The set A does not have to be rectilinear. Similarly, conditional fractiles can be matched using linear programming and concordance probabilities can be matched using quadratic programming.

It is relatively straightforward to generate random vectors that have a patchwork distribution. The basic procedure consists of the following steps:

1. Generate the (random) index (J_1, J_2, J_3) of the cell $C(J_1, J_2, J_3)$ containing the uniform random vector U from the discrete distribution formed by the $q(j_1, j_2, j_3)$s. With some preprocessing, it is possible to do this in constant

time using, e.g., the alias method (Walker 1977). The following description is adapted from Law and Kelton (2000):

- **Alias Method Setup:** Two arrays are calculated from the qs. The array $(AC_j : j = 1, \ldots, n^3)$ contains *cutoff values*, and the array $(AA_j : j = 1, \ldots, n^3)$ contains *aliases*. These arrays can be computed as follows; see Kronmal and Peterson (1979).

  a. Set $AC_j = n^3 q(j_1, j_2, j_3)\ \forall j = 1, \ldots, n^3$, where the index j represents the cell $C(j_1, j_2, j_3)$ via $j = (j_1 - 1)n^2 + (j_2 - 1)n + j_3$.
  b. Define sets $TG = \{j : AC_j \geq 1\}$ and $TS = \{j : AC_j < 1\}$.
  c. Do the following steps until TS becomes empty:

     i. Remove an element k from TG and remove an element m from TS.
     ii. Set $AA_m = k$ and replace AC_k by $AC_k - 1 + AC_m$.
     iii. If $AC_k < 1$, put k into TS; otherwise put k back in TG.

- **Generating Cells:** Once the arrays AA and AC have been calculated, the random cell index (J_1, J_2, J_3) (and equivalently the index J) can be generated as follows:

  a. Generate I from the discrete uniform distribution over $\{1, \ldots, n^3\}$ and $U \sim U(0, 1)$ independent of I.
  b. If $U \leq AC_I$, return $J = I$. Otherwise, return $J = AA_I$.

2. Generate U conditional on $U \in C(J_1, J_2, J_3)$ via (1). Here we need to be able to generate random vectors from the base copula $\mathcal{C}(J_1, J_2, J_3)$, but since we can select the base copula, this should present little difficulty.
3. Generate the components of X via $X_i = F_i^{-1}(U_i)$.

If q is an extreme-point solution to the d-dimensional version of the linear program (4), then there are on the order of nd strictly positive cell probabilities. The exact number of positive values depends on the number of equality constraints in the LP and the degree to which the extreme-point solution is degenerate. On the other hand, there are n^d cells. Therefore, for large d, the fraction of cells receiving positive mass is quite small.

The fact that nd is small relative to n^d can be viewed as an advantage with respect to variate generation since it reduces the setup time required to implement the alias method. However, it can also be viewed as a disadvantage. As the dimension d increases, the fraction of cells receiving positive probabilities is vanishingly small. This means that the set of values that the random vector X can assume is somewhat limited, and so the distributions take a nonintuitive form. As more constraints are added due to the need to match more distributional properties, the problem severity is reduced, but it still remains. Mackenzie (1994) avoids this problem by maximizing the entropy of the discrete distribution q. In this case, all of the cells receive positive probability. However, the problem of maximizing the entropy of q subject to linear constraints is a convex optimization problem that is more difficult to solve than the LPs discussed above. A computationally attractive alternative is to place a lower bound on the cell probabilities. We do not discuss this issue further here as it would lead us too far afield.

3 Modeling Power

In this section we focus on matching covariance matrices. We say that a covariance matrix Σ is *feasible* if a copula exists with that covariance matrix. Let Ω denote the set of feasible covariance matrices. (We suppress the dependence on dimension d.) We view Ω as a subset of the vector space $\mathbb{R}^{d(d-1)/2}$ equipped with the usual inner product, because each $\Sigma \in \Omega$ is symmetric, the elements on the diagonal are all equal to $1/12$, and there are $d(d-1)/2$ elements above the diagonal. (It is therefore also a subset of $[-1/12, 1/12]^{d(d-1)/2}$.) In what follows, the notation Σ will represent both the actual covariance matrix and its vector form in Ω.

One might expect that Ω corresponds with the set of symmetric positive semidefinite matrices with diagonal elements equal to $1/12$. For $d \leq 3$, this is correct (Joe 1997), but for $d > 3$ it is not known whether this is the case or not. It *is* known that in any dimension d, Ω is convex, closed, and full-dimensional (Ghosh and Henderson 2002).

We are now ready to state some results about the modeling power of patchwork distributions. The proofs of these results are, in general, similar to the corresponding results for chessboard distributions (Ghosh and Henderson 2002) and so, for the most part, are omitted. We start with the following lemma, the proof of which is needed later in this paper.

Lemma 1 *Suppose that $\Sigma \in \Omega$. Then the optimal objective value of the linear program* (4) *is at most $d(d-1)/n$.*

Proof Since $\Sigma \in \Omega$, there exists a random vector V with uniform marginals and covariance matrix Σ. We modify the distribution of V as follows. We keep the total mass within each cell constant, but we modify the distribution of V within each cell to conform with the corresponding base copula. This process yields a patchwork copula corresponding to a random vector U, say. The cell probabilities (the qs) for U (which are the same as those for V) constitute a feasible solution to the linear program (4). Furthermore, we can bound the differences in the covariance matrices of V and U, as detailed below. These bounds translate into a bound on the objective value of the solution q. Since the optimal solution of the linear program can do no worse, we obtain a bound on the optimal objective value, thereby proving the result.

For now, assume that $d = 3$. Let $q(j_1, j_2, j_3) = P(V \in C(j_1, j_2, j_3))$ and note that

$$
\begin{aligned}
&\mathrm{Cov}(U_1, U_2) - \Sigma_{12} \\
&= E[U_1 U_2] - E[V_1 V_2] \\
&= \sum_{j_1, j_2, j_3 = 1}^{n} \big(\mu_{12}(j_1, j_2, j_3) - E[V_1 V_2 | V \in C(j_1, j_2, j_3)]\big) q(j_1, j_2, j_3). \qquad (5)
\end{aligned}
$$

But

$$\frac{j_1 - 1}{n}\frac{j_2 - 1}{n} \le E[V_1 V_2 | V \in C(j_1, j_2, j_3)] \le \frac{j_1}{n}\frac{j_2}{n}. \tag{6}$$

Combining (5) with (6) we see that

$$\begin{aligned} &\operatorname{Cov}(U_1, U_2) - \Sigma_{12} \\ &\le \sum_{j_1, j_2, j_3 = 1}^{n} q(j_1, j_2, j_3) \left(\mu_{12}(j_1, j_2, j_3) - \frac{(j_1 - 1)(j_2 - 1)}{n^2} \right) \end{aligned} \tag{7}$$

and

$$\operatorname{Cov}(U_1, U_2) - \Sigma_{12} \ge \sum_{j_1, j_2, j_3 = 1}^{n} q(j_1, j_2, j_3) \left(\mu_{12}(j_1, j_2, j_3) - \frac{j_1 j_2}{n^2} \right). \tag{8}$$

These bounds will prove useful later, but for now we obtain more-explicit bounds. The bounds (6) also apply to $\mu_{12}(j_1, j_2, j_3)$ and so from (7),

$$\begin{aligned} \operatorname{Cov}(U_1, U_2) - \Sigma_{12} &\le \sum_{j_1, j_2, j_3 = 1}^{n} q(j_1, j_2, j_3) \left(\frac{j_1 j_2}{n^2} - \frac{(j_1 - 1)(j_2 - 1)}{n^2} \right) \\ &= n^{-2} \sum_{j_1, j_2, j_3 = 1}^{n} q(j_1, j_2, j_3)(j_1 + j_2 - 1) \\ &\le n^{-2} \sum_{j_1, j_2, j_3 = 1}^{n} q(j_1, j_2, j_3)(2n - 1) \\ &= \frac{2n - 1}{n^2}. \end{aligned}$$

A lower bound follows similarly, so that

$$|\operatorname{Cov}(U_1, U_2) - \Sigma_{12}| \le \frac{2n - 1}{n^2}. \tag{9}$$

The bound (9) was derived assuming $d = 3$, but the same argument and bound hold in higher dimensions. Hence, if Σ^U denotes the covariance matrix of U, we have that

$$r(\Sigma^U, \Sigma) \le \frac{d(d - 1)}{2} \frac{2n - 1}{n^2}$$

and the result follows. □

We can now state the main result of this section. Let A° and ∂A denote the interior and boundary of a set A.

Theorem 2 *Patchwork distributions can get arbitrarily close to any $\Sigma \in \Omega$ and can exactly match any $\Sigma \in \Omega^\circ$ (for sufficiently large n), but cannot necessarily exactly match $\Sigma \in \partial\Omega$. Furthermore, $\Sigma \notin \Omega$ iff the optimal objective value of the linear program* (4) *exceeds $d(d-1)/n$ for some $n \geq 1$.*

Proof This result is proved using the bounds given in Lemma 1. Most of the proof (specifically the $\Sigma \in \Omega^\circ$ and the $\Sigma \notin \Omega$ parts) is very similar to corresponding results in Ghosh and Henderson (2002) and so is omitted. All that needs to be shown is that a patchwork distribution may, or may not, be able to exactly match matrices on the boundary $\partial\Omega$. Ghosh and Henderson (2002) showed that chessboard distributions, which are a special case of patchwork distributions, cannot exactly match *any* matrix on the boundary for a finite n. It remains to show that *some* patchwork distributions can match *some* boundary covariance matrix exactly. This is trivially true, for instance, when the base copulas of the patchwork distribution all have a covariance $\Sigma^b \in \partial\Omega$. Then one can exactly match the boundary covariance matrix Σ^b with $n = 1$ using this base copula. □

Theorem 2 establishes that patchwork distributions can exactly match feasible covariance matrices lying in the interior of Ω, and that infeasible covariance matrices can be proved to be infeasible in finite time by the linear program, but that little can be concluded for covariance matrices lying on the boundary of Ω. The boundary matrices are feasible because Ω is closed, but why are they difficult to match?

Theorem 3 below shows that the joint distribution of a copula with a covariance matrix that lies on the boundary of Ω is a rather strange creature! Recall that any copula F can be decomposed into a singular part F_s and an absolutely continuous part F_{ac} with respect to Lebesgue measure restricted to $(0, 1]^3$. (This is simply the Lebesgue Decomposition; e.g., see Billingsley 1995, p. 414.) Thus, $F = F_{ac} + F_s$. Moreover, the absolutely continuous part has a density f_{ac} with respect to Lebesgue measure.

Theorem 3 *Suppose that f_{ac} is defined as above for a distribution F with covariance matrix $\Sigma \in \partial\Omega$. Then, there cannot exist an open set G such that*

$$f_{ac}(x) \geq \phi > 0 \quad \text{a.e. in } G. \tag{10}$$

(Recall that a property holds *almost everywhere (a.e.)* on the set G if it is true for all $x \in G$ except on a subset of Lebesgue measure 0.)

Proof For notational ease we give a proof in the 3-dimensional case. The general case is virtually identical. Suppose such a G exists. We assume, without loss of generality, that $f_{ac}(x) \geq \phi > 0$ for *all* $x \in G$ and not just a.e. (If not, just redefine f_{ac} on the set of measure 0.) Now, we can choose an open ball $B(x, \epsilon)$ within G and an open cubical region C with faces aligned with the axes within $B(x, \epsilon)$ such that the interior of C is non-empty. Split f_{ac} into two parts, f_C and $f_{\bar{C}}$, defined as:

$$f_C(x) = \begin{cases} \phi & x \in C \\ 0 & \text{elsewhere} \end{cases} \quad \text{and} \quad f_{\bar{C}}(x) = \begin{cases} f_{ac}(x) - \phi & x \in C \\ f_{ac}(x) & \text{elsewhere} \end{cases}.$$

Let u and v be the endpoints that define C, so that

$$C = \{(x_1, x_2, x_3) \in (0, 1]^3 : u_i < x_i \le v_i,\ i = 1, 2, 3\}.$$

Divide the region C into 4 (equal size) subregions,

$$C_{ab} = \left\{(x_1, x_2, x_3) \in C : u_1 + (a-1)\frac{v_1 - u_1}{2} < x_1 \le u_1 + a\frac{v_1 - u_1}{2},\right.$$
$$\left. u_2 + (b-1)\frac{v_2 - u_2}{2} < x_2 \le u_2 + b\frac{v_2 - u_2}{2}\right\},$$

for $1 \le a, b \le 2$.

Define a new distribution H from F as follows. The singular parts H_s and F_s coincide, as do the $h_{\bar{C}}$ and $f_{\bar{C}}$ parts, respectively, of the absolutely continuous density. The density h_C takes the value 2ϕ on C_{11} and C_{22}, and 0 on C_{12} and C_{21}. Then it is straightforward to show that H has uniform marginals, that the $(1, 2)$th covariance is strictly increased, and that the other covariances remain unchanged. Alternatively, if h_C takes the value 0 on C_{11} and C_{22}, and 2ϕ on C_{12} and C_{21}, then the covariance strictly decreases.

The argument above could be repeated for each pair of components. Convexity of Ω then implies that Σ must lie in the interior Ω° which is a contradiction, and the proof is complete. □

One consequence of Theorem 3 is that we cannot hope to exactly match covariance matrices on the boundary of Ω if we use a base copula which has a density component that satisfies (10) for some set G. This gives another explanation for why chessboard distributions cannot match covariance matrices on the boundary of Ω.

We have already seen a singular base copula that can exactly match a covariance matrix on the boundary of Ω. We might ask whether a base copula exists that can match all matrices on the boundary of Ω. We do not have a complete answer to this question, but we will shed further light on it in Section 4.

In summary, the import of the results in this section is that patchwork distributions

- can prove that a given covariance matrix is infeasible in finite time,
- can arbitrarily closely approximate any feasible covariance matrix,
- can exactly match any feasible covariance matrix in the interior of the set of feasible covariance matrices, but
- *might not* exactly match any covariance matrix on the boundary of the set of feasible covariance matrices.

4 Modeling Effort: Theoretical Results

In order to use a patchwork distribution, we need to perform the setup steps outlined in Section 2. The main computational bottleneck there is the solution of the linear programming problem. The time required to solve a linear program typically increases with the size of the linear program, which in turn depends on the discretization level n. So it is of interest to see how large n needs to be to match a given covariance matrix for a fixed dimension d of the random vector, and that is the subject of this section and the next one. Here we focus on theoretical analysis, while the next section performs some computational experiments.

We limit ourselves to the case where the patchwork distribution uses the same base copula in all cells, since this makes the arguments more elegant. Let $\Omega^n(\Sigma^b)$ represent the set of covariance matrices that can be matched by patchwork distributions of size n with a base copula that has Σ^b as its covariance matrix. (In many contexts, the argument Σ^b will be clear and hence shall be dropped.) The set Ω^n shares many of the properties of Ω, namely that it is non-empty, closed, convex, and full-dimensional in $\mathbb{R}^{d(d-1)/2}$ (Ghosh and Henderson 2002). We have shown in Theorem 2 that patchwork distributions can achieve any feasible covariance matrix in the interior of Ω for some finite n. Thus, in a sense the sequence $\{\Omega^n, n \geq 1\}$ grows to cover the whole of Ω as $n \to \infty$; we shall establish this rigorously and provide bounds on the rate of convergence in terms of n. Our results show that, roughly speaking, the set Ω^n is smaller than Ω by a factor that is somewhere between $(1 - \kappa_1/n)$ and $(1 - \kappa_2/n^2)$ for some constants κ_1 and κ_2. In order to state these results precisely we need some definitions.

Let $B(x, \epsilon) = \{y : \|x - y\|_2 < \epsilon\}$ be the (open) ϵ−ball centered at x, defined under the l_2 metric on the space $\mathbb{R}^{m(d)}$, where $m(d) = d(d-1)/2$. The ball $B(0, 1)$, the unit open ball centered at the origin, is denoted by B. Thus, $B(x, \epsilon) = x + \epsilon B$, where the notation vM denotes the set $\{vx : x \in M\}$ for any scalar v, and $y + M = \{y + x : x \in M\}$.

We call any compact, convex set with a non-empty interior a *convex body*. The *Minkowski subtraction* set operation on two convex bodies M and N can be defined (Schneider 1993, Chapter 3) as

$$M \sim N \triangleq \{x \in M : x + N \subset M\}.$$

A convex body E is said to be *centered* if it contains the origin as an interior point. Sangwine-Yager (1988) defines, for an $\epsilon > 0$, the ϵth *relative inner parallel body* of a convex body M with respect to a centered convex body E to be $M \sim \epsilon E$.

The families of sets $\mathcal{U}(\epsilon, \Sigma^b)$ and $\mathcal{L}(\epsilon)$ are indexed by ϵ and defined as

$$\mathcal{U}(\epsilon) \triangleq \Omega \sim \epsilon\Omega,$$

$$\mathcal{U}(\epsilon, \Sigma^b) \triangleq \mathcal{U}(\epsilon) + \epsilon\Sigma^b, \quad \text{and} \tag{11}$$

$$\mathcal{L}(\epsilon) \triangleq \Omega \sim \epsilon B. \tag{12}$$

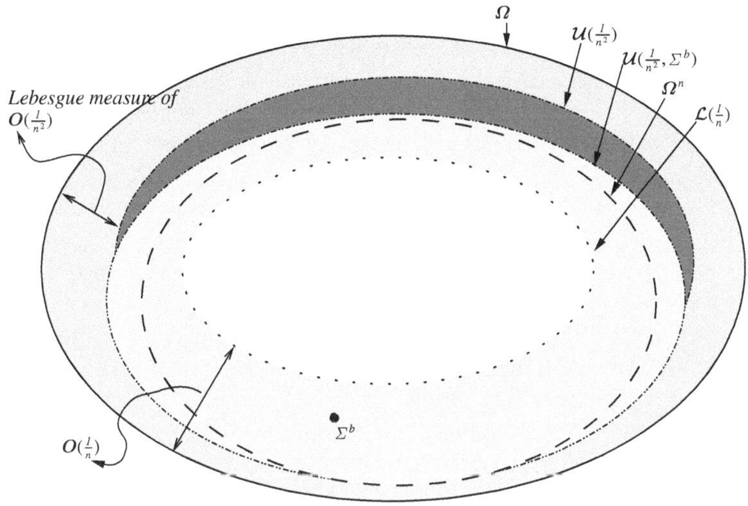

Fig. 1 The sets $\mathcal{U}(\frac{1}{n^2}), \mathcal{U}(\frac{1}{n^2}, \Sigma^b)$, and $\mathcal{L}(\frac{1}{n})$

These definitions are illustrated in Fig. 1.

A matrix z belongs to $\mathcal{U}(\epsilon) \subset \Omega$ if the set $z + \epsilon\Omega$ belongs to Ω. The set $\mathcal{U}(\epsilon)$ has a non-empty interior for all $0 < \epsilon < 1$. The set $\mathcal{U}(\epsilon) + \epsilon\Sigma^b$ is simply the set $\mathcal{U}(\epsilon)$ translated by the matrix $\epsilon\Sigma^b$. Similarly, a matrix z belongs to $\mathcal{L}(\epsilon)$ if the $\epsilon-$ball $B(z, \epsilon) \subset \Omega$. This has a simple interpretation, in that $\mathcal{L}(\epsilon)$ is the subset of points in Ω that are at least an l_2-distance ϵ away from the boundary $\partial\Omega$. Again, the sets $\mathcal{L}(\epsilon)$ can be empty for large ϵ, but are non-empty for sufficiently small $\epsilon > 0$. Note that the lower-bound sets $\mathcal{L}(\epsilon)$ are defined independent of the base covariance Σ^b.

We are now ready to state the main result of this section.

Theorem 4 *Let $\ell = \sqrt{m(d)}$. Then,*

$$a) \qquad \Omega^n(\Sigma^b) \subseteq \mathcal{U}\left(\frac{1}{n^2}, \Sigma^b\right), \text{ and}$$

$$b) \qquad \mathcal{L}\left(\frac{2\ell}{n}\right) \subseteq \Omega^n(\Sigma^b).$$

Theorem 4 establishes that the "gap" between $\Omega^n(\Sigma^b)$ and Ω has a width that is somewhere between $O(n^{-1})$ and $O(n^{-2})$. The following corollary uses that result to obtain bounds on the volume of the set $\Omega^n(\Sigma^b)$ relative to that of Ω. Let $\mathbb{L}$ represent Lebesgue measure on the real vector space $\mathbb{R}^{m(d)}$.

Corollary 1 *There is a constant $K(d)$ that depends on the dimension d such that*

$$\mathbb{L}(\Omega) - \frac{K(d)}{n} \leq \mathbb{L}(\Omega^n(\Sigma^b)) \leq \left(1 + \frac{1}{n^2}\right)^{-m(d)} \mathbb{L}(\Omega).$$

Corollary 1 formalizes the rather imprecise statement we made earlier about the rate at which Ω^n approaches Ω. The rate at which patchwork distributions can cover the set Ω of feasible covariance matrices is at least of the order $1 - K(d)n^{-1}$, but can be no faster than a factor of the order $(1 + n^{-2})^{-m(d)}$ which, in turn, is of the order $1 - m(d)n^{-2}$ when n is large. These results are illustrated in Fig. 1.

We now turn to proving these results.

Proof of Theorem 4(a) For notational ease we prove the result for $d = 3$. The case $d > 3$ is proved similarly. We establish the result by showing that a certain operation on any n-sized patchwork distribution having a covariance matrix $\Sigma \in \Omega^n$ constructs a new distribution with a new covariance matrix in Ω. One can obtain an upper bound on the distance between these matrices, which then gives the result.

Let $\{q(\cdot,\cdot,\cdot)\}$ represent the solution to the LP (4) that exactly matches a covariance matrix $\Sigma \in \Omega^n$. Then

$$\Sigma_{12} = E[U_1U_2] - E[U_1]E[U_2]$$
$$= \sum_{j_1,j_2,j_3=1}^{n} E[U_1U_2|U \in C(j_1,j_2,j_3)]q(j_1,j_2,j_3) - \frac{1}{4}. \tag{13}$$

Let $Z = (Z_1, Z_2, Z_3)$ be a random vector distributed according to the base copula, and let $\Sigma^b \in \Omega$ be its covariance matrix. Since $E[Z_i] = 1/2$, $i = 1, 2, 3$, we see that

$$E[U_1U_2|U \in C(j_1,j_2,j_3)] = E\left[\left(\frac{Z_1}{n} + \frac{j_1 - 1}{n}\right)\left(\frac{Z_2}{n} + \frac{j_2 - 1}{n}\right)\right]$$
$$= \frac{E[Z_1Z_2]}{n^2} + \frac{j_1 + j_2 - 2}{2n^2} + \frac{(j_1 - 1)(j_2 - 1)}{n^2}$$
$$= \frac{E[Z_1Z_2]}{n^2} + t(j_1, j_2), \tag{14}$$

where $t(j_1, j_2)$ is a function only of j_1, j_2, and n.

Suppose now that we replace the base copula in each cell with another copula represented by the random vector Z'. The result is still a valid patchwork copula because of Theorem 1, and represents the distribution of a random vector U', say. If Σ' is the covariance matrix of U', then

$$\Sigma'_{12} = \sum_{j_1,j_2,j_3=1}^{n} \left(\frac{E[Z'_1Z'_2]}{n^2} + t(j_1, j_2)\right) q(j_1,j_2,j_3) - \frac{1}{4}. \tag{15}$$

Let $\Sigma^{b'}$ be the covariance matrix of Z'. The net change in covariance due to the replacement operation is, from (13), (14), and (15),

$$\Sigma'_{12} - \Sigma_{12} = \sum_{j_1,j_2,j_3=1}^{n} \frac{1}{n^2}(E[Z'_1Z'_2] - E[Z_1Z_2])q(j_1,j_2,j_3)$$

$$= \frac{1}{n^2}(\Sigma_{12}^{b'} - \Sigma_{12}^{b}). \tag{16}$$

Equation (16) holds for every component of the covariance matrix. Hence,

$$\Sigma' = \Sigma + \frac{1}{n^2}(\Sigma^{b'} - \Sigma^{b}),$$

and is contained in Ω. We can choose $\Sigma^{b'} \in \Omega$ arbitrarily. Thus,

$$\left(\Sigma + \frac{\Omega}{n^2}\right) - \frac{\Sigma^b}{n^2} \subset \Omega,$$

and we have established that for any $\Sigma \in \Omega^n$, $\Sigma \in \mathcal{U}(n^{-2}, \Sigma^b)$. This gives the result. □

This result is tight in a certain sense. Consider the case where chessboard copulae of size n are used to match a perfectly correlated uniform random vector with pairwise covariances all equal to $1/12$. (This target covariance matrix belongs to $\partial\Omega$.) A chessboard copula can be constructed by equally distributing its mass on the diagonal cells, and all pairwise covariances of this copula are equal to $1/12 - 1/12n^2$. If we perform the transformation described in the proof above, where $\Sigma^{b'}$ is the covariance matrix of a perfectly correlated uniform random vector (so all entries in the covariance matrix are equal to $1/12$), then we obtain the distribution of the perfectly correlated uniform random vector as a result. Thus, we see that Ω^n can have some points in common with the boundary of $\mathcal{U}(n^{-2}, \Sigma^b)$.

We now prove the second part of Theorem 4. First, recall that all norms in a real vector space are equivalent; see, for example, Golub and Van Loan (1996, p. 53). Indeed, for any $x \in \mathbb{R}^{m(d)}$,

$$\|x\|_\infty \leq \|x\|_2 \leq \ell \, \|x\|_\infty. \tag{17}$$

Proof of Theorem 4(b) The result is trivial if $\mathcal{L}(2\ell/n)$ is empty, so assume it is nonempty. The proof of Lemma 1 derived a bound on the optimal objective function of the linear program (4). Specifically, if $\Sigma \in \Omega$ denotes a target covariance matrix and Σ^n is an optimal solution to the linear program then, from (9),

$$|\Sigma(i, j) - \Sigma^n(i, j)| \leq \frac{2}{n} \quad \forall\, 1 \leq i < j \leq 3. \tag{18}$$

Equation (18) shows that we can get within l_∞–distance $2/n$ from any $\Sigma \in \Omega$ using patchwork distributions. From (17), we then have that $\Sigma^n \in B(\Sigma, 2\ell/n)$. Hence, in particular, for any $\Sigma \in \partial\Omega$, we can pick a matrix $\Sigma^n \in \Omega^n$ such that $\Sigma^n \in B(\Sigma, 2\ell/n)$.

Now, suppose the assertion in the theorem is false, and there exists a $\Lambda \in \mathcal{L}(2\ell/n)$ that does not belong to Ω^n. Since Ω^n is convex, the celebrated Separating

Hyperplane Theorem (see, e.g., Luenberger 1969, Theorem 3, Section 5.12) gives us a hyperplane $\mathcal{H}$ through Λ that separates the point Λ from Ω^n.

Consider a line $\mathcal{N}$ passing through Λ that is orthogonal to the hyperplane $\mathcal{H}$. Busemann (1958, Chapter 1) tells us that since Λ is in the interior of Ω, this line intersects the boundary $\partial\Omega$ of the convex set Ω at exactly two points, say Σ^1 and Σ^2. By definition, the point $\Lambda \in \mathcal{L}(2\ell/n)$ does not belong to either of the sets $B(\Sigma^i, 2\ell/n)$, $i = 1, 2$. Thus, $\mathcal{H}$ separates each of the sets $B(\Sigma^i, 2\ell/n)$, $i = 1, 2$, from Λ. Moreover, the sets lie on opposite sides of $\mathcal{H}$, since $\Lambda \in \Omega^\circ$. Thus, at least one ball is separated from Ω^n by the hyperplane $\mathcal{H}$. But this contradicts the earlier observation that one can always choose a point that belongs to Ω^n from each ball $B(\Sigma^i, 2\ell/n)$, $i = 1, 2$. This completes the proof. □

In order to prove Corollary 1, we need the following result. Brannen (1997, Theorem 1) quotes a lower bound from Sangwine-Yager (1988) for the Lebesgue measure of a relative inner parallel body $M \sim \epsilon E$. That result establishes that

$$\mathbb{L}(M \sim \epsilon E) \geq \mathbb{L}(M) - \epsilon S(M; E) + R(m(d), \epsilon),$$

where $S(M; E)$ represents the *relative surface area* of M with respect to E, and the function $R(m(d), \epsilon)$ is nonnegative. They also give conditions under which $S(M; E)$ is finite and positive, and these are satisfied by using the sets Ω and B in the definition of $\mathcal{L}(\epsilon)$ as M and E respectively. Thus, if $\epsilon < 1$, then

$$\mathbb{L}(\mathcal{L}(\epsilon)) \geq \mathbb{L}(\Omega) - k(d)\epsilon \tag{19}$$

for some positive constant $k(d)$ that possibly depends on the dimension $m(d)$ of the sets.

Proof of Corollary 1 From (19), for n large enough that $2\ell/n < 1$,

$$\mathbb{L}(\Omega) - k(d)\left(\frac{2\ell}{n}\right) \leq \mathbb{L}\left(\mathcal{L}\left(\frac{2\ell}{n}\right)\right),$$

where $k(d)$ is a positive value that depends on d. This equation, along with Theorem 4(b), gives the lower bound in the statement of the result with $K(d) = 2k(d)\ell$.

For the upper bound, first note that $\mathcal{U}(n^{-2}, \Sigma^b)$ is a translation of the set $\mathcal{U}(n^{-2})$, and so both sets have the same Lebesgue measure. Also, if $\Lambda \in \mathcal{U}(n^{-2})$, then, by definition, $\Lambda + n^{-2}\Omega \subseteq \Omega$. In particular, $\Lambda + n^{-2}\Lambda \in \Omega$, i.e., $\Lambda \in (1 + n^{-2})^{-1}\Omega$. Hence,

$$\mathcal{U}(n^{-2}) \subseteq (1 + n^{-2})^{-1}\Omega.$$

The Lebesgue measure of the linearly scaled set $(1 + n^{-2})^{-1}\Omega$ is given by $(1 + n^{-2})^{-m(d)}\mathbb{L}(\Omega)$ (see Billingsley 1995, Theorem 12.2). This, along with

Theorem 4(a), establishes the upper bound on the Lebesgue measure of Ω^n and we are done. □

We conclude this section by showing that for any fixed choice of base copula, there will exist covariance matrices in Ω that cannot be exactly matched, no matter how n is chosen. This result shows that it is pointless to attempt to identify a "powerful" base copula that matches all of Ω for some finite n.

Proposition 1 *For any fixed base copula, there exists a covariance matrix $\Sigma \in \Omega$ that cannot be exactly matched for any n.*

Proof On the contrary, suppose that such a base copula exists, and let Σ^b be its covariance matrix. Consider a line $\mathcal{N}$ through Σ^b and the origin. (If Σ^b is equal to the origin, then pick an arbitrary line through the origin.) Since Ω is compact, convex, and the origin is in its interior, this line intersects $\partial\Omega$ at two points. Follow the line from Σ^b through the origin until you reach one of those points. Call that point $\bar{\Sigma}$. By the supposition, a Σ^b-based patchwork copula, of size n, say, can exactly match $\bar{\Sigma}$. Then, by the argument establishing Theorem 4(a), a Σ' of value

$$\Sigma' = \bar{\Sigma} + \frac{\bar{\Sigma} - \Sigma^b}{n^2}$$

can also be achieved by replacing the base copula in each cell with a copula that has covariance matrix $\bar{\Sigma}$. The matrix Σ' is, however, outside Ω, and we have the desired contradiction. □

Whether patchwork copulas with a bounded number of base copulas can match all of Ω for some finite n is an open problem. We conjecture that this is impossible.

5 Modeling Effort: Computational Results

Corollary 1 proves that patchwork distributions with discretization level n can match covariance matrices that are a distance r from the boundary set $\partial\Omega$, where the order of r lies somewhere between n^{-1} and n^{-2}. In this section we describe a computational study that sheds further light on this rate for the special case of chessboard distributions.

Let $\mathcal{S}$ be a collection of $d(d-1)/2$–dimensional vectors that represent the off-diagonal elements of covariance matrices Σ in $\mathbb{R}^{d\times d}$. Consider rays from the origin through each of these vectors. We determine the rate at which each ray is "covered" by chessboard distributions as the discretization level n grows. We populate the set $\mathcal{S}$ by sampling uniformly from the set of all positive semidefinite matrices of size $d \times d$. (Ghosh and Henderson 2003 provides such a sampler.) This allows us to test whether the rate varies in different regions of Ω.

The origin is in the strict interior of the set of all positive semidefinite (PSD) matrices, which implies that there is a finite maximum value $r^{sd}(\Sigma) > 0$ such that $r\Sigma$ is positive semidefinite for all $r^{sd}(\Sigma) \geq r > 0$. We compute $r^{sd}(\Sigma)$ by formulating and solving a semidefinite program. The set of feasible covariance matrices Ω also contains the origin in its strict interior, and so a finite maximum $r^*(\Sigma)$ exists such that $r\Sigma$ is a feasible covariance matrix for uniform marginals for all $r^*(\Sigma) \geq r > 0$. Finally, let $r^*(n, \Sigma)$ represent the point at which the ray $r\Sigma$ intersects the set Ω^n of all covariance matrices that chessboards of size n can exactly match. Fig. 2 illustrates these definitions.

We numerically determine $r^*(n, \Sigma)$ by solving, for each $\Sigma \in \mathcal{S}$, the LP

$$\begin{aligned} r^*(n, \Sigma) = \ & \max r \\ \text{s.t.} \quad & r\Sigma(i, j) = \mathrm{Cov}(X_i, X_j), \ \forall i < j, \\ & r \geq 0, \end{aligned} \tag{20}$$

along with the constraints (2) that ensure that the distribution of X is a copula. The LPs (20) are feasible ($r = 0$, $q = 1/n^d$ is a feasible solution) and terminate with a positive finite optimal solution $r^*(n, \Sigma)$. The rate of change in $r^*(n, \Sigma)$ along rays $r\Sigma, r > 0, \forall\Sigma \in \mathcal{S}$, provides an indication of the rate at which the set Ω^n covers Ω. To see why, recall that two convex closed bodies that differ in their sizes by a small $\epsilon > 0$ also differ in volume (Lebesgue measure) by the same order. (We used this result in the proof of Corollary 1.)

For the $d = 3$ case, the set Ω is known to coincide with the set of all PSD matrices (e.g., Joe 1997) and thus $r^*(\Sigma) = r^{sd}(\Sigma)$, which facilitates the calculation of the exact coverage rate. Figure 3 plots the relative difference $(r^{sd}(\Sigma) - r^*(n, \Sigma))/r^{sd}(\Sigma)$ against n in the log-log scale for 20 different values of $\Sigma \in \mathcal{S}$ and n taking values up to 128. (Here, and in what follows, in the interest of notational

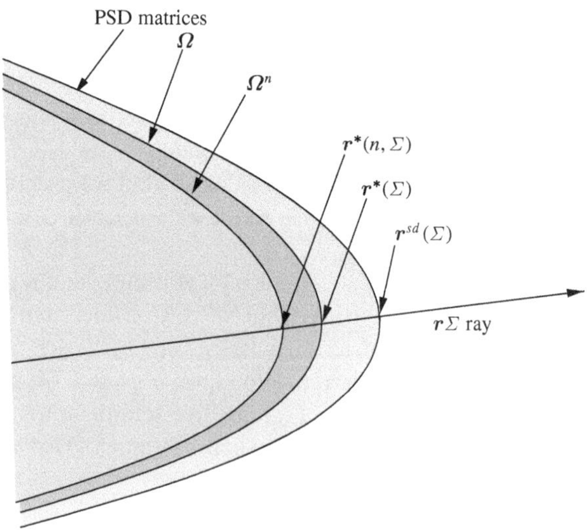

Fig. 2 The points $r^{sd}(\Sigma)$, $r^*(\Sigma)$, and $r^*(n, \Sigma)$

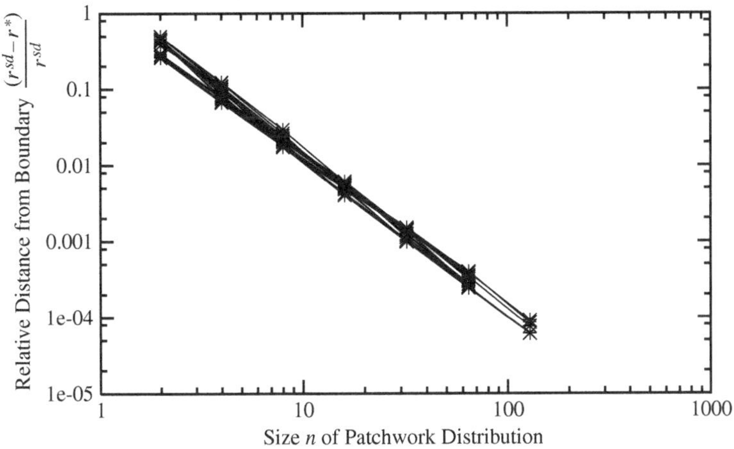

Fig. 3 Log-log plots for 20 covariance rays $r\Sigma$, $r > 0$, in 3 dimensions

brevity we write r^* and r^{sd} in place of $r^*(\Sigma)$ and $r^{sd}(\Sigma)$ when no confusion should arise.) Table 1 provides the estimated slopes of the curves for the 20 illustrated rays. The slope estimates are calculated from only a handful of sample points and are thus noisy, but the slopes hover close to -2 in all the cases tested. Thus, it would seem likely that the coverage rate is closer to n^{-2} than n^{-1} for $d = 3$.

Table 1 Slopes of log-log plots for 20 covariance rays in 3 dimensions. All values reported to 3 decimal places

Correlations			
Σ_{12}	Σ_{13}	Σ_{23}	Calculated Slope
−0.996	−0.055	0.075	−1.902
−0.944	−0.300	−0.136	−2.162
0.912	0.280	0.299	−2.066
−0.773	0.371	0.514	−2.175
−0.731	−0.188	0.656	−2.117
−0.613	0.427	0.664	−2.197
−0.488	−0.070	0.870	−2.136
−0.300	0.223	0.928	−2.130
−0.118	0.142	0.982	−2.035
0.198	0.256	0.946	−2.001
−0.989	−0.118	0.092	−1.900
−0.912	−0.324	−0.251	−2.136
−0.849	−0.122	0.514	−2.135
−0.731	−0.678	0.080	−2.120
−0.713	−0.634	−0.300	−2.171
−0.592	0.493	0.638	−2.200
−0.368	−0.327	0.870	−2.035
−0.236	−0.221	0.946	−1.879
−0.021	0.514	0.857	−2.166
0.336	0.360	0.870	−1.921

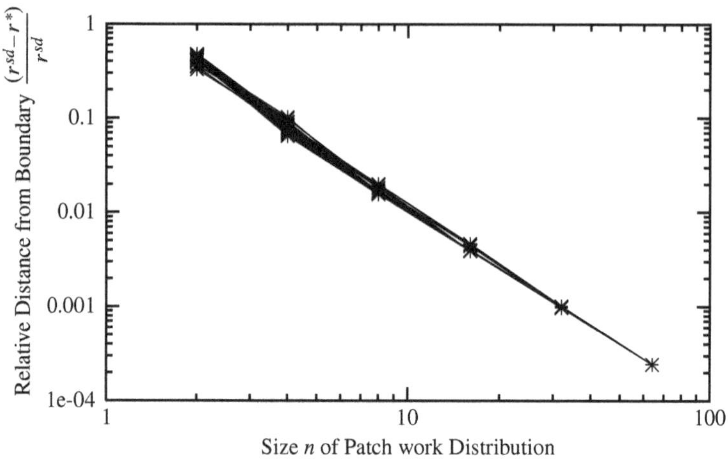

Fig. 4 log-log plots for 20 covariance rays $r\Sigma$, $r > 0$, in 4 dimensions

In the case of higher dimensions $d \geq 4$, the value r^* is not known (in general $r^* \leq r^{sd}$), and the rates of coverage calculated via the same log-scale plot as in the $d = 3$ case will only yield approximate values. Figure 4 plots the relative difference $(r^{sd} - r^*(n, \Sigma))/r^{sd}$ against n in the log-log scale for 20 different values of $\Sigma \in \mathcal{S}$ for dimension $d = 4$. If r^* were strictly less than r^{sd}, then $(r^{sd} - r^*(n, \Sigma))/r^{sd}$ would not drop linearly (in the log scale) to 0 with increasing n. No non-linearity is manifest in the range of values $n = [1, 64]$ plotted. The plots seem fairly linear, with the slopes varying within $[-2.117, -2.013]$.

Our implementation could not solve the linear programs for larger n due to numerical instability (the relative error is below 0.0004 for $n = 64$). It might be the case that the difference $(r^{sd} - r^*)$ is non-zero but too small to be detected by our tests. This indicates that for the $d = 4$ case, $r^* = r^{sd}$ is a good assumption for practical purposes, and that the rate of coverage is again closer to the upper bound in Corollary 1.

6 Conclusions

We have shown that patchwork distributions represent a flexible class of distributions that can match many desired properties of a random vector. They are primarily effective in low dimensions, say up to dimension 5. (We have solved patchwork LPs in dimension 5 for $n = 32$ using a column generation technique, but we will not report in detail on these ideas here.) The primary bottleneck in constructing patchwork distributions is solving a certain linear program, the size of which is related to the discretization level n. The discretization level required to match a given correlation matrix depends on the distance of the correlation matrix to the boundary of the set of all feasible correlation matrices. Our theoretical and computational results give strong evidence that the set of feasible correlation matrices not matched by patchwork distributions using discretization level n diminishes at the rate n^{-2}.

Acknowledgments The authors would like to thank the editors for their superb efforts in improving the presentation of this chapter. This work was partially supported by National Science Foundation grants DMI 0400287 and CMMI 0800688.

References

Avramidis, A. N., N. Channouf, and P. L'Ecuyer. 2009. Efficient correlation matching for fitting discrete multivariate distributions with arbitrary marginals and normal copula dependence. *INFORMS Journal on Computing* 21:88–106.

Avramidis, A. N., A. Deslauriers, and P. L'Ecuyer. 2004. Modeling daily arrivals to a telephone call center. *Management Science* 50:896–908.

Biller, B. 2009. Copula-based multivariate input models for stochastic simulation. *Operations Research* forthcoming.

Biller, B., and S. Ghosh. 2004. Dependence modeling for stochastic simulation. In *Proceedings of the 2004 Winter Simulation Conference*, eds. R. G. Ingalls, M. D. Rossetti, J. S. Smith, and B. A. Peters, pp. 153–161. Piscataway, New Jersey: Institute of Electrical and Electronics Engineers.

Biller, B., and S. Ghosh. 2006. Multivariate input processes. In *Handbooks in Operations Research and Management Science: Simulation*, eds. B. L. Nelson and S. G. Henderson, pp. 123–154. Amsterdam: Elsevier Science.

Billingsley, P. 1995. *Probability and Measure*, 3rd ed. New York: Wiley.

Brannen, N. S. 1997. The Wills conjecture. *Transactions of the American Mathematical Society* 349:3977–3987.

Busemann, H. 1958. *Convex Surfaces*. Interscience Tracts in Pure and Applied Mathematics, No. 6. New York: Interscience.

Clemen, R. T., G. W. Fischer, and R. L. Winkler. 2000. Assessing dependence: Some experimental results. *Management Science* 46:1100–1115.

Devroye, L. 1986. *Non-Uniform Random Variate Generation*. New York: Springer-Verlag.

Fishman, G. S. 2001. *Discrete-event Simulation: Modeling, Programming and Analysis*. Springer Series in Operations Research. New York: Springer.

Ghosh, S., and S. G. Henderson. 2001. Chessboard distributions. In *Proceedings of the 2001 Winter Simulation Conference*, eds. B. A. Peters, J. S. Smith, D. J. Medeiros, and M. W. Rohrer, pp. 385–393. Piscataway, New Jersey: Institute of Electrical and Electronics Engineers.

Ghosh, S., and S. G. Henderson. 2002. Chessboard distributions and random vectors with specified marginals and covariance matrix. *Operations Research* 50:820–834.

Ghosh, S., and S. G. Henderson. 2003. Behaviour of the NORTA method for correlated random vector generation as the dimension increases. *ACM Transactions on Modeling and Computer Simulation* 13:276–294.

Golub, G. H., and C. F. Van Loan. 1996. *Matrix Computations*, 3rd ed. Baltimore, Maryland: The Johns Hopkins University Press.

Henderson, S. G. 2005. Should we model dependence and nonstationarity, and if so how? In *Proceedings of the 2005 Winter Simulation Conference*, eds. M. E. Kuhl, N. M. Steiger, F. B. Armstrong, and J. A. Joines, pp. 120–129. Piscataway, New Jersey: Institute of Electrical and Electronics Engineers.

Hill, R. R., and C. H. Reilly. 2000. The effects of coefficient correlation structure in two-dimensional knapsack problems on solution procedure performance. *Management Science* 46:302–317.

Joe, H. 1997. *Multivariate Models and Dependence Concepts*. London: Chapman & Hall.

Johnson, M. E. 1987. *Multivariate Statistical Simulation*. New York: Wiley.
Johnson, N. L., and S. Kotz. 2004. Cloning of distributions. Preprint.
Kronmal, R. A., and A. V. Peterson. 1979. On the alias method for generating random variables from a discrete distribution. *American Statistician* 33:214–218.
Law, A. M., and W. D. Kelton. 2000. *Simulation Modeling and Analysis*, 3rd ed. New York: McGraw-Hill.
Luenberger, D. G. 1969. *Optimization by Vector Space Methods*. New York: John Wiley and Sons.
Lurie, P. M., and M. S. Goldberg. 1998. An approximate method for sampling correlated random variables from partially-specified distributions. *Management Science* 44:203–218.
Mackenzie, G. R. 1994. Approximately maximum-entropy multivariate distributions with specified marginals and pairwise correlations. PhD dissertation, Department of Decision Sciences, University of Oregon, Eugene, Oregon.
Nelsen, R. B. 1999. *An Introduction to Copulas*. Lecture Notes in Statistics, vol. 139. New York: Springer-Verlag.
Nelson, B. L. 2004. Personal communication.
Sangwine-Yager, J. R. 1988. Bonnesen-style inequalities for Minkowski relative geometry. *Transactions of the American Mathematical Society* 307:373–382.
Schneider, R. 1993. *Convex Bodies: The Brunn-Minkowski Theory*. Cambridge, UK: Cambridge University Press.
Sklar, A. 1959. Fonctions de répartition à n dimensions et leurs marges. *Publications de l'Institut Statistique de l'Université de Paris* 8:229–231.
Walker, A. J. 1977. An efficient method for generating discrete random variables with general distributions. *ACM Transactions on Mathematical Software* 3:253–256.

Asymptotic Validity of Batch Means Steady-State Confidence Intervals

Peter W. Glynn and Eunji Lim

Abstract The method of batch means is a widely applied procedure for constructing steady-state confidence intervals. The traditional theoretical support for the method of batch means has rested on the assumption of a functional central limit theorem for the underlying process. We establish here that the method of batch means is valid for Harris recurrent Markov processes whenever the associated process satisfies a simple (non–functional) central limit theorem. This weaker condition for validity of the method of batch means is also shown to hold in the setting of one-dependent regenerative processes.

1 Introduction

Consider a real-valued stochastic process $Y = (Y(t) : t \geq 0)$, with right-continuous paths having left limits, that represents the output of a simulation. Assume that the goal of the simulation is to compute the steady-state mean of Y. More precisely, suppose that Y obeys a law of large numbers (LLN), so that there exists a (deterministic) constant α for which

$$\overline{Y}(t) \triangleq \frac{1}{t}\int_0^t Y(s)\,\mathrm{d}s \Rightarrow \alpha \tag{1}$$

as $t \to \infty$, where $\Rightarrow$ denotes weak convergence (also known as "convergence in distribution"). The quantity α is called the *steady-state mean* of Y, and the problem of computing α is known in the literature as the *steady-state simulation problem*.

The LLN (1) immediately suggests a sampling-based (Monte Carlo) algorithm for computing α. In particular, simulate Y for t units of simulated time, and return $\overline{Y}(t)$ as the estimator for α. As with all numerical procedures, a key issue is the

P.W. Glynn (✉)
Department of Management Science and Engineering, Stanford University,
Stanford, CA 94305, USA
e-mail: glynn@stanford.edu

C. Alexopoulos et al. (eds.), *Advancing the Frontiers of Simulation: A Festschrift in Honor of George Samuel Fishman*, International Series in Operations Research & Management Science 133, DOI 10.1007/b110059_5,

question of how fast the algorithm converges to the answer. This issue is generally addressed by the central limit theorem (CLT).

In the presence of (1), it is typically the case that Y also enjoys a CLT, so that there exists a further (deterministic) constant σ such that

$$t^{1/2}\left(\overline{Y}(t)-\alpha\right) \Rightarrow \sigma N(0,1) \tag{2}$$

as $t \to \infty$, where $N(0,1)$ is a normal random variable (rv) with mean zero and unit variance. The quantity σ^2 is called the *time-average variance constant* (TAVC) of Y. Because (2) asserts that $\overline{Y}(t)$ has an error of order $t^{-1/2}$, it is evident that the above simulation-based estimator has a slow rate of convergence. (In particular, to add one additional significant figure of accuracy requires increasing the length t of the simulation run by a factor of 100.) As a consequence, it is important to provide the user of such a procedure with some quantitative assessment of accuracy.

Such an accuracy assessment usually comes in the form of a confidence interval. Suppose that the CLT (2) can be generalized to a multivariate CLT of the form

$$t^{1/2}\left(\overline{Y}_1(t)-\alpha,\ldots,\overline{Y}_m(t)-\alpha\right) \Rightarrow \left(\sigma\sqrt{m}N_1(0,1),\ldots,\sigma\sqrt{m}N_m(0,1)\right) \tag{3}$$

as $t \to \infty$, where

$$\overline{Y}_i(t) = \frac{m}{t}\int_{(i-1)t/m}^{it/m} Y(s)\,\mathrm{d}s$$

is the ith *batch mean* for $1 \le i \le m$, and $N_1(0,1),\ldots,N_m(0,1)$ is a collection of m independent and identically distributed (iid) normal rv's with mean zero and unit variance. (Note that (2) is the special case in which $m = 1$, so that (3) is indeed a generalization of (2).) The continuous mapping principle (see, for example, Billingsley 1999, p. 16) then guarantees that if $\sigma^2 > 0$, then

$$\frac{m^{1/2}\left(\frac{1}{m}\sum_{i=1}^{m}\overline{Y}_i(t)-\alpha\right)}{\sqrt{\frac{1}{m-1}\sum_{i=1}^{m}\left(\overline{Y}_i(t)-\frac{1}{m}\sum_{j=1}^{m}\overline{Y}_j(t)\right)^2}} \Rightarrow t_{m-1} \tag{4}$$

as $t \to \infty$, where t_{m-1} is a Student t rv with $m-1$ degrees of freedom. Hence, if one selects z so that $P(-z \le t_{m-1} \le z) = 1-\delta$, then

$$P\left(-z \le \frac{m^{1/2}(\overline{Y}(t)-\alpha)}{s_m(t)} \le z\right) \to 1-\delta$$

as $t \to \infty$, where

$$s_m(t) = \left(\frac{1}{m-1} \sum_{i=1}^{m} \left(\overline{Y}_i(t) - \frac{1}{m} \sum_{j=1}^{m} \overline{Y}_j(t) \right)^2 \right)^{1/2}.$$

In other words,

$$P\left(\alpha \in \left[\overline{Y}(t) - z\frac{s_m(t)}{\sqrt{m}},\ \overline{Y}(t) + z\frac{s_m(t)}{\sqrt{m}} \right] \right) \to 1 - \delta \tag{5}$$

as $t \to \infty$, so that

$$\left[\overline{Y}(t) - z\frac{s_m(t)}{\sqrt{m}},\ \overline{Y}(t) + z\frac{s_m(t)}{\sqrt{m}} \right]$$

is an asymptotic $100(1-\delta)\%$ confidence interval for α. This confidence interval procedure is, not surprisingly, known as the *method of batch means* and is a widely used method for constructing steady-state confidence intervals; see Fishman (1978) for an early textbook discussion of the method of batch means. The above argument establishes that the method of batch means is asymptotically valid whenever (3) holds with $\sigma^2 > 0$.

This paper is devoted to developing general conditions for the validity of the method of batch means. In Section 2, we review the existing state of the literature, and discuss the validity of the method of batch means under a functional central limit theorem hypothesis. Section 3 shows that when Y shift-couples to a stationary version Y^*, the validity of batch means for Y can be reduced to verifying validity for Y^*. Turning next to validity for Y^*, Theorem 3 of Section 4 proves that the method of batch means holds under a (non-functional) central limit theorem hypothesis, provided that Y^* satisfies a condition closely related to the requirement of ergodicity. Finally, Section 5 applies the theory of Section 3 and 4 to prove the main result of this paper, Theorem 5. Specifically, Theorem 5 proves that when Y is a one-dependent regenerative process, then the method of batch means holds under a simple (non-functional) central limit theorem.

2 Validity Based on a FCLT Hypothesis

Note that the TAVC σ appears in (3), whereas the unknown σ is not present in (4). The reason, of course, is that the common factor σ appears in the limit distribution for both the numerator and denominator of the left-hand side of (4), so that σ is "cancelled out." Steady-state confidence interval procedures that are based on cancellation of the TAVC σ are, not surprisingly, called "cancellation methods." In particular, steady-state confidence intervals based on "standardized time series" (as introduced by Schruben 1983) form a general class of cancellation methods, of which batch means is a special case; see Glynn and Iglehart (1990) for a discussion of the theory of standardized series.

As noted in Schruben (1983) and Glynn and Iglehart (1990), standardized time series steady-state confidence intervals (and hence batch means) are asymptotically valid when Y is presumed to satisfy a functional CLT (FCLT) with $\sigma^2 > 0$.

Definition 1 The process $Y = (Y(t) : t \geq 0)$ is said to satisfy a FCLT if there exist (deterministic) constants α and σ such that

$$\varepsilon^{-1/2} Z_\varepsilon \Rightarrow \sigma B \tag{6}$$

as $\varepsilon \downarrow 0$ (where the weak convergence in (6) is with respect to the Skorohod topology on the function space $D[0, \infty)$ of right-continuous functions with left limits), with Z_ε defined as

$$Z_\varepsilon(t) = \varepsilon \left(\int_0^{t/\varepsilon} Y(s)\,\mathrm{d}s - \alpha t/\varepsilon \right)$$

and where B is standard Brownian motion.

As noted above, the method of batch means provides an asymptotically valid confidence interval (as specified via the limit theorem (5)) whenever Y satisfies a FCLT with $\sigma^2 > 0$. (A direct proof of validity is also straightforward. Note that $g(x) = (x(1/m), x(2/m), \ldots, x(1))$ is a continuous mapping from $D[0, \infty)$ to R^m, so that the continuous mapping principle immediately establishes (3)).

A large class of discrete-event simulations can be represented as Markov process simulations. Specifically, it is typically the case that $Y(t)$ can then be represented as $Y(t) = f(X(t))$, where $X = (X(t) : t \geq 0)$ is Markov and $f : S \to R$ is a given performance measure. In the discrete-event setting, the Markov state $X(t)$ at time t must incorporate both the "physical state" of the system (e.g. the vector number-in-system process) and the state of the future-event schedule (e.g. the remaining time to the next scheduled event for each event type in the system). Under modest conditions on the discrete-event simulation, the Markov process X is positive Harris recurrent, and contains embedded regenerative structure; see Glynn and Haas (2006) for definitions and details. In view of this fact, it is of significant interest to know when a regenerative process satisfies a FCLT. (This easily translates into conditions on the function f appearing in the representation $Y(t) = f(X(t))$.)

Definition 2 Let $Y = (Y(t) : t \geq 0)$ be a real-valued stochastic process. Then, Y is *(classically) regenerative* if there exist random times $0 \leq T(0) < T(1) < \cdots$ such that:

(i) $W_1, W_2, \ldots$ is a sequence of identically distributed random elements; and
(ii) $W_0, W_1, \ldots$ is a sequence of independent random elements,

where $T(-1) = 0$ and for $i \geq 0$,

$$W_i(t) = \begin{cases} Y(T(i-1)+t)\,, & \text{if } 0 \le t < \tau_i \triangleq T(i) - T(i-1)\,, \\ \Delta\,, & \text{if } t \ge \tau_i\,, \end{cases} \tag{7}$$

with Δ chosen as a point not in R (for example, we can set $\Delta = (0, 0)$). The regenerative process Y is said to be *positive recurrent* if $E\tau_1 < \infty$.

In the setting of such regenerative processes, we can provide a necessary and sufficient condition for validity of the FCLT; see Theorem 1 of Glynn and Whitt (1993). We assume throughout the remainder of this paper that:

Assumption A $\int_0^t |Y(s)|\, \mathrm{d}s < \infty$ a.s. for each $t \ge 0$.

Theorem 1 *Suppose that $Y = (Y(t) : t \ge 0)$ is a real-valued positive recurrent classically regenerative process satisfying Assumption A. Then, there exists α and σ such that*

$$\varepsilon^{-1/2} Z_\varepsilon \Rightarrow \sigma B$$

as $\varepsilon \downarrow 0$ (in the sense of weak convergence on $D[0, \infty)$) if and only if

$$E\left(\int_0^{\tau_1} [Y(T(0)+s) - \alpha]\, \mathrm{d}s\right)^2 < \infty \tag{8}$$

and

$$t^2 P\left(\sup_{0 \le s \le \tau_1} \left|\int_0^s [Y(T(0)) + u) - \alpha]\, \mathrm{d}u\right| > t\right) \to 0 \quad \text{as } t \to \infty\,. \tag{9}$$

It follows that if Y is a positive recurrent classically regenerative process satisfying A, then (3) is valid if and only if (8) and (9) hold, in which case the method of batch means provides asymptotically valid confidence intervals.

3 Validity Based on a CLT Hypothesis: Reduction to the Stationary Setting

In the next two sections, we prove that the method of batch means can be valid even in situations in which the FCLT fails to be satisfied. In particular, we will prove that the method of batch means is asymptotically valid under a CLT hypothesis, plus a modest additional regularity condition.

We establish in this section that the validity of (3) can typically be reduced to the setting in which Y is a stationary stochastic process. Let $\|\cdot\|$ be the *total variation norm* defined by

$$\|P_1 - P_2\| \triangleq \sup\{|P_1(A) - P_2(A)| : A \text{ is measurable}\}\,.$$

Theorem 2 *(i) Suppose that Y satisfies Assumption A and that there exists $Y^* = (Y^*(t) : t \geq 0)$ for which*

$$\left\| \frac{1}{t} \int_0^t P((Y(u+s) : u \geq 0) \in \cdot\,)\,\mathrm{d}s - P((Y^*(u) : u \geq 0) \in \cdot\,) \right\| \longrightarrow 0 \tag{10}$$

as $t \to \infty$. Then Y^* satisfies Assumption A, so that

$$\int_0^t |Y^*(s)|\,\mathrm{d}s < \infty \quad a.s.$$

for each $t \geq 0$.

(ii) If Y satisfies Assumption A, (10), and there exists α and σ such that

$$\begin{aligned} t^{1/2} &\left(\frac{m}{t} \int_0^{t/m} [Y^*(s) - \alpha]\,\mathrm{d}s, \ldots, \frac{m}{t} \int_{(m-1)t/m}^t [Y^*(s) - \alpha]\,\mathrm{d}s \right) \\ &\Rightarrow (\sigma \sqrt{m} N_1(0,1), \ldots, \sigma \sqrt{m} N_m(0,1)) \end{aligned} \tag{11}$$

as $t \to \infty$, then $Y = (Y(t) : t \geq 0)$ satisfies (3).

Proof For part (i), note that Assumption A guarantees that

$$P\left(\int_0^t |Y(s)|\,\mathrm{d}s < \infty, t \in Q_+ \right) = 1 \tag{12}$$

where Q_+ is the set of nonnegative rational numbers. Since

$$\int_0^t |Y(s)|\,\mathrm{d}s$$

is clearly a nondecreasing function of t, it follows from (12) that

$$P\left(\int_0^t |Y(s)|\,\mathrm{d}s < \infty,\ t \in R_+ \right) = 1\,, \tag{13}$$

where $R_+ \triangleq [0, \infty)$.

We next establish that the limit Y^* appearing in (10) must necessarily be stationary. To see this, we observe that for each $\gamma \geq 0$,

$$\begin{aligned} P((Y^*(u) : u \geq 0) \in \cdot) &= \lim_{t \to \infty} \frac{1}{t} \int_0^t P((Y(u+s) : u \geq 0) \in \cdot\,)\,\mathrm{d}s \\ &= \lim_{t \to \infty} \frac{1}{t+\gamma} \int_0^{t+\gamma} P((Y(u+s) : u \geq 0) \in \cdot\,)\,\mathrm{d}s \end{aligned}$$

$$= \lim_{t\to\infty} \frac{1}{t} \int_{\gamma}^{t+\gamma} P((Y(u+s) : u \ge 0) \in \cdot\,)\, ds$$
$$= \lim_{t\to\infty} \frac{1}{t} \int_{0}^{t} P((Y(u+s+\gamma) : u \ge 0) \in \cdot\,)\, ds$$
$$= P((Y^*(u+\gamma) : u \ge 0) \in \cdot\,)\,.$$

Assumption (10) therefore ensures that Y shift-couples to Y^*; see pp. 162 and 167 of Thorisson (2000). In particular, we can presume that Y and Y^* are defined on a common probability space upon which two finite-valued rv's T_1 and T_2 can be defined, and for which

$$Y(T_1+u) = Y^*(T_2+u) \tag{14}$$

for $u \ge 0$. Because Y^* is stationary,

$$P\left(\int_0^t |Y^*(s)|\, ds > c\right) = P\left(\int_{\gamma}^{\gamma+t} |Y^*(s)|\, ds > c\right)$$

for $\gamma \ge 0$. Hence,

$$P\left(\int_0^t |Y^*(s)|\, ds > c\right) \le P\left(\int_{\gamma}^{\gamma+t} |Y^*(s)|\, ds > c,\, T_2 \le \gamma\right) + P(T_2 > \gamma)$$
$$\le P\left(\int_{T_2}^{T_2+\gamma+t} |Y^*(s)|\, ds > c\right) + P(T_2 > \gamma)$$
$$= P\left(\int_{T_1}^{T_1+\gamma+t} |Y(s)|\, ds > c\right) + P(T_2 > \gamma) \quad \text{(due to (14))}$$
$$\le P\left(\int_{0}^{T_1+\gamma+t} |Y(s)|\, ds > c\right) + P(T_2 > \gamma)\,.$$

It is therefore evident from (13) that

$$\limsup_{c\to\infty} P\left(\int_0^t |Y^*(s)|\, ds > c\right) \le P(T_2 > \gamma)\,.$$

Since $\gamma \ge 0$ was arbitrary, we can now send $\gamma \downarrow 0$ to conclude that

$$\limsup_{c\to\infty} P\left(\int_0^t |Y^*(s)|\, ds > c\right) = 0\,. \tag{15}$$

So, Assumption A is satisfied by Y^*.

Turning next to part (*ii*), we shall prove (3) when $m = 1$; the proof for general m is essentially identical (but with a higher notational burden). Put $Y_c(t) = Y(t) - \alpha$, $Y_c^*(t) = Y^*(t) - \alpha$, and $a \wedge b \triangleq \min(a, b)$ for $a, b \in R$. Then,

$$
\begin{aligned}
t^{-1/2}\int_0^t Y_c(s)\,\mathrm{d}s &= t^{-1/2}\int_0^{t\wedge T_1} Y_c(d)\,\mathrm{d}s + t^{-1/2}\int_{T_1}^t Y_c(s)\,\mathrm{d}s\,I(T_1 \le t) \\
&= t^{-1/2}\int_0^{t\wedge T_1} Y_c(d)\,\mathrm{d}s + t^{-1/2}\int_0^{t-T_1} Y_c(T_1+s)\,\mathrm{d}s\,I(T_1 \le t) \\
&= t^{-1/2}\int_0^{t\wedge T_1} Y_c(d)\,\mathrm{d}s + t^{-1/2}\int_0^{t-T_1} Y_c^*(T_2+s)\,\mathrm{d}s\,I(T_1 \le t) \\
&= t^{-1/2}\int_0^{t\wedge T_1} Y_c(d)\,\mathrm{d}s + t^{-1/2}\int_{T_2}^{t+T_2-T_1} Y_c^*(s)\,\mathrm{d}s\,I(T_1 \le t) \\
&= t^{-1/2}\int_0^{t\wedge T_1} Y_c(d)\,\mathrm{d}s + t^{-1/2}I(T_1 \le t)\times\left(\int_0^t Y_c^*(s)\,\mathrm{d}s\right. \\
&\quad \left. - \int_0^{T_2} Y_c^*(s)\,\mathrm{d}s + \int_t^{t+T_2-T_1} Y_c^*(s)\,\mathrm{d}s\right).
\end{aligned}
$$

Clearly,

$$
\left|t^{-1/2}\int_0^{t\wedge T_1} Y_c(s)\,\mathrm{d}s\right| \le t^{-1/2}\int_0^{T_1} |Y_c(s)|\ \mathrm{d}s \Rightarrow 0 \tag{16}
$$

as $t \to \infty$, because (13) implies that $\int_0^{T_1} |Y_c(s)|\,\mathrm{d}s < \infty$ a.s. in view of the finiteness of T_1.

Furthermore, on $\{T_1 \le r, T_2 \le r\}$,

$$
\begin{aligned}
\left|t^{-1/2}\int_0^{t+T_2-T_1} Y_c^*(s)\,\mathrm{d}s\right| &\le t^{-1/2}\int_{t-r}^{t+r} |Y_c^*(u)|\,\mathrm{d}u \\
&\overset{\mathcal{D}}{=} t^{-1/2}\int_0^{2r} |Y_c^*(u)|\,\mathrm{d}u \Rightarrow 0\,,
\end{aligned} \tag{17}
$$

where $\overset{\mathcal{D}}{=}$ denotes "equality in distribution" and (15) was used in the final step. So,

$$
\begin{aligned}
&P\left(t^{-1/2}\int_0^t Y_c(s)\,\mathrm{d}s \le x\right) \\
&= P\left(t^{-1/2}\int_0^t Y_c(s)\,\mathrm{d}s \le x,\, T_1 \le r,\, T_2 \le r\right) + P(T_1 > r) + P(T_2 > r) \\
&\le P\left(t^{-1/2}\int_0^{t\wedge T_1} Y_c(s)\,\mathrm{d}s + t^{-1/2}\int_0^t Y_c^*(s)\,\mathrm{d}s\,I(T_1 \le t)\right. \\
&\quad \left. - t^{-1/2}\int_{t-r}^{t+r} |Y_c^*(u)|\,\mathrm{d}u\,I(T_1 \le t) \le x\right) + P(T_1 > r) + P(T_2 > r)\,.
\end{aligned}
$$

Exploiting (11), (16), and (17) yields

$$\limsup_{t\to\infty} P\left(t^{-1/2}\int_0^t Y_c(s)\,ds \le x\right) \\ \le P(\sigma N(0,1) \le x) + P(T_1 > r) + P(T_2 > r)\,.$$

Since $r \ge 0$ was arbitrary, we can send $r \downarrow 0$ to conclude that

$$\limsup_{t\to\infty} P\left(t^{-1/2}\int_0^t Y_c(s)\,ds \le x\right) \le P(\sigma N(0,1) \le x)\,.$$

A similar argument proves that

$$\liminf_{t\to\infty} P\left(t^{-1/2}\int_0^t Y_c(s)\,ds \le x\right) \ge P(\sigma N(0,1) \le x)\,,$$

proving part (*ii*) of the theorem for $m = 1$. □

We note that the proof establishes that the limit process Y^* appearing in (10) must necessarily be a stationary process. We conclude this section by showing that in typical discrete-event simulation contexts, we can expect (10) to be a relatively benign hypothesis.

Any one-dependent regenerative process for which $E\tau_1 < \infty$ automatically satisfies (10); see Glynn and Sigman (1992) for details. Since all Markov processes that are positive recurrent in the sense of Harris fall into this class (see Sigman 1990), (10) also holds for this important class of processes. As noted earlier, such Harris recurrence broadly applies to discrete-event simulations; see Glynn and Haas (2006).

4 Validity Based on a CLT Hypothesis: The Main Result

The discussion of Section 3 permits us to presume that we can reduce our analysis to that of a stationary process $Y^* = (Y^*(t) : t \ge 0)$. Without loss of generality, we can extend Y^* to a two-sided version $(Y^*(t) : -\infty < t < \infty)$. Our next result is the main theorem in this paper. It shows that the method of batch means is valid when Y^* satisfies the CLT, plus a modest additional regularity condition (namely (18)).

Theorem 3 *Suppose that Y^* satisfies Assumption A and that*

$$\sup\left\{\left|\frac{1}{t}\int_0^t P\left((Y^*(r) : r \le 0) \in B_1, (Y^*(s+u) : u \ge 0) \in B_2\right) ds \right.\right. \\ \left.\left. - P\left((Y^*(r) : r \le 0) \in B_1\right) P\left((Y^*(u) : u \ge 0) \in B_2\right)\right| : B_1, B_2 \textit{ measurable}\right\} \\ \to 0 \textit{ as } t \to \infty\,. \qquad (18)$$

If Y^* satisfies the CLT (2), then Y^* also automatically satisfies (3) for each $m \geq 1$.

Proof To simplify the proof notationally, we specialize to the setting where $m = 2$; the proof for general m is essentially identical. Note that the stationarity of Y^* implies that

$$P\left(t^{-1/2}\int_0^t Y_c^*(s)\,ds \leq x,\ t^{-1/2}\int_t^{2t} Y_c^*(s)\,ds \leq y\right)$$
$$= P\left(t^{-1/2}\int_{-t}^0 Y_c^*(s)\,ds \leq x,\ t^{-1/2}\int_0^t Y_c^*(s)\,ds \leq y\right). \tag{19}$$

Furthermore, according to (18), for each $\varepsilon > 0$, there exists r_0 such that

$$\Bigg|\frac{1}{r_0}\int_0^{r_0} P\left(t^{-1/2}\int_{-t}^0 Y_c^*(u)\,du \leq x,\ t^{-1/2}\int_s^{s+t} Y_c^*(u)\,du \leq y\right) ds$$
$$-P\left(t^{-1/2}\int_{-t}^0 Y_c^*(u)\,du \leq x\right) P\left(t^{-1/2}\int_0^t Y_c^*(u)\,du \leq y\right)\Bigg| < \varepsilon \tag{20}$$

uniformly in $t > 0$ and $x, y \in R$.

In addition, the following inequalities hold uniformly in $s \in [0, r_0]$ and $\delta > 0$:

$$P\left(t^{-1/2}\int_{-t}^0 Y_c^*(u)\,du \leq x,\ t^{-1/2}\int_0^t Y_c^*(u)\,du \leq y\right)$$
$$= P\left(t^{-1/2}\int_{-t}^0 Y_c^*(u)\,du \leq x,\ t^{-1/2}\int_s^{s+t} Y_c^*(u)\,du + t^{-1/2}\int_0^s Y_c^*(u)\,du\right.$$
$$\left. - t^{-1/2}\int_t^{t+s} Y_c^*(u)\,du \leq y\right)$$
$$\leq P\left(t^{-1/2}\int_{-t}^0 Y_c^*(u)\,du \leq x,\ t^{-1/2}\int_s^{s+t} Y_c^*(u)\,du \leq y + 2\delta,\right.$$
$$\left. t^{-1/2}\int_0^s Y_c^*(u)\,du \leq \delta,\ t^{-1/2}\int_t^{t+s} Y_c^*(u)\,du \geq -\delta\right)$$
$$+P\left(t^{-1/2}\int_0^s Y_c^*(u)\,du > \delta\right) + P\left(t^{-1/2}\int_t^{t+s} Y_c^*(u)\,du < -\delta\right)$$
$$\leq P\left(t^{-1/2}\int_{-t}^0 Y_c^*(u)\,du \leq x,\ t^{-1/2}\int_s^{s+t} Y_c^*(u)\,du \leq y + 2\delta\right)$$
$$+P\left(t^{-1/2}\int_0^{r_0} \left|Y_c^*(u)\right|\,du > \delta\right) + P\left(t^{-1/2}\int_t^{t+r_0} \left|Y_c^*(u)\right|\,du > \delta\right)$$
$$\leq P\left(t^{-1/2}\int_{-t}^0 Y_c^*(u)\,du \leq x,\ t^{-1/2}\int_s^{s+t} Y_c^*(u)\,du \leq y + 2\delta\right)$$
$$+2P\left(t^{-1/2}\int_0^{r_0} \left|Y_c^*(u)\right|\,du > \delta\right) \quad \text{(using the stationarity of } Y\text{)}. \tag{21}$$

Hence,

$$
\begin{aligned}
&P\left(t^{-1/2}\int_{-t}^{0} Y_c^*(u)\,du \le x,\ t^{-1/2}\int_0^t Y_c^*(u)\,du \le y\right)\\
&= \frac{1}{r_0}\int_0^{r_0} P\left(t^{-1/2}\int_{-t}^{0} Y_c^*(u)\,du \le x,\ t^{-1/2}\int_0^t Y_c^*(u)\,du \le y\right) ds\\
&\le \frac{1}{r_0}\int_0^{r_0} P\left(t^{-1/2}\int_{-t}^{0} Y_c^*(u)\,du \le x,\ t^{-1/2}\int_s^{s+t} Y_c^*(u)\,du \le y+2\delta\right) ds\\
&\quad +2P\left(t^{-1/2}\int_0^{r_0} \left|Y_c^*(u)\right|\,du > \delta\right) \quad \text{(using (21))}\\
&\le P\left(t^{-1/2}\int_{-t}^{0} Y_c^*(u)\,du \le x\right) P\left(t^{-1/2}\int_0^t Y_c^*(u)\,du \le y+2\delta\right) + \varepsilon\\
&\quad +2P\left(t^{-1/2}\int_0^{r_0} \left|Y_c^*(u)\right|\,du > \delta\right) \quad \text{(using (20))}\\
&= P\left(t^{-1/2}\int_0^t Y_c^*(u)\,du \le x\right) P\left(t^{-1/2}\int_0^t Y_c^*(u)\,du \le y+2\delta\right) + \varepsilon\\
&\quad +2P\left(t^{-1/2}\int_0^{r_0} \left|Y_c^*(u)\right|\,du > \delta\right) \quad \text{(using the stationarity of } Y\text{)}\,. \qquad (22)
\end{aligned}
$$

It follows from (19), (22), and Assumption A that

$$
\begin{aligned}
&\limsup_{t\to\infty} P\left(t^{-1/2}\int_0^t Y_c^*(u)\,du \le x,\ t^{-1/2}\int_t^{2t} Y_c^*(u)\,du \le y\right)\\
&\quad \le \limsup_{t\to\infty} P\left(t^{-1/2}\int_0^t Y_c(u)\,du \le x\right) P\left(t^{-1/2}\int_0^t Y_c^*(u)\,du \le y+2\delta\right)\\
&\qquad + \varepsilon + 2P\left(t^{-1/2}\int_0^{r_0} \left|Y_c^*(u)\right|\,du > \delta\right) \quad \text{(using (22))}\\
&\quad \le P\left(\sigma N(0,1) \le x\right) P\left(\sigma N(0,1) \le y+2\delta\right) + \varepsilon \quad \text{(using (2))}\,.
\end{aligned}
$$

Sending $\delta \downarrow 0$ and then $\varepsilon \downarrow 0$ provides the inequality

$$
\begin{aligned}
&\limsup_{t\to\infty} P\left(t^{-1/2}\int_0^t Y_c^*(u)\,du \le x,\ t^{-1/2}\int_t^{2t} Y_c^*(u)\,du \le y\right)\\
&\qquad\qquad \le P\left(\sigma N(0,1) \le x\right) P(\sigma N(0,1) \le y)\,.
\end{aligned}
$$

A similar argument establishes that

$$\liminf_{t\to\infty} P\left(t^{-1/2}\int_0^t Y_c^*(u)\,du \le x,\ t^{-1/2}\int_t^{2t} Y_c^*(u)\,du \le y\right)$$
$$\ge P(\sigma N(0,1) \le x)\,P(\sigma N(0,1) \le y)\,,$$

proving the theorem. □

We note that (18) implies that

$$\sup\left\{\left|\frac{1}{t}\int_0^t P\left((Y^*(r): r\le 0)\in B_1, (Y^*(s+u): u\ge 0)\in B_2\right)\,ds \right.\right. \tag{23}$$
$$\left.\left. - P\left((Y^*(r): r\le 0)\in B_1, (Y^*(u): u\ge 0)\in B_2\right)\right| \ :\ B_2 \text{ measurable}\right\} \to 0$$

as $t \to \infty$, for each fixed (measurable) B_1, so that (18) implies Cesaro mixing; see p. 199 of Thorisson (2000) for the definition (and the remark concerning the need to verify (23) only for finite-dimensional sets B_1). As a consequence of Theorem 2.2 of Chapter 6 of Thorisson (2000), this is equivalent to asserting that the invariant σ-algebra of Y^* is trivial. Of course, since Y^* is stationary, triviality of the invariant σ-algebra is equivalent to ergodicity of Y^*. In other words, (23) (which is a slight weakening of (18)) is equivalent to asserting that Y^* is an ergodic stationary stochastic process.

We further note that the condition (18) is weaker than requiring that Y^* is mixing. All mixing conditions that are present in the literature minimally require that

$$P\left((Y^*(r): r\le 0)\in B_1, (Y^*(s+u): u\ge 0)\in B_2\right)$$
$$\to P((Y^*(r): r\le 0)\in B_1)P((Y^*(u): u\ge 0)\in B_2) \tag{24}$$

as $s \to \infty$ for each (measurable) B_1, B_2. But (24) precludes stationary stochastic processes that exhibit periodic behavior (e.g. all periodic positive recurrent Harris chains violate (24)). Our formulation of (18) as a (slight) strengthening of Cesaro mixing (which is equivalent to ergodicity) is intended to permit easy application of our results to generic stationary processes, regardless of whether the process is periodic or not.

5 Validity Based on a CLT Hypothesis: Specializing the Results to the Regenerative Setting

As argued in Section 2, the typical discrete-event simulation can be viewed as the simulation of a corresponding Markov process. Under modest conditions on the simulation, the Markov process is positive recurrent in the sense of Harris. Such a Markov process contains embedded regenerative structure under which Y is a one-dependent regenerative process. Specifically, the cycle variables $W_0, W_1, \ldots$ then satisfy Definition 2, part (i); and Definition 2, part (ii) is replaced by:

(ii′) $W_0, W_1, \ldots$ is a one-dependent sequence of random elements (so that $(W_0, W_1, \ldots, W_{i-1})$ is independent of $(W_{i+1}, W_{i+2}, \ldots)$ for $i \geq 1$).

Theorem 4 *Let $Y^* = (Y^*(t) : -\infty < t < \infty)$ be a stationary one-dependent positive recurrent regenerative process. Then (18) is automatically satisfied. Furthermore, if*

$$\int_0^{\tau_1} |Y^*(\tau(0) + s)| \, ds < \infty \quad \text{a.s.}, \tag{25}$$

then Assumption A is satisfied by Y^*.

Proof Let f and g be two nonnegative (measurable) functions that are bounded by one in absolute value. It is a standard fact in the theory of Palm processes that

$$Eg(Y^*(u) : u \geq 0) = \lambda E\Gamma_1(g)$$

where $\lambda = 1/E\tau_1$ and

$$\Gamma_j(g) \triangleq \int_0^{\tau_j} g(Y^*(T(j-1) + s)) \, ds$$

for $j \geq 1$; see for example, Thorisson (2000).

Furthermore, because the sequence $(W_n : n \geq 0)$ is one-dependent, it follows from the strong law of large numbers for iid sequences that

$$\frac{1}{n} \sum_{j=0}^{n-1} \tau_{2j} \to E\tau_1 \quad \text{a.s.}$$

and

$$\frac{1}{n} \sum_{j=0}^{n-1} \tau_{2j+1} \to E\tau_1 \quad \text{a.s.}$$

as $n \to \infty$ so that

$$\frac{1}{n} \sum_{j=0}^{n-1} \tau_j \to E\tau_1 \quad \text{a.s.}$$

as $n \to \infty$. Hence, if $N(t) = \max\{n \geq -1 : T(n) \leq t\}$, the inequality

$$\frac{T(N(t))}{N(t)} \le t \le \frac{T(N(t)+1)}{N(t)+1} \cdot \frac{N(t)+1}{N(t)}$$

implies that

$$\frac{N(t)}{t} \to \lambda \quad \text{a.s.} \tag{26}$$

as $t \to \infty$.

It follows that for each $\varepsilon > 0$,

$$\begin{aligned}
&Ef(Y^*(r) : r \le 0)\frac{1}{t}\int_0^t g(Y^*(s+u) : u \ge 0)\,\mathrm{d}s \\
&\quad - Ef(Y^*(r) : r \le 0)Eg(Y^*(u) : u \ge 0) \\
&\le Ef(Y^*(r) : r \le 0)\frac{1}{t}\int_0^{T(0)\wedge t} g(Y^*(s+u) : u \ge 0)\,\mathrm{d}s \\
&\quad + Ef(Y^*(r) : r \le 0)\frac{1}{t}\sum_{j=1}^{N(t)+1} \Gamma_j(g) - \lambda Ef(Y^*(r) : r \le 0)E\Gamma_1(g) \\
&\quad \text{(since } g \text{ is nonnegative)} \\
&\le E(T(0) \wedge t)\frac{1}{t} + Ef(Y(r) : r \le 0)\frac{1}{t}\sum_{j=1}^{\lceil(\lambda+\varepsilon)t\rceil} \Gamma_j(g) \\
&\quad - \lambda Ef(Y^*(r) : r \le 0)E\Gamma_1(g) + P(N(t) \ge (\lambda+\varepsilon)t) \\
&\quad \text{(since } f \text{ and } g \text{ are bounded by one)} \\
&\le E(T(0) \wedge t)\frac{1}{t} + Ef(Y^*(r) : r \le 0)\Gamma_1(g)\frac{1}{t} \\
&\quad + \lceil(\lambda+\varepsilon)t\rceil\frac{1}{t}Ef(Y^*(r) : r \le 0)E\Gamma_1(g) \\
&\quad - \lambda Ef(Y^*(r) : r \le 0)E\Gamma_1(g) + P(N(t) \ge (\lambda+\varepsilon)t) \\
&\le E(T(0) \wedge t)\frac{1}{t} + E\tau_1 \cdot \frac{1}{t} \\
&\quad \text{(using the one-dependence and identical distribution property of the cycles)} \\
&\quad + \left(\varepsilon + \frac{1}{t}\right)\lambda Ef(Y^*(r) : r \le 0)E\Gamma_1(g) + P(N(t) \ge (\lambda+\varepsilon)t) \\
&\le E(T(0) \wedge t)\frac{1}{t} + \left(\varepsilon + \frac{1}{t}\right) + P(N(t) \ge (\lambda+\varepsilon)t) + \frac{1}{t}E\tau_1 \\
&\quad \text{(since } f \text{ and } g \text{ are bounded by one).}
\end{aligned}$$

Note that $(T(0) \wedge t)/t$ converges a.s. to zero and is bounded by one, so that the Bounded Convergence Theorem ensures that $t^{-1}E(T(0) \wedge t) \to 0$. Also, (26) guarantees that $P(N(t) \geq (\lambda + \varepsilon)t) \to 0$ as $t \to \infty$. Hence, by sending $t \to \infty$ and then $\varepsilon \downarrow 0$, we conclude that

$$\limsup_{t\to\infty} \sup \Big\{ Ef(Y^*(r) : r \leq 0) \frac{1}{t} \int_0^t g(Y^*(s+u) : u \geq 0)\, ds \\ - Ef(Y^*(r) : r \leq 0) Eg(Y^*(u) : u \geq 0) : \\ f, g \text{ are (measurable) functions bounded by one} \Big\} \leq 0\,.$$

A similar argument proves that

$$\liminf_{t\to\infty} \inf \Big\{ Ef(Y^*(r) : r \leq 0) \frac{1}{t} \int_0^t g(Y^*(s+u) : u \geq 0)\, ds \\ - Ef(Y^*(r) : r \leq 0) Eg(Y^*(u) : u \geq 0) : \\ f, g \text{ are (measurable) functions bounded by one} \Big\} \geq 0\,,$$

thereby proving (18).

To prove Assumption A, note that for any $c \geq 0$,

$$\begin{aligned} P&\left(\int_0^{T(0)} |Y^*(s)|\, ds > c \right) \\ &= \lambda E \int_0^{\tau_1} I\left(\int_s^{\tau_1} |Y^*(T(0)+u)|\, du > c \right) ds \\ &\quad \text{(using Palm theory)} \\ &\leq \lambda E \int_0^{\tau_1} I\left(\int_0^{\tau_1} |Y^*(T(0)+u)|\, du > c \right) ds \\ &= \lambda E \tau_1 I\left(\int_0^{\tau_1} |Y^*(T(0)+u)|\, du > c \right) . \end{aligned}$$

In light of (25), the Dominated Convergence Theorem therefore proves that

$$\lim_{c\to\infty} P\left(\int_0^{T(0)} |Y^*(s)|\, ds > c \right) = 0\,,$$

so that

$$\int_0^{T(0)} |Y^*(s)|\, ds < \infty \quad \text{a.s.}$$

Since (25) also guarantees that

$$\int_{T(k-1)}^{T(k)} |Y^*(u)|\, du < \infty \quad \text{a.s.}\,,$$

it is evident that

$$\int_0^{T(k)} |Y^*(u)|\, du < \infty$$

for $k \geq 0$, proving Assumption A. □

We now focus on the implications of our theory for one-dependent regenerative processes.

Proposition 1 *(i) Let $Y^* = (Y^*(t) : t \geq 0)$ be a stationary one-dependent positive recurrent regenerative process. Then, Y^* satisfies Assumption A if and only if*

$$\int_0^{\tau_1} |Y^*(T(0) + s)|\, ds < \infty \quad \text{a.s.} \tag{27}$$

(ii) Let $Y = (Y(t) : t \geq 0)$ be a one-dependent positive recurrent regenerative process. Then, Y satisfies Assumption A if and only if

$$\int_0^{T(0)} |Y(s)|\, ds < \infty \quad \text{a.s.} \tag{28}$$

and

$$\int_0^{\tau_1} |Y(T(0) + s)|\, ds < \infty\,. \tag{29}$$

Proof We start with (*ii*). If Assumption A is in force, then the proof of Theorem 2 shows that

$$P\left(\int_0^t |Y(s)|\, ds < \infty \text{ for each } t \geq 0\right) = 1\,.$$

Since the $T(j)$'s are finite-valued, it follows that

$$\int_0^{T(1)} |Y(s)|\, ds < \infty \quad \text{a.s.}\,,$$

proving (28) and (29). Conversely, if (29) holds, then

$$\int_{T(j-1)}^{T(j)} |Y(s)|\, ds < \infty \quad \text{a.s.}$$

for each $j \geq 1$, so that (28) shows that

$$\int_0^{T(n)} |Y(s)| \, ds < \infty \quad \text{a.s.}$$

for each $n \geq 1$. Since $T(n) \to \infty$ a.s., we may conclude that Assumption A is valid.

As for part (*i*), Theorem 4 shows that (27) implies (28) for the process Y^*. Furthermore,

$$\int_0^{\tau_1} |Y^*(T(0) + s)| \, ds \stackrel{\mathcal{D}}{=} \int_0^{\tau_1} |Y(T(0) + s)| \, ds \; ,$$

so (29) is also validated. Hence, part (*ii*) can be applied to prove the result. □

Implicit in assuming the CLT (2) for the process Y is the presumption that Y is a.s. integrable over finite intervals $[0, t]$. In view of Theorems 2, 3, and 4, we have therefore proved the following result.

Theorem 5 *Let $Y = (Y(t) : t \geq 0)$ be a positive recurrent one-dependent regenerative process. Then, (2) holds if and only if (3) is valid.*

Corollary 1 *Suppose that $Y(t) = f(X(t))$, where $X = (X(t) : t \geq 0)$ is an S-valued Markov process that is positive recurrent in the sense of Harris and where $f : S \to R$. Then (2) holds for $Y = (Y(t) : t \geq 0)$ if and only if (3) is valid.*

Proposition 2 and Corollary 1 show that when $\sigma^2 > 0$, the method of batch means provides asymptotically valid confidence intervals for positive recurrent one-dependent processes and for positive recurrent Harris processes whenever Y satisfies a CLT. In particular, the FCLT is not required to establish validity of the method of batch means.

It has been shown in Glynn and Whitt (2002) that the conditions required for validity of a CLT in the classically regenerative context are strictly weaker than those associated with the FCLT. Specifically, Y satisfies the CLT (2) under Assumption A if and only if (8) holds. So condition (9) is the extra condition required to ensure a FCLT. Examples of processes satisfying (8) but not (9) are easy to construct; see Glynn and Whitt (2002).

Our discussion therefore establishes that the method of batch means is valid under the weakest possible conditions (namely, the assumption of a CLT). In contrast, it can easily be shown that the validity of general standardized time series confidence interval methods (specifically those that look at the maximum of the standardized time series) fundamentally relies on the FCLT assumption. Hence, we may conclude that the method of batch means is the standardized time series confidence interval procedure that is most generally applicable.

Acknowledgments The authors wish to thank the reviewer for the suggestions made to improve the readability of this paper.

References

Billingsley, P. 1999. *Convergence of Probability Measures*, 2nd ed. New York: John Wiley.

Fishman, G. S. 1978. *Principles of Discrete Event Simulation*. New York: John Wiley.

Glynn, P. W., and P. Haas. 2006. Laws of large numbers and functional central limit theorems for generalized semi-Markov processes. *Stochastic Models* 22:201–231.

Glynn, P. W., and D. L. Iglehart. 1990. Simulation output analysis using standardized time series. *Mathematics of Operations Research* 15(1):1–16.

Glynn, P. W., and K. Sigman. 1992. Uniform Cesaro limit theorems for synchronous processes with applications to queues. *Stochastic Processes and their Applications* 40:29–43.

Glynn, P. W., and W. Whitt. 1993. Limit theorems for cumulative processes. *Stochastic Processes and Their Applications* 47:299–314.

Glynn, P. W., and W. Whitt. 2002. Necessary conditions in limit theorems for cumulative processes. *Stochastic Processes and Their Applications* 98:199–209.

Schruben, L. 1983. Confidence interval estimation using standardized time series. *Operations Research* 31:1090–1108.

Sigman, K. 1990. One-dependent regenerative processes and queues in continuous time. *Mathematics of Operations Research* 15(1):175–189.

Thorisson, H. 2000. *Coupling, Stationarity, and Regeneration*. New York: Springer-Verlag.

Efficient Modeling of Delays in Discrete-Event Simulation

James O. Henriksen

Abstract Efficient modeling of delay conditions is a critical requirement of modern discrete-event simulation software. In this article four forms of delay are discussed: time-based, state-based, compound combinations of time-based and state-based, and user-managed. Most of the algorithms described are those used by the SLX simulation language. An informal survey of SLX users provides many examples of the demands placed on delay mechanisms and the resultant performance of the SLX models.

1 Introduction

This paper describes efficient modeling of delay conditions in discrete-event simulation software. Four forms of delay are discussed: time-based, state-based, compound combinations of time-based and state-based, and user-managed. Of these forms, only time-based delays have received significant attention in the discrete-event simulation literature. Time-based delays are modeled using event lists, and the number of papers that have been written about event list algorithms is quite large. Furthermore, an event list is a specialized form of priority queue, and the number of papers published on priority queue algorithms is vast. The remaining three forms of delay have received scant attention. Notable exceptions include Schriber (1990), Ståhl (1998), Crain and Brunner (1990), and Henriksen (1981).

Much of the material in this paper is drawn from experience with the SLX simulation language (Henriksen 2000). Most of the algorithms described are those used by SLX, and many of the examples of demands placed on delay mechanisms and resultant performance have been gathered through an informal survey of SLX users.

In Section 2, an overview of the four forms of delays is presented. Section 3 considers the impact of modeling world-view and simulation-tool architecture on demands placed on delay mechanisms, and Section 4 considers the impact of

J.O. Henriksen (✉)
Wolverine Software Corporation, 3131 Mount Vernon Avenue, Alexandria, VA 22305, USA
e-mail: mail@wolverinesoftware.com

C. Alexopoulos et al. (eds.), *Advancing the Frontiers of Simulation: A Festschrift in Honor of George Samuel Fishman*, International Series in Operations Research & Management Science 133, DOI 10.1007/b110059_6, © Springer Science+Business Media, LLC 2009

application characteristics and modeling style. Sections 5–8 describe the design and implementation of each of the four forms of delay. Finally, Section 9 presents striking examples of the use of delay mechanisms, followed by conclusions in Section 10.

2 The Four Forms of Delays

Discrete-event simulation models describe the behavior of components of a system over a sequence of instants of time. System components are usually modeled using some form of object, with the available forms of objects highly dependent on the software used to do the modeling. In the remainder of this paper, we will simply use the term *object* to refer to all these forms. Objects can be *active* or *passive*, depending on the world-view of the modeling tool and, within the world-view, modeler preference. Active objects flow through a model, interacting with one another and competing for control of passive objects. Active objects have executable behavior patterns, while passive objects are only *acted upon* by other active objects. The earliest widely used simulation language that incorporated the use of active objects was GPSS (Schriber 1990). In GPSS, active objects have a single, generic form and are called *transactions*. The transaction-flow world-view has been widely adopted by a number of simulation tools.

In a simplistic model of a grocery store, customers could naturally be described as active objects, while a butcher could naturally be modeled as a passive object, responding only to requests for service. In a more realistic model, a butcher would not only respond to requests for service, but carry out a number of activities, such as cutting meat, rearranging display cases, cooperating with the deli, etc. The more realistic depiction of the butcher is far more easily represented using an active object.

Simulation *languages* that include active objects contain *statements* that request or imply delays in the execution of the active objects' behavior patterns, while graphically-based simulation tools contain pictorial building blocks for the same purpose. When an active object is delayed, it is suspended at the current point in its behavior pattern, and it resumes execution at a later point in simulated time. Resumption can be implicitly triggered by the simulation executive or explicitly triggered by other active objects. During the execution of its behavior pattern, an active object can undergo exactly four forms of delay. These forms are discussed in Sections 2.1–2.4.

Passive objects have no executable behavior patterns, so suspension and resumption of their activities that take place over time must be carried out by other, *active* objects. For example, if jobs in a jobshop are modeled as passive objects, an active object executing an arrival process could create passive job objects and place them into a queue, and active objects executing server processes could remove the jobs from the queue. Delays such as "wait for queue nonempty" are incurred by the active objects that manipulate the passive objects.

In the event-scheduling world-view, there are no active objects in the sense we have defined, so delays must be modeled in a different fashion, described in Section 3.2.

2.1 Scheduled Delays

Scheduled (or *time-based*) delays take the form "wait Δt," where Δt is often determined by sampling from a probability distribution. Scheduled delays are prototypically used to model service time intervals and interarrival times. Resumption of scheduled delays is managed by a *simulation executive*. (We use the term *simulation executive* to describe the software under whose control simulation applications run.) The set of all currently scheduled delays is called an event list or event set.

2.2 State-Based Delays

State-based delays take the form "wait until Boolean expression is true." Delays of this form are commonly used to control access to servers, e.g., "wait until server is idle." The variables used in the Boolean expression are *state variables*, which can be visible, user-declared variables, or hidden, implicit variables, or a mixture thereof, depending on the simulation tool used. Resumption of the execution of an active object undergoing a state-based delay can be accomplished in a variety of ways. The simplest and least efficient approach is *polling*, in which the Boolean condition is reevaluated one or more times during each instant of simulated time until the condition becomes true ("Can you hear me now?"), at which point the simulation executive allows the active object to resume. A near-optimal algorithm for managing state-based delays is presented in Section 6.

2.3 Compound Delays

Compound delays include both a time-based delay and a state-based delay. Such delays are completed either as the result of the passage of time (the time-based portion) or as the result of a condition becoming true (the state-based portion). Compound delays are extremely useful for representing a wide range of system component behavior. For example, reneging can be described as a compound delay, e.g. "wait until a server is available or the maximum tolerable queuing time has elapsed." Oddly enough, although compound delays are a very convenient modeling construct, very few simulation tools directly provide generic support for them. Compound delays are useful in modeling material-handling systems. For example, if an object is placed on a conveyor belt that moves at constant speed, we can easily compute the time required for the object to travel from point A to point B and express the delay as "wait until time = calculated arrival time at B *or* some 'special' event occurs." The special event that most easily comes to mind is a conveyor stoppage.

2.4 User-Managed Delays

User-managed delays for active objects take the form "put an active object (almost always the currently executing active object) to sleep in a manner such that it can be subsequently *awakened* by another active object." Delays of this form are often used to describe system operating rules in situations where time-based, state-based, and compound delays cannot describe the required behavior, or if using them would result in inefficient model execution. For example, in a jobshop model in which the scheduling discipline is "most imminent deadline," server queues could be implemented as sets of active objects representing jobs ranked by ascending deadline. Arriving jobs could be modeled as active objects that place themselves into a ranked set and "sleep" until awakened by a server process. When a user-managed delay is completed, resumption of an active object's behavior pattern is triggered by an explicit "awaken" verb executed by another active object.

3 Demands Imposed by World-View and Simulation Tool Architecture

3.1 Introduction

The majority of this paper focuses on implementation of delay mechanisms for active objects; however, the relevance of implementation quality depends on which mechanisms are actually used in a model and the extent of their usage. These in turn depend on the characteristics of the system being modeled, the world-view of the modeling tool, and the style of modeling embraced within the world-view. These properties collectively determine the *demands* placed on delay mechanisms. This paper carefully considers such demands.

The world-view embodied in a simulation tool and the architecture of the *implementation* of the world-view can have a profound impact on the demands placed on mechanisms for expressing and implementing delays in a simulation software package. In the next three sections, we examine the impacts of three different world-views/architectures on delay management.

3.2 The Event-Scheduling World-View

The simplest and oldest simulation world-view is the event-scheduling world-view. In this world-view, systems are described using *event routines* that are called by a central *timing routine*. The hallmark of this world-view is that all event routines execute in zero simulated time. Execution cannot be suspended in the middle of an event routine (or anywhere else, for that matter) and resumed later. Any activity that takes place over simulated time must be described by *scheduling* "start" and "end" events. Processes that take place over time must be described as a collection of event routines, each of which models one or more changes in the state of one or more system components. For complex processes, process flow can become obscured, because the *sequence* in which events occur is not obvious.

For the implementer of an event-scheduling simulation package, implementation is simplified by the absence of any suspend-and-resume constructs. Their absence implies all model execution can be accomplished by function calls, each of which models behavior at a particular instant in time. No matter how many levels of function call may take place as the result of calling an event routine, execution always returns to the timing routine in zero simulated time. Virtually any procedural language can be used to implement the event-scheduling world-view. In the early days of discrete-event simulation, software such as GASP and the original SIMSCRIPT were popular event-scheduling tools.

Consider how one would model a single-waiting-line, single-server queuing system under the event-scheduling world-view. In this simple queuing system, there are five types of events:

- Customer Arrival
- Start-of-Service
- End-of-Service
- Customer Exit
- End-of-Simulation

Upon further examination, the start-of-service and customer exit events can be subsumed by the customer arrival and end-of-service events. The logic of the model would be as follows:

Main Program

- Schedule the first arrival event.
- Schedule the end-of-simulation event.
- Turn control over to the timing routine.
- When the timing routine returns, exit.

Arrival Event

- Schedule a successor arrival event.
- If the server is idle, mark the server as busy and schedule an end-of-service event; otherwise, place the current customer into a data structure representing the server queue.
- Return to the timing routine.

End-of-Service Event

- If the queue is nonempty, remove the first customer from the queue and schedule an end-of-service event for him/her; otherwise, mark the server as idle.

- Return to the timing routine.

End-of-Simulation Event

- Print statistics and tell the timing routine that execution is over.

For the implementer of an event-based simulation package, the only challenges are implementing a "schedule" verb and an efficient event list algorithm to support it, and to provide adequate capabilities for constructing and managing data structures in which pending, but currently unscheduled, events are stored.

3.3 The Transaction-Flow World-View

In the transaction-flow world-view, active objects (transactions) travel through a program, competing for passive objects. In its lifetime, a transaction can undergo time-based (scheduled), state-based, or user-managed delays. Many simulation tools embodying the transaction flow world-view have no direct provision for compound delays; i.e., they have no "wait until" statements. This is not due to any inherent weakness in the world-view; rather it reflects the manner in which "traditional" *implementations* of the world-view have been constructed. Modeling compound delays can entail additional end-user programming. For example, reneging (as discussed in Section 2.3) can be modeled by *cloning* a transaction, having the original transaction undergo a time delay and its clone undergo a state-based delay, and implementing logic that distinguishes which of the two transactions completes its delay first and allows the "winner" to proceed and the "loser" to exit gracefully.

In the transaction-flow world-view, time-based delays are modeled using a single statement whose form is "advance Δt." State-based delays are provided in two forms: so-called *unique* blocking conditions and *nonunique* blocking conditions.

The prototypical transaction-flow language GPSS provides a multiplicity of statements for predefined unique blocking conditions. For example, two different statements, "seize" and "enter," are used to attempt to capture servers with unit capacity and capacity N, respectively. The internal logic of such statements, hidden from users, is quite similar to the user-coded logic for the queuing model sketched for the event-scheduling world-view in Section 3.1. In GPSS parlance, transactions waiting for *unique* blocking conditions are placed into *delay chains*. When a state change affecting transactions experiencing a unique blocking condition takes place, all transactions on the delay chain associated with the blocking condition are removed from the delay chain and placed on the *current events chain*. For example, the GPSS "release" and "leave" statements, which relinquish a single server and relinquish N (often 1) units of a multiple server, respectively, effectively transfer any transactions waiting on a delay chain associated with a server to the current events chain. Reactivated transactions are subsequently "picked up" by the GPSS simulation executive, and they retest their blocking conditions.

GPSS also provides several forms of *nonunique* blocking conditions. Each of these is implemented by polling. Delayed transactions remain on GPSS's *current events chain*, and their delay conditions are reevaluated every time they are encountered by the GPSS simulation executive. This will occur at least once per instant of simulated time. Since many state changes in GPSS force rescanning of the current events chain (in order to make sure that all relieved blocking conditions are recognized at the same instant they are relieved), it is common for transactions in nonunique blocking conditions to be processed more than once per instant of simulated time.

From the user's standpoint, the transaction flow world-view is vastly superior to the event-scheduling world-view. In contrast to the event-scheduling world-view, process flow is readily apparent. Since delays occur only at well-defined points (statements that can delay transaction flow), a process's, wait, execute, wait, execute, ... sequence is also readily apparent. Furthermore, since the management of delay chains is performed by the simulation executive, users are relieved of the burden of having to supply the kind of logic shown in Section 3.1.

The flip side of the transaction-flow world-view is that while easy to use, it can make it easy to unknowingly construct models that execute very inefficiently. For example, consider a GPSS queuing model in which up to 1,000 transactions compete for a single server simultaneously. If one transaction owns a server, and 999 transactions are on a delay chain for that server, when the server is freed, GPSS allows all 999 delayed transactions to reattempt to acquire the server. One of these transactions will succeed in acquiring the server, and 998 will fail, resulting in horrendous overhead. This characteristic of GPSS is well known and has been described by many authors (Schriber 1990, Ståhl 1998, Crain and Brunner 1990, Henriksen 1981). For nonunique blocking conditions, the situation is potentially much worse, since such conditions are in many cases retested more than once per instant of simulated time, as opposed to being tested only when a blocking condition changes.

For both designers and users of a simulation package, ease-of-use is a double-edged sword. While it is convenient to have easy-to-use features, features that "silently" lead to inefficient execution when abused are a dirty trick. SLX, described in the next section, provides feedback that makes it easy to detect such inefficiencies.

In the transaction-flow world-view, the cure for inefficient execution of state-based delays is resorting to user-managed delays, as exemplified by GPSS's *user chains*. When built-in algorithms for managing state-based delays result in poor execution, more intelligent delay management is required. Unfortunately, that intelligence must be provided by the end user.

User-managed delays can also be used for reasons other than improving efficiency. For example, complex queuing disciplines can be implemented as user-managed delays. GPSS's user chains are sets of transactions that can be ordered by a single attribute common to all transactions on the chain. Such a user chain could be used for modeling "most imminent deadline" scheduling by using deadline time as a ranking attribute.

3.4 The SLX World-View

The SLX world-view is the end result of generalization of, and simplification of, the transaction flow world-view. The following similarities and differences between SLX and traditional transaction-flow simulation tools are noteworthy:

- All time-based delays are specified using a statements of the form "advance Δt," exactly as in GPSS.
- State-based and compound delays are specified using a single statement of the form "wait until (expression)."
- All blocking conditions are unique.
- Each variable used in a state-based based "wait until" or in the state-based portion of a compound "wait until" must be declared as a *control* variable. "Control" is a prefix that can be applied to any data type, e.g., it is possible to have control integer variables, control floating point variables, control character strings, etc.
- User-managed waits are specified using statements of the form "wait" and "reactivate x," where x designates a delayed active object. Active objects that are subject to user-managed delays are often managed by placing them into SLX sets. SLX provides sets that can be ranked according to one or more attributes of a single object type, with each attribute ranked in ascending or descending order.
- SLX has an option for enabling collection of performance statistics for a model's "wait until" statements. When monitoring is enabled, statistics are displayed at the end of a run, showing the number of "wait untils" attempted and the ratio of attempts to successes. High ratios of attempts to successes are called *control variable abuse* in SLX.
- In traditional implementations of the transaction flow world-view, transactions have two different types of properties. Some properties are user-defined and user-modifiable. For example, an active object representing a ship in a model of a harbor might have properties representing the ship's type, tonnage, destination, etc. All such properties can be directly manipulated by end users. Other properties are defined by the underlying simulation software. Some of these properties are user-accessible and user-modifiable, while others are not. For example, GPSS transactions have a priority that users can modify directly, but they also have internal pointers used for event list management that users cannot modify or even examine. In SLX, an active object takes the form of a traditional object and a second object pointing to it called a *puck*. SLX pucks are the central repository of information pertaining to an active object's scheduling/delaying. In SLX, it is possible for an active object to have multiple pucks. For example, a machine capable of executing three simultaneous operations would often be modeled in SLX as a single object with three pucks. Ramifications of this architecture are discussed in the next section.

3.5 Ramifications of World-Views

The above characteristics have profound impact on modeling of delays. As we have seen, in the event-scheduling world-view, all delays are user-managed and are primarily modeled by scheduling events. Typical implementations of the transaction-flow world view lack the capability to directly express compound wait conditions. In contrast, SLX's "wait until" is a single, general statement capable of handling all state-based and compound waits. Once an end user masters the use of "wait until," he/she can model *any* state-based or compound delay. Having a single, general construct enables language implementers to focus their development resources on implementation of a single problem.

SLX's "control" declaration prefix makes it easy to designate any variable of any type to be a state variable that can be used in state-based or compound delays. For designated variables, the SLX compiler inserts checks for pending delays at all points at which the variables are modified. The compiler never misses state changes. Other architectures that depend on end user recognition of, and reaction to, state changes can be error-prone.

The ability to attach multiple pucks to a single object makes it easy to describe parallel activities within a single object; however, if multiple pucks are used for large numbers of active objects, then the demands on delay management functions can be greatly increased. This topic is explored further in the next section.

4 Demands Imposed by Application Characteristics and Modeling Style

The demands placed on delay mechanisms are highly application-dependent. For example, telecommunications models often contain very large numbers of simultaneously active packets of information. Massive parallelism and complex synchronization rules place extreme demands on delay-handling mechanisms. For example, I have experimented with one long-running SLX telecommunications model that executes more than 10^{12} "wait until" statements in a single run. For such models, every nanosecond spent handling the execution of a "wait until" adds 1,000 seconds of execution time.

In air traffic control applications, tracking thousands of simultaneous active airplanes is commonplace. In the U.S. airspace alone, the maximum number of simultaneously active flights controlled by the FAA under peak conditions exceeds 5,000. This results in large event lists for time-based delays and places heavy demands on mechanisms for handling state-based delays.

For a given application, modeling style can also have a profound impact on the demands imposed on delay handling mechanisms. For example, consider the impact of using modeling constructs for expressing state-based delays that are implemented using polling. I once conducted a test using the telecommunications model described above, in which I replaced SLX's implementation of "wait until" with a polled implementation, and I added code to detect the ratio of failures to successes for all state-based delay evaluations. The resultant ratio was on the order of 10^4. Running the model to completion was infeasible, as that would have required over 10^{16} evaluations of state-based blocking conditions.

As another example, consider how the activities of an airplane might be described in a model of the U.S. airspace. For every active flight, a large number of state variables must be continuously monitored, some by pilots and others by air traffic controllers. In an SLX model, using multiple pucks for each flight, with each puck monitoring a set of closely-related state variables would probably result in greater model clarity than would be possible in a monolithic description of the monitoring of all state variables. If every flight has ten pucks, delay handling mechanisms must contend with upwards of 50,000 simultaneously active pucks.

5 Modeling Time-Based Delays

5.1 Introduction

The principal challenge in modeling time-based delays is implementing a good *event list algorithm*. A tutorial on event list algorithms is beyond the scope of this paper. Instead, we briefly catalog some of the approaches that have been used and illustrate some of the difficulties in implementing a good algorithm by describing a new algorithm currently under development.

5.2 History

A poor event list algorithm can be the Achilles heel in an otherwise well-implemented simulation. Papers describing novel event list algorithms first appeared in the mid 1970s, and new papers describing event list algorithms are still being written. For a well-written review of the many classes of algorithms that have been proposed, see Chung et al. (1993). At least two Ph.D. theses, McCormack (1979) and Kingston (1984), have explored aspects of event list algorithms.

5.3 Required and Desired Properties of Event List Algorithms

- An event list algorithm must resolve time ties (one or more events scheduled to occur at the same time) in a predictable manner, e.g., FIFO within priority. Consider the use of a tree-based algorithm. Tree-based algorithms typically incorporate complicated rules for keeping an event tree approximately balanced. The inclusion of a tree-balancing strategy implies that a single change in a model can drastically affect the form of the event tree. If time ties are not dealt with explicitly, then the sequence in which they are processed is subject to perturbation at the hands of the tree-balancing algorithm.
- An event list algorithm should be insensitive to the mixture of distributions used for generating random event times.
- An event list algorithm should be able to handle bursts of activity (periods of time with large numbers of near-term events).
- An event list algorithm should handle arbitrary deletions (event cancellations) efficiently.

- An event list algorithm should not require user input to specify the characteristics of the event list for an application. Algorithm parameters should be set automatically and changed when necessary by an internal adaptive process.
- Some applications may need to examine the event list. While this is relatively rare, the need to examine the next scheduled event is much more commonplace. Some event list algorithms do not easily lend themselves to traversing predecessor/successor events. For example, a tree-based algorithm might contain parent, left child, and right child pointers connecting scheduled events, but not incorporate explicit time predecessor/successor pointers. If this is the case, finding an event's predecessor/successor requires recursive tree traversal. If traversing the event list in sequence is important, an algorithm with explicit successor/predecessor pointers is preferable. Later in this paper, we consider the adaptation of event list algorithms to serve as algorithms for managing ranked sets used in user-managed waits. Traversal of ranked sets is a very common operation. For example, a server may need to examine all consumers in its queue in order to decide which one to serve next. If an event list algorithm is to be adapted to serve this purpose, explicit representation of successor/predecessor relationships is required.

5.4 An Event List Algorithm Is a Highly Specialized Form of Priority Queue

Event list algorithms are frequently considered equivalent to priority queue algorithms. A priority queue is a very general construct that can take many forms. Consider the Wikipedia definition of a priority queue:

> A **priority queue** is an abstract data type in computer programming that supports the following three operations:
> **InsertWithPriority**: add an element to the queue with an associated priority.
> **GetNext**: remove the element from the queue that has the *highest priority*, and return it (also known as "PopElement(Off)", or "GetMinimum").
> **PeekAtNext** (optional): look at the element with *highest priority* without removing it.

For example, a telephone book is a form of priority queue (ranked by last name, first name). An event list has the following special properties:

- Event times have a lower bound equal to the value of the simulation clock and an upper bound that is a hardware-dependent maximum representable floating point value. While we can probably safely assume that entries in a telephone book fall in the interval ("aaaa, aaaa", "zzzz zzzz"), there are no conveniently expressed, guaranteed bounds; and the highest priority (first) entry in a telephone book rarely changes.
- Because event times are numeric values, an event list algorithm can include *computations* on the primary ranking attribute, scheduled event time. In Section 5.8, we discuss calendar queue algorithms, in which division is a key operation. In a priority queue, ranking attributes may be nonnumeric, as with a telephone book.

Division cannot be performed on character strings, so a calendar queue algorithm could not be used for a telephone book.

- Event times inexorably increase over the duration of a simulation run. Consider the operation of Henriksen's Algorithm (Henriksen 1977). This algorithm features an event tree that is dynamically adjusted in a manner biased to reflect the inexorable increasing of event times. Using Henriksen's Algorithm for a priority queue that does not have this property may result in less-than-desirable performance.
- Event times are typically sampled from a diverse collection of statistical distributions, while priority queues are often used with only a single distribution.
- The times between events can fall into ranges that differ by many orders of magnitude. For example, consider a telecommunications model. Such a model can track large numbers of events with microsecond or even nanosecond resolution, but also include statistics collection events that occur at intervals measured in minutes or hours. A system shutdown event may have a time measured in days.

5.5 Event List Implementation Approaches

A large number of classes of algorithms have been proposed for use in event list management, including linear lists, indexed lists, multiple lists, heaps, leftist trees, pagodas, top-down and bottom-up skew heaps, splay trees, pairing heaps, Henriksen's algorithm, and calendar queues.

5.6 Measuring Event List Algorithm Performance

The most important measure of performance for the simulation user is model execution time. While analysis of algorithm complexity provides instructive insights for algorithm designers, users have at best casual interest in such measures, and theoretical results are often misleading. For example, an algorithm can possess logarithmic complexity in theory, but if event list sizes are very large, their memory footprint can easily cause degradation of hardware memory cache performance, yielding a linear component to complexity as a function of event list size.

The oldest approach for measuring the efficiency of an event list algorithm is called the *hold model*. This approach was first described in Vaucher and Duval (1975) and operates as follows:

- The event list is initially populated with N scheduled events.
- N remains constant throughout the test.
- At each step of the test, the lowest time event is removed form the event list and reinserted at a randomly sampled time. Franta and Maly (1978) used a family of six event time distributions. Each of their tests used a single distribution.

The hold model is deficient in many ways; however, notwithstanding its shortcomings, it has been very widely used in comparisons of alternative event list algorithms. Among the deficiencies of the hold model are the following:

- Simulation models rarely use a single distribution for scheduling events. Even the single-waiting-line, single-server queuing model uses different distributions for interarrival times and service times. Real-world models typically contain a rich mixture of distributions.
- The distributions in real-world models often have means that differ by orders of magnitude. For example, in a detailed model of a call center, the mean time between questions asked by a caller might be 20 seconds, but if a question requires that an operator look something up on a computer, a burst of events whose interarrival times are measured in milliseconds or microseconds is likely to occur.
- Many real-world systems are prone to bursts of activity. For example, hub-and-spoke routes employed by many airlines result in spikes of activity at airports.

A number of approaches have been taken to improve on the hold model. The simplest of these is called the *interaction hold model* (a phrase coined by McCormack and Sargent 1981) and permits a mixture of distributions. *Markovian* descriptions of model behavior (Chung et al. 1993) are capable (in theory, at least) of realistically representing event scheduling behavior by characterizing a model as a collection of N states, each of which uses a given distribution of event times, and an $N \times N$ matrix of transition probabilities from each state to all others. Other approaches have included procedures for inducing transient behavior (runs up and runs down in the size of the event list).

5.7 What Is the Best We Can Do?

The discussion that follows assumes that no a priori knowledge exists of event scheduling distributions or other special properties.

Linear search algorithms have $O(n)$ complexity for insertions into lists of size n; i.e., insertion time is proportional to the length of the list.

Tree-based algorithms tend to have logarithmic performance. For example, if the event list is maintained as a nearly balanced binary tree, we can expect insertion times to be $O(\log_2(n))$.

Henriksen's algorithm, which utilizes a binary tree whose leaves are short sublists of the full event list and are examined by linear search, has been shown to exhibit $O\left(\sqrt{n}\right)$ worst-case performance, although in my 30 years' experience with this algorithm, I have found that its performance is nearly always much closer to logarithmic.

The newest widely used class of event list algorithms is the calendar queue, first described in Brown (1988). Under ideal operating conditions, these algorithms exhibit $O(1)$ complexity for event list insertions; i.e., insertion times are constant.

This is, of course, the nirvana of event list algorithms. The difficulty is how to assure "ideal operating conditions" in the general case. Under less than ideal operating conditions, calendar queue algorithms can fail spectacularly. A number of algorithms have been proposed for dynamically adjusting calendar queue algorithm parameters in order to maintain reasonable operating conditions. One such algorithm, H2, is described in Section 5.9.

5.8 Basics of Calendar Queues

A calendar queue is an array of *buckets*, each of which contains a linked list. We use the following notation to describe calendar queues:

NB The number of buckets
BW The width (in time) of each bucket
YL The length of one calendar year (YL = NB × BW)
CY The starting time of the current year
B_j The jth bucket
N_j The number of events in the jth bucket, where $0 \leq j \leq \text{NB} - 1$
k The index of the currently active bucket

To insert an event scheduled to occur at time t into a calendar queue, an integer bucket number $j = t \mod \text{BW}$ is calculated; and the event is inserted into a sorted list anchored in B_j. Under ideal conditions (described below), the lists for all buckets are very short, resulting in nearly constant CPU search times. Since the CPU time required to calculate j is constant, the CPU time required to perform an insertion is nearly constant.

At time zero, $k = 0$ and $\text{CY} = 0.0$. To delete the first event in the calendar queue, B_k through $B_{\text{NB}-1}$ are examined until a bucket is found with a first event that falls within the current calendar year. Thus, the search interval is $\text{CY}, \ldots, \text{CY}+\text{YL}$. If no such event is found, CY is incremented by YL, k is set to zero, and the test is repeated for the next calendar year. If large jumps in simulated time are possible, it is advantageous to skip a number of calendar years by scanning the entire bucket list for the minimum event time, rather than incrementing one year at a time, resulting in fruitless searches.

Proper selection of calendar queue algorithm parameters is critical to its success:

- If BW is too large, buckets will contain large numbers of events, and the CPU time required to maintain the buckets' sorted lists will become large.
- If BW is too small, many buckets will contain no events, and the CPU time spent skipping over empty buckets when finding the next scheduled event will become large.
- If NB is too small, buckets will contain large numbers of events. In the degenerate case of NB = 1, *all* events will fall in a single bucket, and large amounts of CPU time will be spent inserting events into the single sorted list. Increasing

NB "unreasonably" results in increased memory requirements and little gain in performance. The upper bound for NB is determined by limits on available memory. Thus the challenge in choosing NB is to avoid the large penalty for choosing a value that is too small, and to determine what value is "unreasonably" large.

A number of strategies for dynamically modifying calendar queue parameters have been proposed. See Oh and Ahn (1999), Tan and Thng (2000), and Ronngren et al. (1997) for examples. Under many of these approaches, parameter modification is triggered by changes in the size of the event list. For example, modification might be triggered every time the size of the event list grows or shrinks by a factor of two or more. While the *size* of the event list is an important influence on the performance of the calendar queue, the *distribution* of events over total time is of even greater importance.

Since there are no guarantees on the properties of the collection of events to be scheduled, achieving $O(1)$ insertion times depends on dynamically determining acceptable values of NB and BW.

5.9 H2, a Calendar Queue Algorithm under Development

This section describes H2, a novel form of calendar queue algorithm currently being developed. We present it as an illustration of a strategy for maintaining "reasonable" operating conditions in a calendar queue. Insertions into H2's event list are performed as follows:

- The calendar queue, per se, comprises only one year. Events whose scheduled times fall within the current year are inserted into binary trees anchored in the bucket determined by event time. Trees are used as a safety measure to avoid excessive cost during intervals in which parameters of the calendar queue may have less than desirable settings. No attempt is made to balance the binary trees. When an event falls in a nonempty bucket, a "search count" variable is incremented, and for each comparison required in the search, a "comparison count" variable is incremented. The ratio of comparisons to searches is used to alter NB when necessary (described below).
- Events whose scheduled times fall outside the current year are placed into either a "near overflow" list or a "distant overflow" list. Both lists are simple LIFO lists.
- At the start of a simulation, the calendar year is set to end at time -1, and the distant overflow interval threshold is set to infinity, so all events are placed into the near overflow list up until the time the first event is removed from the event list.
- The sum, sum-of-squares, and minimum of the event times (expressed as distances beyond the end of the current year) are accumulated on-the-fly for the near overflow list.
- No statistics are collected for events placed into the distant overflow list.

Removing the first event from H2's event list is considerably more complicated than inserting an event. Such deletions are handled as follows:

- Calendar queue buckets are examined in sequence. Since each bucket is either empty or contains one or more events that fall within the current year, removing the first event from a nonempty calendar queue is very straightforward.
- When the current year is exhausted, the size of the near overflow list is examined. If the near overflow list contains exactly one event, then that event is removed from the near overflow list and becomes the next event. If the overflow list is empty, then the entire distant overflow list is transferred to the overflow list, accumulating the sum, sum-of-squares, and minimum of event times.
- When the calendar queue is empty, and the size of the near overflow list is greater than one, a new calendar year is defined. The start of the new year is always the minimum near overflow time. The end of the next calendar year is determined by a somewhat involved, but nevertheless rapid procedure.
- The sum and sum-of-squares are used to compute the mean, standard deviation, and coefficient of variation (CV) of overflow times. The degenerate case in which the standard deviation is zero is treated as a CV of 1.0.

The H2 algorithm is in essence CV-driven. It uses CVs to determine the characteristics of a new calendar year as follows:

- The start of the next calendar year is set to the minimum overflow list event time.
- CVs that are "too large," i.e., CVs that exceed an algorithm threshold parameter, indicate that events in the overflow list are widely dispersed. When large CVs are encountered, events whose times exceed a threshold multiple (an algorithm parameter) of standard deviations above the mean are removed from the current overflow list and placed into the distant overflow list. The mean and CV of the reduced overflow list are computed on-the-fly as the list is thinned, and the resultant CV is reexamined. If the CV of the reduced overflow list is still too large, then the reduction process is repeated. The H2 algorithm uses an experimentally determined maximum of three such iterations. In practice, more than one iteration is rarely needed.
- CVs that are "too small" indicate densely clustered times. If a CV is less than an algorithm parameter, then two possibilities are considered. If the mean overflow time is "close" (an algorithm parameter) to the start of the next calendar year, then the overflow list is regarded as a "near burst," and the end of the next calendar year is set to the mean of the overflow times that lie a certain number of deviations (an algorithm parameter) above the mean. If the mean overflow time is "too far" (an algorithm parameter) from the start of the next calendar year, then the overflow list is regarded as a "far burst," and the all events whose times are greater than a number of standard deviations beyond the start of the next calendar year are transferred to the distant overflow list, and the end of the new calendar year is set to the maximum event time in the reduced near overflow list.
- If the CV is neither too small or too large, then the end of the new calendar year is set to a multiple of standard deviations (an algorithm parameter) above the mean overflow time.

Once the start and end of the new calendar year have been determined, the possibility exists that the end of the new calendar year may overlap the start of the distant overflow list. This is easily determined by examining the distant overflow threshold time. When overlap exists, the distant overflow list must be examined. Events that fall within the new calendar year are transferred to the overflow list where they will be subsequently transferred to the new calendar year. In the process of examining the distant overflow list, we calculate a new distant overflow threshold value equal to the minimum event time in the pruned distant overflow list.

5.10 Observations on H2

- H2 is an extremely complex algorithm with a considerable number of parameters; however, in early experimentation, it has proven possible to develop parameter settings that work well across a variety of event time distributions. H2 has been compared to Henriksen's Algorithm and a splay tree algorithm. Tests conducted to date have used only the hold model. H2 has proven to be faster than the splay tree algorithm and much faster than Henriksen's Algorithm. For both algorithms, the larger the event set, the greater H2's advantage. H2 comes close to $O(1)$ complexity for event lists of up to 10,000 events. For example, I have observed slopes on the order of 10^{-4} nanoseconds per event list insertion/deletion per unit size of the event list. Under such circumstances, if an insertion/deletion cycle for a given distribution takes 40 nanoseconds when the event list size is 1,000, increasing the size to 2,000 will increase the cycle time by a time on the order of 10^{-1} nanoseconds.
- While H2 was not explicitly designed to perform well under the hold model, it is ideally suited for doing so. Most implementations of the hold model populate the event list with a large number of events before moving into time-advance, event-rescheduling mode. Since H2 defers setting of its parameters until the first deletion from the event set, it has the distinct advantage of observing a large number of sample event times before making any commitments.
- When transferring event notices among H2's various lists, great care must be exercised to maintain FIFO order for time ties. For example, if the near overflow list is maintained as a LIFO list, events must be transferred from the near overflow list to the calendar queue in LIFO order. Recall that LIFO(LIFO) = FIFO.
- H2's adaptation strategy is based on statistical properties of event times. Clearly, if computation time were not a consideration, a calendar queue algorithm whose adaptation strategy is statistically based ought to be superior to one that is not. The performance of H2 depends on the effectiveness of its strategy and the speed with which the required computations can be performed. The latter considerations are discussed in Section 5.11, which follows.
- Computing means and standard deviations by using sums and sums-of-squares is well-known to be computationally unstable. In H2, computational instability is mitigated in part by expressing times not as absolute values, but as offsets from the end of the current calendar year. H2 uses these statistics in a very primitive manner, so high precision is not necessary. One could, of course, use much more

sophisticated techniques to capture statistical properties to be used in setting calendar queue properties; however, spending a great deal of time doing so in order to save a little time elsewhere is counterproductive.

5.11 Priority Queue Algorithm Computational Costs

On modern CPUs, accumulation of sums and sums-of-squares can be completely pipelined. Since these values are not used until much later, the cost of collecting them is determined by the rate at which instructions can be *issued*, not the time that it takes to *complete* them.

The primary advantage of a calendar queue algorithm is maximizing the extent to which *indexing* operations are used and minimizing the extent to which *search* operations are used. While the speed of indexing operations is in theory superior to search operations, hardware characteristics cloud the issue to an extent. For example, performing an insertion into a standard calendar queue requires performing a floating point modulo division and converting the result to an integer. Floating point division is the most expensive arithmetic operation in a computer's instruction set. For example, on my computer, division takes 3.5 times as long as addition. To be used as an index, the floating point result must be converted to an integer. How is floating point-to-integer conversion performed? The normal operating mode for IEEE-compliant floating point hardware is to perform calculations that round to the nearest bit. Conversion from floating point to integer is customarily performed by truncation, which requires changing the operating mode of IEEE-compliant hardware from "round" to "chop" and restoring "round" mode after conversion is complete. Hence, many language compilers generate code that performs conversion from floating point to integer by *function call*! Computers that have extended versions of the X86 instruction set may offer alternative instructions for performing floating point to integer conversion. For example, machines with the SSE2 instruction set, introduced by Intel in 2001 and implemented by AMD in 2003, can perform a truncated floating point to integer conversion in a single instruction. Software that exploits this SSE2 instruction will run faster than it will when constrained to use older instructions.

A similar situation exists for floating point comparisons. In early PCs, floating point instructions were implemented using coprocessor chips. Floating point comparisons set status bits in the coprocessor, so before the CPU could execute a conditional branch following a floating point comparison, status bits had to be transferred from the coprocessor to the CPU. Intel's Pentium Pro processors, introduced in 1995, were the first to include floating point comparison instructions that set CPU status bits directly.

Calendar queue indices can be calculated either by rounding or by truncation. It does not matter which form of conversion is used, as long as all such conversions are performed consistently. The problem is that no higher-level programming languages provide mechanisms for telling their compilers to operate in "nonstandard" modes. The consequence of all of the above is that the skilled assembly language programmer still has advantages over compilers when the compilers are constrained to generate code that will run on "old" machines.

Most simulation models schedule one or more distant events at time zero. Examples of such events include events for model shutdown, statistics collection, shift changes, etc. Such events usually end up in H2's distant overflow list. Such outlier events are excluded from the process by which parameters of the calendar queue are determined. Thus, the fact that H2's operation is so ideally suited for the hold model is a fortuitous coincidence that arises for well-motivated reasons.

When event list sizes are considerably larger than 10,000, hardware cache performance becomes a factor. For event lists of size 100,000, H2 is up to 10 times faster than Henriksen's Algorithm and typically 2–3 times faster than the splay tree algorithm, depending on the event time distributions used.

6 Modeling State-Based Delays

6.1 Introduction

Although management of time-based delays has long been recognized as a great source of overhead, very little has been published on efficient algorithms for handling state-based delays. This section presents SLX's near-optimal algorithm for handling state-based delays of arbitrary complexity. We use SLX parlance. State variables are called *control variables*, and active objects involved in state-based wait conditions are referred to by their SLX *pucks*. We use "&&" to represent Boolean *and* operations and "||" to represent Boolean *or* operations. State-based delays are portrayed using SLX's "*wait until*" statement.

6.2 The Objective

The objective of this algorithm is to achieve the fastest possible implementation of the generalized "wait until" statement for discrete event simulation:

```
wait until (arbitrary Boolean expression);
```

First, following modern language convention, the Boolean expression must be evaluated using an "earliest possible exit" strategy. This allows statements such as

```
wait until (p != NULL && p -> attrib > 10);

wait until (p == NULL || p -> attrib < 10);
```

The symbolism "`p -> attrib`" is used to specify an attribute named "`attrib`" in an object pointed to by pointer variable "`p`." In both of the above statements, the second terms of the Boolean expressions are "guarded" by the first terms, eliminating the possibility that a NULL pointer is used to reference an object.

Second, when a Boolean expression evaluates to false, the puck executing a "wait until" waits for value changes only for those control variables that contribute to the falsity of the expression. Consider the following statement:

```
Control Boolean a, b;
Int c, d, e;
wait until ((a && b) || c + d > e);
```

In order for the above expression to be false, `c + d` *must* be less than or equal to `e`, and either `a` or `b` must be false; however, if `a` is false, then `b` is not examined. Thus, when waiting takes place, `c`, `d`, `e`, and either `a` or `b` must be placed into *state change queues*.

Statements that modify a state variable must process the variable's state change queue when it is found to be nonempty. For example, a statement of the form

```
e = 12.0;
```

must trigger reevaluation of the above "wait until" statement.

6.3 Run-Time Data Structures

State change queues are organized as a double-ring structure. Consider the following example:

State variables:

```
control Boolean a, b, c;
```

"wait until" statements:

```
1. wait until (a);
2. wait until (a || b);
3. wait until (c && b);
```

If `a`, `b`, and `c` are all FALSE, the double-ring structure would appear as in Fig. 1.

Vertical rings link each control variable to the instances in which a puck is waiting for a change in the variable's state. Horizontal rings link all the control variables that contribute to the falsity of a given "wait until" statement. Black dots represent control blocks used to store object/set connections.

Assume that `a` is set to TRUE. By traversing `a`'s vertical ring, we can identify all the pucks that are waiting for `a` to change (in this case, P1 and P2). By

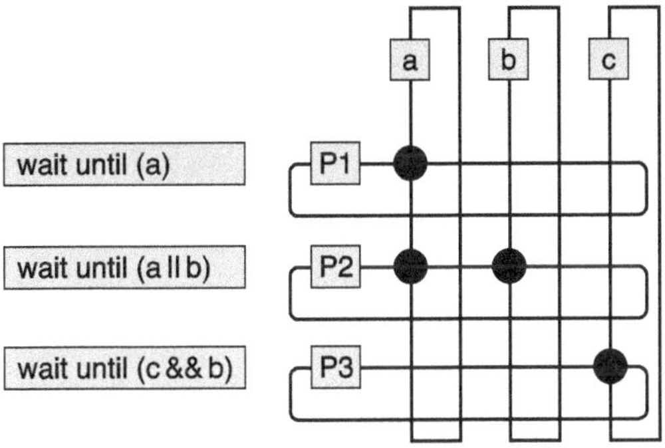

Fig. 1 The "wait until" double ring data structure

traversing P1's horizontal ring, we can remove the connection between P1 and a, and by traversing P2's horizontal ring, we can remove the connections between P2 and a and b. If b is set to TRUE, only P2 is affected, and if c is set to TRUE, only P3 is affected.

6.4 Implementation Strategy

Each control variable is prefixed with "first" and "last" pointers that anchor the variable's vertical ring. At all points at which a control variable is modified, a test is inserted to detect a nonempty ring. The overhead incurred when the ring is empty is only 2–3 machine instructions (roughly a nanosecond or so on current hardware).

The execution of "wait until" statements is carried out in two phases. In Phase I, the Boolean expression is evaluated, keeping track of those variables that contribute to falsity in the event the expression is false. If the expression is false, in Phase II, the control variables actually contributing to falsity are stacked and passed to a run-time support function.

Postponing the issue of *when* we must remember control variables contributing to falsity, let us consider *how* we remember control variables. To accomplish this task, *jump words* are employed. In contexts that require us to remember a control variable, a jump word is used to store labels (memory addresses) that are used as branch targets in Phase II. There are other alternatives to the use of jump words. For example, control variables could be remembered by pushing them onto the run-time stack. The problem with this approach is that if variables already pushed onto the stack are later discovered to be irrelevant, they must be popped off the stack. Management of such pushes and pops would be extremely difficult. Jump words enable clean, low-overhead execution. Consider the following example:

```
wait until (a && b || c);
```

```
L6:  MOV    RootESP,ESP    Start Phase I; Remember stack top
     CMP    a,0
    *MOV    EAX,L1            Label used to enqueue a in Phase II
     JZ     L2
     CMP    b,0
    *MOV    EAX,L3            Label used to enqueue b in Phase II
     JNZ    L4               Jump ahead if the expression is true
L2: *MOV    JW1,EAX           Store label value in a jump word
     CMP    c,0
     JNZ    L4             Jump ahead if the expression is true

     LEA    ESI,queue(c)      Start Phase II; c is known to be false
     PUSH   ESI
     JMP    JW1               Jump => enqueue a or enqueue b
L3:  LEA    ESI,queue(b)
     PUSH   ESI
     JMP    L5
L1:  LEA    ESI,queue(a)
     PUSH   ESI
L5:  CALL   Wait_UntilN    Run-time support
     JMP    L6             Retest upon eventual resumption
L4:  ----------
```

Fig. 2 A simple "wait until" example

The pseudocode in Fig. 2 shows how this statement executes.

With the exception of jump word processing and special handling for control variables whose locations are computed, e.g., array references, Phase I uses standard techniques for evaluation of the Boolean expression. Only the *-flagged instructions are peculiar to "wait until" processing. The instructions that move label values into register EAX are carefully sandwiched between test instructions and their corresponding conditional branches. CPUs usually stall for a cycle or two when a test instruction is immediately followed by a conditional branch, so the MOV instructions typically add *no* overhead.

When we encounter a control variable whose location is dynamically computed, the address of the selected location, e.g., an array element or an attribute of an object accessed by means of a pointer, is saved in a temporary location for subsequent use in Phase II. Consider the use of a random variable as an array subscript. Phase II cannot recalculate the address of a randomly selected array element evaluated as false in Phase I, because reevaluation will almost certainly yield a different subscript value. While such usage is fairly rare, we are solving a very general problem, and it is easier to solve the problem in general than to develop an elaborate set of restrictions on expression contents.

In Phase II, the tree representing the Boolean expression is traversed in top-to-bottom, right-to-left order. Enqueuing control variables for an OR operator is easy: if the OR is false, then both its sides are false and must be enqueued. Enqueuing control variables used in AND operators is accomplished by means of indirect jumps that enqueue either the left or right side of the AND.

There are subtle asymmetries involved in the forms of Boolean expressions used in "wait until" statements. Consider the following example, which is very similar to the preceding example:

```
wait until (a || (b && c));
```

The pseudocode in Fig. 3 shows how this statement executes.

In the above example, the AND operator is the right child of the OR operator, so if either of a or b is false, then execution can jump directly to the respective enqueuing instruction sequences for a and b without the need for an intervening step to remember which was false.

6.5 A Complex "wait until" Example

Consider the sample code shown in Fig. 4.

The tree representation of this program fragment is shown in Fig. 5.

```
L4:  MOV    RootESP,ESP    Start Phase I; remember stack top
     CMP    a,0
     JNZ    L1             If a is true, the entire expression is true
     CMP    b,0
     JZ     L2             If b is false, the expression is false
     CMP    c,0
     JNZ    L1             If c is true, the entire expression is true

     LEA    ESI,queue(c)      Phase II: fall-through from above => enqueue c
     PUSH   ESI
     JMP    L3
L2:  LEA    ESI,queue(b)      Enqueue b (direct branch to L2 above)
     PUSH   ESI
L3:  LEA    ESI,queue(a)      Always enqueue a
     PUSH   ESI
     CALL   Wait_UntilN    Run-time support
     JMP    L4             Retest upon eventual resumption
L1:  ----------
```

Fig. 3 An example of "wait until" asymmetry

```
control boolean      a, b, c, d, e, f, g, h, i;
control double            x[10], z;
control int          ix, jx;

wait until (((a && b || c && d) || (e && f || g && h)) && i || x[ix+jx] > z);
```

Fig. 4 A complex "wait until" example

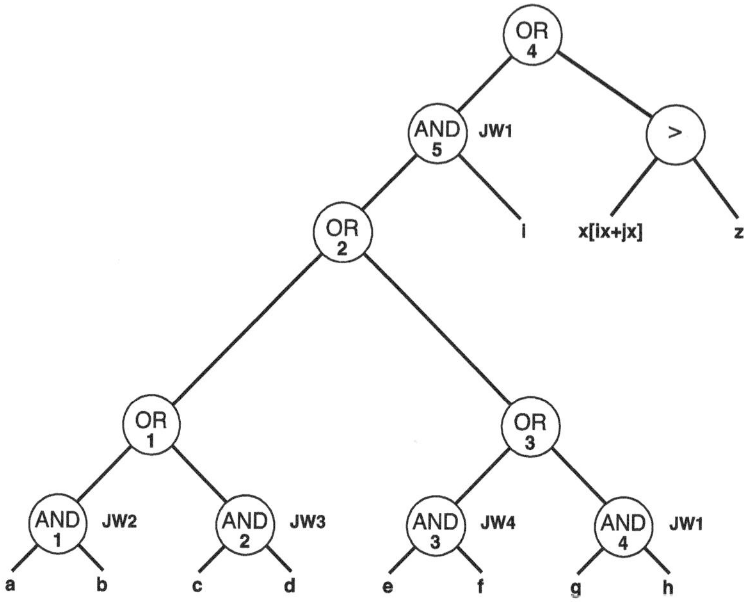

Fig. 5 The expression tree for a complex "wait until"

The Objective

Assume the following values:

```
a          true
b          false
c          false
d          true
e          true
f          false
g          false
h          true
i          true
x[ix+jx]   0.0
z          1.0
```

Let us walk through the execution of the statement.

1. Since `b` is false, "`a` AND_1 `b`" is false, so "`c` AND_2 `d`" must be examined.

2. Since `c` is false, "`c` AND_2 `d`" is false, but `d` is never evaluated, due to the falsity of `c`.
3. Since both its subtrees are false, OR_1 is false, the left subtree of OR_2 is false, so OR_2's right subtree must be evaluated.
4. Since `e` is true, but `f` is false, so AND_3 is false, and OR_3's right subtree must be examined.
5. Since `g` is false, `h` is never evaluated; OR_3 is false; OR_2 is false; and AND_5 is false, so OR_4's right subtree must be examined.
6. Since `x[ix+jx]` is less than `z`, the entire expression is false.

Upon completion of the statement, the following control variables must be inserted into state-change queues: `b`, `c`, `f`, `g`, `x[ix+jx]`, `ix`, `jx`, and `z`.

The pseudocode in Fig. 6 shows exactly how our example is executed.

In the code shown above, for jump word branches, the variable actually enqueued is underscored.

6.6 Compiling "wait until" Statements

Phase I is carried out by recursively traversing the tree structure representing the "wait until" expression from top-to-bottom, left-to-right. As the tree is traversed, jump words are allocated as follows:

1. Jump words are required for AND operators that lie on the left branch of a higher-level OR operator. In Fig. 5, AND_1 is the right child of OR_1, and AND_3 is the left child of OR_3, so their need for jump words is obvious. The operator AND_2 is reached via the left branch of OR_1, and AND_4 is reached via the left branch of OR_4, so their need for jump words is a little less obvious.
2. We could simply allocate a new jump word for each qualifying AND, but we can achieve improved efficiency by allowing consecutive AND operators to share the same jump word. Consider the expression "`a && b && c && d || e`". If any of `a`, `b`, `c`, or `d` are false, evaluation continues with `e`. Clearly, a single jump word can be used to select distinguish the first of `a`, `b`, `c`, and `d` that was found false, in the event `e` is also false.
3. A new jump word is allocated only when one is required and sharable jump word does not exist.

Phase II consists of a second recursive traversal of the expression tree. This traversal is performed top-to-bottom, right-to-left (*not* left-to-right). Traversal in this manner results in enqueuing control variables in the reverse order of their occurrence in the "wait until" expression. A quick glance at Fig. 6 verifies that this is indeed the case.

In Phase II, when an AND node that has a jump word is encountered, an indirect jump to the address contained in the jump word is generated. When a node is encountered that has no jump word, it is either a fall-through node or is the subject of a "branch on false" conditional jump. For conditional branches, the branch target

Phase I

```
L17:  MOV     RootESP,ESP  start Phase I; save the current stack pointer for
                           use by run-time support
      CMP     a,0
      MOV     EAX,L1       jump word target for a
      JZ      L2
      CMP     b,0
      MOV     EAX,L3       jump word target for b
      JNZ     L4
L2:   MOV     JW2,EAX      a or b
      CMP     c,0
      MOV     EAX,L5       jump word target for c
      JZ      L6
     *CMP     d,0
     *MOV     EAX,L7       jump word target for d
     *JNZ     L4
L6:   MOV     JW3,EAX      c or d
      CMP     e,0
      MOV     EAX,L8       jump word target for e
      JZ      L9
      CMP     f,0
      MOV     EAX,L10      jump word target for f
      JNZ     L4
L9:   MOV     JW4,EAX      e or f
      CMP     g,0
      MOV     EAX,L11      jump word target for g
      JZ      L12
     *CMP     h,0
     *MOV     EAX,L13      jump word target for h
     *JZ      L12
L4:  *CMP     i,0
     *MOV     EAX,L14      jump word target for i
     *JNZ     L15
L12:  MOV     JW1,EAX      g, h, or i
      ...                  calculate and verify address of x[ix+jx]
      MOV     T1,ECX       save array element address
      FLD     z
      FLD     [ECX+8]
      FCOMIP  ST(0),ST(1)
      FSTP    ST(0)
      JA      L15

      LEA     ESI,queue(z)        start Phase II; enqueue z
      PUSH    ESI
      PUSH    T1           enqueue x[ix+jx]
      LEA     ESI,queue(ix) enqueue ix
      PUSH    ESI
      LEA     ESI,queue(jx) enqueue jx
      PUSH    ESI
      JMP     JW1          enqueue i, h, or g
L14: *LEA     ESI,queue(i)
     *PUSH    ESI
```

Fig. 6 Compiled instructions for a complex "wait until"

```
      *JMP     L16
L13:  *LEA     ESI,queue(h)
      *PUSH    ESI
      *JMP     JW4             enqueue f or e
L11:   LEA     ESI,queue(g)
       PUSH    ESI
       JMP     JW4             enqueue f or e
L10:   LEA     ESI,queue(f)
       PUSH    ESI
       JMP     JW3             enqueue d or c
L8:   *LEA     ESI,queue(e)
      *PUSH    ESI
      *JMP     JW3             enqueue d or c
L7:   *LEA     ESI,queue(d)
      *PUSH    ESI
      *JMP     JW2             enqueue b or a
L5:    LEA     ESI,queue(c)
       PUSH    ESI
       JMP     JW2             enqueue b or a
L3:    LEA     ESI,queue(b)
       PUSH    ESI
       JMP     L16
L1:   *LEA     ESI,queue(a)
      *PUSH    ESI
L16:   CALL    Wait_UntilN     run-time support
       JMP     L17             retest upon eventual resumption
L15:   --------------------
```

* Instructions not executed

Fig. 6 (continued)

is resolved to the current location. In either case, all of the node's control variables are pushed onto the stack of control variables for which state-change queues must be constructed. A slight improvement can be obtained by recognizing consecutive AND operators that do not share the same jump word. Normally, when the right child of an AND operator is enqueued, a jump around the enqueuing of the left child is required; however, if the next operator for which operand enqueuing must be performed is another AND that has its own jump word, the second AND will be implemented using a jump word jump instruction. If we fail to recognize this situation, a jump to a jump will result. Under these circumstances, the preferred approach is to duplicate the jump word–style jumping of the second AND instead of generating a jump to it. In Fig. 6, this optimization occurs for jump words `JW2`, `JW3`, and `JW4`.

This example contains quite a few subtleties. For example, if the "wait until" expression evaluates to false, execution always falls through to the enqueueing of `z`, `x[ix+jx]`, `ix`, and `jx`, i.e., the right subtree of OR_4 is enqueued. Next, the left subtree of OR_4 is enqueued. For this subtree to be false, exactly one of `i`, `h`, or `g` must be false. Code generated in part 1 has deposited the address of the appropriate label L_i in JW_1, so an indirect branch to the contents of JW_1 is generated. (The three possible labels are L11, L13, and L14.)

Careful study of Fig. 6 will reveal the correctness of the generated code.

We have consistently stated that our approach is "near-optimal." One case that our algorithm does not detect is duplicate enqueuing. For example, in the statement

```
wait until (a < b || a < c);
```

where a, b, and c are all arithmetic control variables, if the condition is false, then a will be enqueued twice. In the general case, we could even go so far as to detect enqueuing of identical subexpressions. Since determination of what is to be enqueued is performed dynamically, detecting duplicate enqueuing for the general case would be extremely difficult. Even though such an approach would lead to faster run times, we have taken the simpler approach. Duplicate enqueuing can always be avoided manually. For example, we can rewrite the previous statements as follows:

```
if (b > c)
        d = c;
else
        d = b;
wait until (a < d);
```

6.7 Historical Perspectives

The utility of "wait until" statements for expressing state-based delays in simulation models was first recognized long ago. One of the earliest, if not the first, implementations of "wait until" in discrete-event simulation was done for SOL (Simulation-Oriented Language) in the 1960s (Knuth and McNeley 1964a,b). SOL was a language that compiled into pseudocode that was executed by an interpreter. Knuth and McNeley offer the following descriptions:

> [active objects] waiting for a condition such as "wait until A = 0," for some global variable A, are kept in a list associated with A; this list is interrogated only when the value of A has been changed.
>
> [when an active object ceases moving,] the simulator examines all other [active objects] that are stopped because of a wait-until statement involving global quantities changed by [the current active object]. If [their wait condition] is now true, those [active objects] become free to move.

Details of the design and implementation of SOL have by now faded into obscurity, but the following characteristics are known:

1. SOL's "wait until" expression could contain AND, OR, and NOT operators, and array references were allowed.
2. No special designations were required for defining state variables. Apparently SOL's compiler was able to draw such inferences.
3. State variables were restricted to global integer and floating point values.

There is no record as to whether only those state variables contributing to falsity were enqueued. This may have been contemplated at one time, but probably was not implemented. According to Knuth (personal communication), SOL *may* have waited only for the first state variable contributing to falsity.

Vaucher (1973) described an implementation of "wait until" that was created as an extension to Simula. The implementation was done in Simula source code. No changes were made to the Simula compiler. Expressions of type "wait until" having arbitrary complexity were allowed. They could include function calls and time-based conditions. Simula processes that were delayed as the result of "wait untils" were placed into a single, common queue and *passivated* (deactivated). A monitor process was responsible for (re)activating delayed processes so they could retest their blocking conditions. The monitor was itself activated at each clock update. Thus, the implementation utilized a polled approach. This architecture placed severe limits on the accuracy of time-based delays, because the monitor process could not control the simulator clock. For example, assume that at time 10, the following statements are executed:

```
wait until (time = 15);

wait until (time > 15);
```

If the next event is scheduled for time 20, the first statement will never succeed, but the second statement will succeed at time 20. To deal with these deficiencies, an "alarm" construct was added, allowing the scheduling of dummy events at user-specified times. Proper execution of the first "wait until" statement shown above required setting alarms for time 15 and time 15. Vaucher does not describe the handling of statements such as

```
wait until (time >= 15);
```

In SLX, "`time > 15`" is interpreted as "the first time that occurs after time 15," and no intervention is taken to force an event at time 15, while "`time >= 15`" forces an event to occur at time 15.

7 Compound Delays

SLX supports the use of at most one time-based condition in a "wait until" statement. Six cases must be handled:

```
wait until (time == t ...);
wait until (time != t ...);
wait until (time > t ...);
wait until (time >= t ...);
wait until (time < t ...);
wait until (time <= t ...);
```

Without loss of generality, we ignore the six additional cases that arise when the operands of the comparison operators are reversed. The six cases are handled as follows:

- "`time == t`" is handled by scheduling the active puck to move at time `t`. State-based components in the wait until expression may cause additional delay beyond time `t`. If this occurs, it is treated as an execution error, since the clock increases monotonically, and once time is greater than `t`, the "`time == t`" condition can never succeed.
- "`time != t`" is handled by treating the simulator clock as a control variable.
- "`time > t`" is handled in two stages. First, the active puck is scheduled to move at time `t`. When time `t` is reached, the active puck waits for the simulator clock to change by treating the clock as a control variable.
- "`time >= t`" is handled exactly like "`time == t`," except that going beyond time `t` is not treated as an error.

One might question the utility or even the validity of the last two forms, "`time < t`" and "`time <= t`." Consider the following example. Suppose `t` is an expression that represents the time at which a corrective action could be completed. If the current time is already greater than `t`, there is not enough time to perform the corrective action. Such usage is rare, but it does occur. SLX handles by requiring that the expression for `t` contain one or more control variables, and makes state change queue insertions only for the expression `t`.

In each of the above cases that require scheduling events, if state-based portions of a compound delay are such that retesting a "wait until" expression is required when a control variable is changed before the scheduled time is reached, then the scheduled events are canceled.

8 User-Managed Delays

The final class of delays that we will consider is user-managed delays. User-managed delays are useful in at least two contexts. First, in the presence of heavy queuing, "wait until" statements can result in high ratios of attempts-to-successes. In such circumstances, user-managed queues offer an attractive alternative. Second, user-managed delays can be used to implement complicated queuing disciplines. Consider the processing of jobs in a jobshop model in which jobs are represented as active objects. If scheduling rules are complex, and there are multiple sources of jobs, each having distinct characteristics, then having job objects determine for themselves whether they should be delayed can be very difficult. An alternative approach is to use another active object as a "queue manager." Incoming jobs can then establish their own unique properties, join a queue, and sleep until awakened by the queue-manager object. The queue-manager object has a centralized view of the queue and can examine the queue in its entirety in order to make an informed decision on when any given job is allowed to continue moving through the system.

The key requirements for user-managed delays are having (1) a "sleep" verb, whereby an active object can suspend its own execution; (2) an "awaken" verb, whereby another active object can cause a sleeping object to resume its execution; and (3) data structures that are rich enough to conveniently store delayed objects.

SLX provides "wait" (no "until" clause) and "reactivate" statements that satisfy requirements (1) and (2). The most commonly used repository for delayed objects in SLX is a *set*. The remainder of this section describes the design and implementation of sets, which fulfill requirement (3).

SLX provides FIFO and LIFO sets and sets ranked on one or more attributes of the objects stored in the sets, with each attribute ranked in ascending or descending order. Generalized sets are convenient for uses other than managing delays. For example, one of the challenges in developing large-scale logistics models is simply "keeping track of where everything is." In such models, data management can be a bigger problem than time management. It is not unusual for such models to spend 25% or more of their time performing set operations.

One of the nasty little problems in set management is determining whether an object is already in a set. (SLX allows a given object to be included in a given set at most once.) In early implementations of SLX, every object had a list of the sets to which it belonged, if any. The theory was that most objects would belong to a small number of sets, so determining whether an object was in a given set could be accomplished by a short, linear search. This approach worked well for a long time, until a user came along with an application containing a relatively small number of objects, each of which belonged to a large number of sets. Search times were excessive. For this application, it would have been more efficient to search the lists comprising the sets' members. The solution that accommodated both extremes was to implement a set membership hash table. Entries in the hash table contain pointers to (set, object) pairs, and hashing is accomplished by using shifting and masking operations to form a hash table index from a (set, object) pair whose membership is to be queried. SLX attempts to keep the size of the hash table around one fourth of the total number of (set, object) pairs that exist in a model.

Ranked sets are very useful. Consider the following description of jobs in a jobshop model:

```
class job
  {
  int job_priority;
  double job_deadline;
  };
set(job) ranked(descending job_priority, ascending job_deadline)
job_queue;
```

Insertion of jobs into the job queue in their proper positions is accomplished automatically by executing statements of the form

```
place j into job_queue;
```

Sets can be traversed as follows:

```
for (j = each job in job_queue)

      ...
```

If more detailed traversal is required, lower-level operators can be used:

```
j = first job in job_queue;
j = last job in job_queue;
j = successor(i) in job_queue;
j = predecessor(i) in job_queue;
```

Operators for querying set membership are provided:

```
if (j is_in set1)
if (k is_not_in set2)
```

Ranked sets are priority queues, so good priority queue algorithms are candidates for use in managing ranked sets. If one already has an event list algorithm, it is tempting to extend the event list algorithm for use with general sets. This is, indeed the approach that was taken with SLX (using, not surprisingly, Henriksen's Algorithm). Some event list algorithms are unsuitable for use in ranked set management. For example, the H2 algorithm uses unordered overflow lists, and it does not maintain explicit successor and predecessor pointers. Among the algorithms that have been used for event lists, the splay tree is one attractive possibility for use in ranked set management. Most implementations of splay trees do not maintain explicit successor and predecessor pointers. To determine a set member's successor or predecessor under these circumstances requires moving up and down the splay tree. In a test implementation of a splay tree algorithm adding successor and predecessor pointers increased insertion times by approximately 10%.

One of the earliest examples of general-purpose sets for use in simulation was SIMSCRIPT II.5 (Kiviat et al. 1973). Many of the capabilities described above were first implemented in SIMSCRIPT over 30 years ago. SIMSCRIPT's implementation of sets suffered in several ways:

1. The pointers for constructing lists comprising sets were stored in the objects comprising a set. This had several consequences. First, all potential set memberships for an object had to be declared in advance in order to have the pointers allocated. Second, if more than one type of object could belong to the same set, the pointers had to be at the same offsets within their respective objects, and accomplishing this was a user responsibility. Third, it was impossible to have different dynamically allocated sets with the same name. For example, in a jobshop model, it might be convenient to have multiple instances of a job queue set; however, this was impossible, because only one instance of the set pointers was allocated per set *name*.
2. The rules by which the compiler generated the names of set pointers were documented, so the variables could be modified directly by an end user. It was easy to corrupt the structure of a set.

The solutions to these problems are (1) connecting set members to each other and to their set by using control blocks that are exogenous to both, and (2) preventing direct user access to such data structures. The former approach facilitates implementation of nonhomogeneous sets and eliminates the need to declare potential set memberships. The latter approach forces users to use built-in first, last, successor, predecessor, is_in, and is_not_in operators to perform low-level set manipulations, thereby assuring the integrity of all sets.

9 Examples of "wait until" Performance

In my preparations for writing this paper, I conducted an informal survey of a number of SLX users. Each user who responded provided statistics gathered by turning on run-time statistics collection for wait until statements. The results of the survey were as follows:

Telecommunications Model 1
9,130,589 Wait Untils Completed
Average Control Variables/Expression: 3.08
Average AND Operators/Expression: 0.00
Average OR Operators/Expression: 2.08
Average Attempts/Completion: 2.01
Average Comparisons/Attempt: 2.54
Average Queue Insertions/Delay: 2.75
Average Time Advances/Delay: 0.00

Telecommunications Model 2
16.12G Wait Untils Completed
Average Control Variables/Expression: 1.47
Average AND Operators/Expression: 0.00

Average OR Operators/Expression: 0.47
Average Attempts/Completion: 1.26
Average Comparisons/Attempt: 1.21
Average Queue Insertions/Delay: 1.46
Average Time Advances/Delay: 0.66

Homeland Security Airport Model
87,547 Wait Untils Completed
Average Control Variables/Expression: 1.04
Average AND Operators/Expression: 0.02
Average OR Operators/Expression: 0.02
Average Attempts/Completion: 1.55
Average Comparisons/Attempt: 1.00
Average Queue Insertions/Delay: 1.00
Average Time Advances/Delay: 0.34

Air Traffic Control Model 1
1,375,823 Wait Untils Completed
Average Control Variables/Expression: 2.55
Average AND Operators/Expression: 0.13
Average OR Operators/Expression: 1.13
Average Attempts/Completion: 2.01
Average Comparisons/Attempt: 2.17
Average Queue Insertions/Delay: 2.31
Average Time Advances/Delay: 0.03

Air Traffic Control Model 2
276,219 Wait Untils Completed
Average Control Variables/Expression: 3.33
Average AND Operators/Expression: 0.46
Average OR Operators/Expression: 0.48
Average Attempts/Completion: 1.97
Average Comparisons/Attempt: 1.95
Average Queue Insertions/Delay: 2.44
Average Time Advances/Delay: 0.49

Transportation Model
19,325 Wait Untils Completed
Average Control Variables/Expression: 1.59
Average AND Operators/Expression: 0.18
Average OR Operators/Expression: 0.41
Average Attempts/Completion: 1.20
Average Comparisons/Attempt: 1.58

Average Queue Insertions/Delay: 1.90 (not counting the simulator clock)
Average Time Advances/Delay: 1.25

Of greatest interest are the ratios of attempts-to-completions. As discussed at many points in this paper, high ratios imply high overhead, and excessive ratios indicate the need to move from state-based and compound delays to user-managed delays. Prior to conducting the survey, I would not have been surprised to see that ratios typically ran in the 5–10 range. I was very pleasantly surprised to see low ratios across the board, indicative of well-constructed models.

Other statistics of interest include the average number of control variables and the average number of state-change queue insertions per delay. The extent to which the latter is less than the former is indicative of the efficacy of enqueuing only control variables that actually contribute to falsity. The results shown above are somewhat disappointing, given the degree of difficulty implementing minimal enqueueing. I have seen other models that utilize very complex "wait untils." For example, in one traffic model, a 19-term "wait until" was used to describe the conditions under which a left turn could be made in one particularly complex environment. The "wait until" expression began

```
wait until (light != RED && ...
```

whenever the traffic light was red, only a single state change was enqueued, notwithstanding the complexity of the remainder of the "wait until" expression. Alas, performance statistics for this model are unavailable.

The approach that yields best performance when constructing "wait until" statements is as follows:

1. In an AND or cascaded sequence of ANDs, supply expression terms in increasing order of probability of being true. In other words, place the least likely alternative first. This approach is best because ANDs exit on the first false condition encountered.
2. In an OR or cascaded sequence of ORs, supply expression terms in decreasing order of probability of being true. In other words, place the likeliest alternative first. This approach is best because ORs exit on the first true condition encountered.

10 Conclusions

We have extensively reviewed the four forms of delay used in discrete event simulation models: time-based, state-based, compound, and user-managed delays. The latter three forms have received scant attention in simulation literature. Poor management of delays is a sure-fire path to long-running models. The techniques presented above offer efficient solutions to delay modeling.

First-rate implementation of delay mechanisms requires compile-time and run-time support. Having a compiler recognize all points at which a state variable is changed is vastly preferable to requiring simulation users to do so. For such reasons, well-crafted simulation-specific compilers will always have an advantage over general-purpose language compilers. Large, long-running models can execute truly vast numbers of delays, placing the quality with which delays are handled in sharp focus.

Acknowledgments I thank Dr. Donald Knuth for kindly sharing design notes from SOL.

References

Brown, R. 1988. Calendar queues: a fast $O(1)$ priority queue implementation for the simulation event set problem. *Communications of the ACM* 31(10):1220–1227.

Chung, K., J. Sang, and V. Rego. 1993. A performance comparison of event calendar algorithms: an empirical approach. *Software, Practice & Experience* 23(10):1107–1138.

Crain, R. C., and D. T. Brunner. 1990. Modeling efficiently with GPSS/H. In *Proceedings of the 1990 Winter Simulation Conference*, eds. O. Balci, R. Sadowski, and R. Nance, pp. 89–93. Piscataway, New Jersey: Institute of Electrical and Electronics Engineers.

Franta, W. R., and K. Maly. 1978. A comparison of heaps and the TL structure for simulation event set. *Communications of the ACM* 21(10):873–875.

Henriksen, J. O. 1977. An improved event list algorithm. In *Proceedings of the 1977 Winter Simulation Conference*, eds. H. J. Highland, R. G. Sargent, and J. W. Schmidt, pp. 547-557. Piscataway, New Jersey: Institute of Electrical and Electronics Engineers.

Henriksen, J. O. 1981. GPSS—Finding the appropriate world-view. In *Proceedings of the 1981 Winter Simulation Conference*, eds. T. Ören, C. Delfosse, and C. Shub, pp. 505–516. Piscataway, New Jersey: Institute of Electrical and Electronics Engineers.

Henriksen, J. O. 2000. SLX: the X is for extensibility. In *Proceedings of the 2000 Winter Simulation Conference*, eds. R. R. Barton, J. A. Joines, P. A. Fishwick, and K. Kang, pp. 183–190. Piscataway, New Jersey: Institute of Electrical and Electronics Engineers. Available online as `www.informs-sim.org/wsc00papers/027.PDF` [accessed February 6, 2008].

Kingston, J. 1984. Analysis of algorithms for the simulation event list. Ph.D. diss., Department of Computer Science, University of Sydney.

Kiviat, P. J., R. Villanueva, and H. M. Markowitz. 1973. *The SIMSCRIPT II.5 programming language*. Los Angeles: Consolidated Analysis Centers, Inc.

Knuth, D. E., and J. L. McNeley. 1964a. SOL—A symbolic language for general-purpose systems simulation. *IEEE Transactions on Electronic Computers* EC-13(4):401–408.

Knuth, D. E., and J. L. McNeley. 1964b. A formal definition of SOL. *IEEE Transactions on Electronic Computers* EC-13(4):409–414.

McCormack, W. M. 1979. Analysis of future event set algorithms for discrete event simulation. Ph.D. diss., Syracuse University.

McCormack, W. M., and R. G. Sargent. 1981. Analysis of future event set algorithms for discrete event simulation. *Communications of the ACM* 24(12):801–812.

Oh, S., and J. Ahn. 1999. Dynamic calendar queue. In *Proceedings of the 32nd Annual Simulation Symposium*, pp. 20–25. Washington, DC: IEEE Computer Society.

Rönngren, R., and R. Ayani. 1997. Lazy queue: an efficient implementation of the pending event set. In *Proceedings of the 24th Annual Simulation Symposium*, pp. 194–204. Los Alamitos, CA: IEEE Computer Society Press.

Schriber, T. J. 1990. *An Introduction to Simulation using GPSS/H*. New York: Wiley.

Ståhl, I. 1998. The efficient implementation of wait statements. In *Proceedings of the 1998 Winter Simulation Conference*, eds. D. J. Medeiros, E. F. Watson, J. S. Carson, and M. S. Manivannan, pp. 523–530. Piscataway, New Jersey: Institute of Electrical and Electronics Engineers. Available online as `www.informs-sim.org/wsc98papers/070.PDF` [accessed February 8, 2009].

Tan, K. L., and L.-J. Thng. 2000. SNOOPy calendar queue. In *Proceedings of the 2000 Winter Simulation Conference*, eds. R. R. Barton, J. A. Joines, P. A. Fishwick, and K. Kang, pp. 487–495. Piscataway, New Jersey: Institute of Electrical and Electronics Engineers. Available online as `www.informs-sim.org/wsc00papers/068.PDF` [accessed February 8, 2009].

Vaucher, J. G. 1973. A "wait until" algorithm for general purpose simulation languages. In *Proceedings of the 1973 Winter Simulation Conference*, eds. A. C. Hoggatt and J. Sussman, pp. 77–83. Piscataway, New Jersey: Institute of Electrical and Electronics Engineers.

Vaucher, J. G., and P. Duval. 1975. A comparison of simulation event list algorithms. *Communications of the ACM* 18(4):223–230.

Sampling from Linear Multivariate Densities

Wolfgang Hörmann and Josef Leydold

Abstract It is well known that the generation of random vectors with non-independent components is difficult. Nevertheless, we propose a new and very simple generation algorithm for multivariate linear densities over point-symmetric domains. Among other applications it can be used to design a simple decomposition-rejection algorithm for multivariate concave distributions.

1 Introduction

Exact sampling from (arbitrary) multivariate distributions over (subsets of) $\mathbb{R}^d$ is a challenging task. Only sampling from a distribution with independent components is simple. As we can generate each component independently, the sampling time for such random vectors scales linearly with dimension; see Hörmann et al. (2004) or Devroye (1986) for surveys of generation methods for univariate distributions. It is thus also very easy to sample from a distribution with constant density over a hyperrectangular domain (i.e., sampling uniformly from a box). For distributions with dependent components the situation is much more difficult. Only for the multinormal and the multi-t distributions well known generation algorithms are available that scale quadratically with the dimension. But even these distributions are not easily generated when their domain is restricted to a subset of $\mathbb{R}^d$.

There exist two general approaches for generating random vectors with a given probability density function (PDF) (see Hörmann et al. 2004): The rejection method and the conditional distribution method. The latter is only applicable in very simple situations as marginal distributions are required. Rejection on the other hand works for arbitrary dimension but its applicability is limited by the fact that the rejection constant often grows exponentially with the dimension.

When designing new rejection algorithms for multivariate distributions we experimented with linear densities and were astonished to realize that they can be generated very easily. We were even more astonished by the fact that we did not find

W. Hörmann (✉)
Department of Industrial Engineering, Boğaziçi University, 34342 Bebek-Istanbul, Turkey
e-mail: hormannw@boun.edu.tr

C. Alexopoulos et al. (eds.), *Advancing the Frontiers of Simulation: A Festschrift in Honor of George Samuel Fishman*, International Series in Operations Research & Management Science 133, DOI 10.1007/b110059_7, © Springer Science+Business Media, LLC 2009

any hints to such methods in the literature. We therefore present our new method for generating random vectors with multivariate linear density over a bounded point-symmetric domain D and some of its applications.

This chapter is organized as follows: Section 2 describes the new idea to sample from a multivariate linear density over point-symmetric domains. In Section 3 an improved rejection algorithm for multivariate concave densities over point-symmetric domains is introduced. Applications of the new algorithm are presented in Section 4.

2 Linear Densities over Point-symmetric Domains

A domain $D \subset \mathbb{R}^d$ is called *point-symmetric with center* $\mathbf{c}$ if $\mathbf{x} \in D$ implies that $\mathbf{x}^* = \mathbf{c} - (\mathbf{x} - \mathbf{c}) \in D$. The quantity $\mathbf{x}^*$ is the reflection of $\mathbf{x}$ in $\mathbf{c}$. For our purposes hyperrectangles $[x_{1l}, x_{1r}] \times \cdots \times [x_{dl}, x_{dr}]$ are the most important case of point-symmetric domains with center $\mathbf{c} = \frac{1}{2}\big((x_{1l}, \dots, x_{dl})' + (x_{1r}, \dots, x_{dr})'\big)$.

Let $\ell\colon D \subset \mathbb{R}^d \to [0, \infty)$, $\mathbf{x} \mapsto \ell(\mathbf{x}) = \mathbf{a}'(\mathbf{x} - \mathbf{c}) + f_c$ be a linear PDF with parameter $\mathbf{a}$, some constant f_c, and point-symmetric domain D. We denote the region below the graph of density ℓ by $\mathcal{F} = \{(\mathbf{x}, u) \in \mathbb{R}^{d+1}\colon \mathbf{x} \in D \text{ and } 0 \le u \le \ell(\mathbf{x})\}$. Notice that $\ell(\mathbf{c}) = f_c$. For such distributions we can easily show the following property.

Theorem 1 *Let* $\mathbf{x} \in D$ *and* $u \in [0, 2f_c]$. *Then* $u < \ell(\mathbf{x})$ *if and only if* $u^* > \ell(\mathbf{x}^*)$, *where* $(\mathbf{x}^*, u^*) = 2(\mathbf{c}, f_c) - (\mathbf{x}, u)$, *i.e.,* $(\mathbf{x}, u)$ *reflected in* $(\mathbf{c}, f_c)$.

Proof Notice that $(\ell(\mathbf{x}^*) - f_c) = -(\ell(\mathbf{x}) - f_c)$ and $(u^* - f_c) = -(u - f_c)$. Hence $u - f_c \le \ell(\mathbf{x}) - f_c$ if and only if $u^* - f_c \ge \ell(\mathbf{x}^*) - f_c$. Thus the statement follows. □

Figure 1 sketches the situation for one dimension. As an immediate consequence we find that $\mathrm{Vol}_{d+1}(\mathcal{F}) = f_c \cdot \mathrm{Vol}_d(D)$, where $\mathrm{Vol}_d(\cdot)$ denotes the d-dimensional volume. Moreover, the reflection $(\mathbf{x}, u) \mapsto (\mathbf{x}^*, u^*)$ is a volume-preserving transformation that maps $\mathcal{F} \setminus (D \times [0, f_c])$ one-to-one onto $(D \times [0, f_c]) \setminus \mathcal{F}$. Algorithm 1 (`linearPDF-reflect`) compiles the relevant steps to sample from a linear density $\ell(\mathbf{x})$ using this property. Notice that Steps 2, 5, and 6 ("squeeze") reduce the average number of evaluations of the density and can speed up the algorithm in higher dimensions. They can also be entirely omitted.

Remark 1 Step 3 in Algorithm 1 is crucial and can be difficult for point-symmetric but irregular shaped domains.[1] Nevertheless, for a hyperrectangle $[x_{1l}, x_{1r}] \times \cdots \times [x_{dl}, x_{dr}]$ this is quite simple: $\mathbf{X} = \big(U_1 x_{1l} + (1 - U_1)x_{1r}, \dots, U_d x_{dl} + (1 - U_d)x_{dr}\big)$, where $U_1, \dots, U_d$ are i.i.d. (0, 1) uniform random numbers.

Let us look at the performance gain of this approach compared to Algorithm 2, which describes the simplest method, rejection from a constant hat. Both the setup

[1] In fact, *every* non-uniform generation problem can be reduced to sampling uniformly from some domain in $\mathbb{R}^{d+1}$.

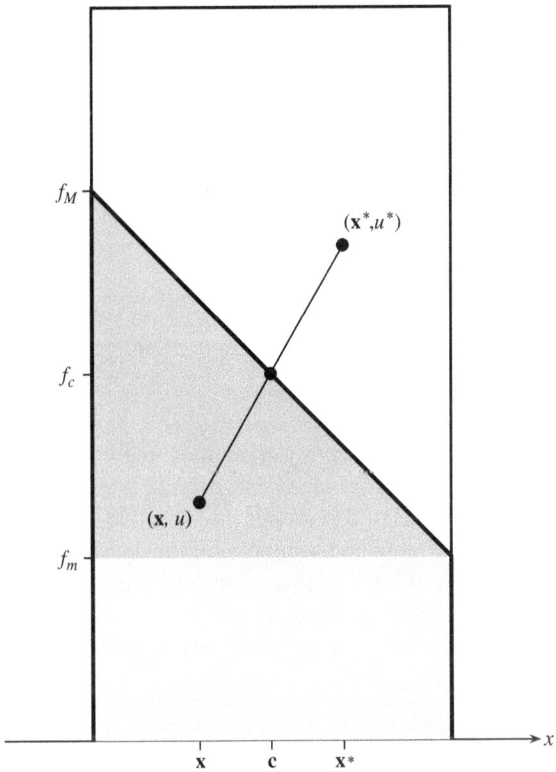

Fig. 1 Linear density on point-symmetric set D with center $\mathbf{c}$ (in one dimension)

Algorithm 1 `linearPDF-reflect`

Input: Linear density $\ell(\mathbf{x})$ on point-symmetric domain D with center $\mathbf{c}$
($\ell(\mathbf{x}) \geq 0$ for all $\mathbf{x} \in D$).
Output: Random vector $\mathbf{X}$ with density ℓ.

/* Setup */
Compute $f_c \leftarrow \ell(\mathbf{c})$. /* squeeze */
Compute $f_m \leftarrow \min_{\mathbf{x}\in D} \ell(\mathbf{x})$.
/* Generator */
Generate $\mathbf{X}$ uniformly in D.
Generate U uniformly in $[0, f_c]$.
if $U \leq f_m$ **then** /* below squeeze */
    return $\mathbf{X}$.
else if $U \leq \ell(\mathbf{X})$ **then** /* below density */
    return $\mathbf{X}$.
else /* reflect point on center */
    return $\mathbf{X}^* = 2\mathbf{c} - \mathbf{X}$.

Algorithm 2 `rejection`

Input: Linear density $\ell(\mathbf{x})$ on domain D.
Output: Random vector $\mathbf{X}$ with density ℓ.

/* Setup */
Compute $f_M \leftarrow \max_{\mathbf{x}\in D} \ell(\mathbf{x})$.
Compute $f_m \leftarrow \min_{\mathbf{x}\in D} \ell(\mathbf{x})$.
/* Generator */
loop
    Generate $\mathbf{X}$ uniformly in D.
    Generate U uniformly in $[0, f_M]$.
    if $U \le f_m$ **then** /* below squeeze */
        return $\mathbf{X}$.
    else if $U \le \ell(\mathbf{X})$ **then** /* below density */
        return $\mathbf{X}$.

and one iteration of the acceptance/rejection loop require the same number of density evaluations and uniform random numbers as Algorithm 1. However, the expected number of repetitions of this loop is $f_M/f_c = 2f_M/(f_M + f_m) \le 2$, where f_M and f_m denote the maximum and minimum of the density ℓ on D, respectively. Hence, on average we save at most one iteration. In the best case the new approach saves 50% of the marginal generation time; in the worst case it is not slower than simple rejection with squeeze. Note that the new algorithm scales linearly with dimension.

2.1 An Extension

Algorithm 1 also works when $\mathbf{c} \notin D$ as long as $\ell(\mathbf{X})$ can be extended to a linear function ℓ' on some point-symmetric superset $D' \supset D$ with center $\mathbf{c}$. A simple example for this situation is a linear density on a ball restricted to its boundary (a sphere). Points $\mathbf{X}$ are still sampled uniformly in D in Step 4 (and not in the superset D').

3 Improved Rejection

Algorithm 2 (`rejection`) also works for distributions where the density is some linear function $\ell(\mathbf{x})$ restricted to its nonnegative part, i.e., $\max(0, \ell(\mathbf{x}))$. Algorithm 1 is not directly applicable for such densities but it can be easily adapted. We only have to add a rejection step to eliminate points with negative U-coordinates. Algorithm 3 (`linearPDF-general`) shows the details. Notice that there is no squeeze when $\min_{\mathbf{x}\in D} \ell(\mathbf{x}) \le 0$.

This algorithm is based on the following modification of Theorem 1.

Theorem 2 *Let* $\mathbf{x} \in D$ *and* $u \in [f_m, f_M] = [\min_{\mathbf{x}\in D} \ell(\mathbf{x}), \max_{\mathbf{x}\in D} \ell(\mathbf{x})]$. *Then* $u < \ell(\mathbf{x})$ *if and only if* $u^* > \ell(\mathbf{x}^*)$, *where* $(\mathbf{x}^*, u^*) = 2(\mathbf{c}, f_c) - (\mathbf{x}, u)$.

Algorithm 3 `linearPDF-general`

Input: Linear function $\ell(\mathbf{x})$ on point-symmetric domain D with center $\mathbf{c}$.
Output: Random vector $\mathbf{X}$ with density $\max(0, \ell(\mathbf{x}))$.

/* Setup */
Compute $f_c \leftarrow \ell(\mathbf{c})$.
Compute $f_m \leftarrow \min_{\mathbf{x}\in D} \ell(\mathbf{x})$ and $f'_m \leftarrow \min(0, f_m)$.
/* Generator */
loop
  Generate $\mathbf{X}$ uniformly in D.
  Generate U uniformly in $[f'_m, f_c]$.
  if $U > \ell(\mathbf{X})$ **then** /* above $\ell \to$ reflect point on center */
    $\mathbf{X} \leftarrow \mathbf{X}^* = 2\mathbf{c} - \mathbf{X}, \quad U \leftarrow U^* = 2\,f_c - U$.
  if $U \geq 0$ **then** /* density must be nonnegative */
    return $\mathbf{X}$.

It is easy to see that the rejection constant of Algorithm 3 is smaller than that of Algorithm 2 (`rejection`) if and only if $f_c = \ell(\mathbf{c}) > 0$. The performance gain is again limited by a factor of 2. We have to note here that the rejection constants of both algorithms can be arbitrarily large. Even when we restrict the domain of the density to its smallest bounding hyperrectangle, the rejection constant grows exponentially with dimension d if f is not linear on its entire domain.

3.1 Concave Densities

Let us look at a concave differentiable density f over a point-symmetric domain D with center $\mathbf{c}$. We can then construct a linear hat function ℓ over D by means of a tangent at some construction point $\mathbf{p}$. Thus we find

$$\ell(\mathbf{x}) = \nabla f(\mathbf{p})(\mathbf{x} - \mathbf{c}) + \big(\nabla f(\mathbf{p})(\mathbf{c} - \mathbf{p}) + f(\mathbf{p})\big) .$$

By the concavity of f, $\ell(\mathbf{x}) \geq 0$ and we can apply Algorithm 1 (`linearPDF-reflect`) for sampling from the majorizing density, Algorithm 4 (`concavePDF`). The acceptance probability of the rejection step is maximized if the area below the hat is minimized, i.e., when $\ell(\mathbf{c})$ is minimized. Such a construction point $\mathbf{p}$ can be easily found.

Theorem 3 *Let D be a point-symmetric domain with center $\mathbf{c}$ and let f be a density that is the restriction of some concave function to D. Then the rejection constant of a rejection algorithm based on the linear hat $\ell(\mathbf{x})$ is minimized if we choose center $\mathbf{c}$ as the construction point of ℓ.*

Proof By Theorem 1 the volume of $\mathcal{F} = \{(\mathbf{x}, u) \in \mathbb{R}^{d+1} \colon \mathbf{x} \in D \text{ and } 0 \leq u \leq \ell(\mathbf{x})\}$ is given by $\mathrm{Vol}_{d+1}(\mathcal{F}) = \ell(\mathbf{c}) \cdot \mathrm{Vol}_d(D)$. Thus the rejection constant is minimized if $\ell(\mathbf{c})$ is minimized. By the concavity of f, $\ell(\mathbf{c}) \geq f(\mathbf{c})$ where equality holds for $\mathbf{p} = \mathbf{c}$. Thus the statement follows. □

Algorithm 4 `concavePDF`

Input: Concave density $f(\mathbf{x})$ on point-symmetric domain D with center $\mathbf{c}$.
Output: Random vector $\mathbf{X}$ with density f.

/* Setup */
Compute $\mathbf{a} \leftarrow \nabla f(\mathbf{c})$ and $f_c \leftarrow f(\mathbf{c})$.
Compute $f_m \leftarrow \min_{\mathbf{x}\in D} f(\mathbf{x})$. /* use concavity of f */
/* Generator */
loop
  Generate $\mathbf{X}$ uniformly in D.
  Generate U uniformly in $[0, f_c]$.
  if $U \le f_m$ **then** /* below squeeze */
    return $\mathbf{X}$.
  if $U > \mathbf{a}\cdot(\mathbf{X}-\mathbf{c}) + f_c$ **then** /* above $\ell \rightarrow$ reflect point on center */
    $\mathbf{X} \leftarrow \mathbf{X}^* = 2\mathbf{c} - \mathbf{X}, \quad U \leftarrow U^* = 2f_c - U$.
  if $U \le f(\mathbf{X})$ **then** /* accept */
    return $\mathbf{X}$.

3.2 General Densities

The concavity property of f above is only necessary to guarantee a simple set-up, which is even simpler and faster than the set-up of the Ahrens algorithm (see next paragraph) as no minimization procedure is required. The concept of Algorithm 4 also works for arbitrary densities. To use it in practice it is necessary to construct a linear upper bound to the density that is nonnegative for every point of the domain D. For example, secants can be used as hat functions for a convex density in one dimension.

4 Applications

4.1 Concave Densities and Ahrens Method

The simplest method for sampling from arbitrary multivariate distributions with given density are multigrid methods, which we call "multivariate Ahrens methods" as they generalize a method for univariate distributions proposed by Ahrens (1993). For this approach the domain of the distribution is partitioned into hyperrectangles. (If necessary the domain has to be extended to a union of proper hyperrectangles.) On each of these the maximum of the given density is estimated and a piecewise constant hat is computed. Thus the region below the density is covered by a union of bars and hence it is extremely simple to draw a random sample by rejection using Algorithm 2 for each of the hyperrectangles. This simple method has again and again attracted the interest of researchers who had to solve multivariate generation problems; see Jadach (2003) and Karawatzki (2006) for two recent examples. However, there are significant drawbacks associated with this approach. First, finding the maximum in each hyperrectangle requires either strong constraints on the given target distribution (e.g., uniorthomodal in Karawatzki 2006) or it is very time consuming when arbitrary densities are assumed (e.g., Jadach 2003).

The most prohibitive downside however is the slow convergence of the rejection constant. For a distribution in d dimensions with a bounded gradient, the rejection constant tends to 1 with rate $1 + O(N^{-1/d})$ for increasing number N of hyperrectangles. Thus this leads to a large number N, and consequently large memory requirements and slow setup times even when computing the maximum is fast. Nevertheless, acceptance probabilities usually remain very small even for a moderate number of dimensions; see Karawatzki (2006). For distributions with concave density functions f we can formulate an alternative to the Ahrens method. It avoids the computation of the maximum in each hyperrectangle by using tangents in lieu of a constant hat and applies Algorithm 4 on each of the hyperrectangles. Then we find that the convergence rate is $1 + O(N^{-2/d})$. The following theorem summarizes this observation.

Theorem 4 (Hörmann et al. 2004) *Let f be a bounded two times differentiable concave density f with bounded domain $D \subset \mathbb{R}^d$.*

(i) Construct a hat function h_1 using constant hats on N subintervals of equal size and shape. Then the volume between the hat and density tends to 0 with rate $O(N^{-1/d})$, i.e., we find for the rejection constant $1 + O(N^{-1/d})$.
(ii) Likewise, when we construct a hat function h_2 using tangents in the center of each of N subintervals of equal size and shape, we find for the rejection constant $1 + O(N^{-2/d})$.

Proof (ii) As h_2 and f are both two times differentiable functions with the same first-order Taylor expansion at the center $\mathbf{c}$, we have $|h_2(\mathbf{x}) - f(\mathbf{x})| = O(r^2)$ around each $\mathbf{c}$, where $r = \|\mathbf{x} - \mathbf{c}\|$ is the distance from the center. Since we have N design points on a regular grid, the average radius is $r = O(N^{-1/d})$, which implies that the average distance $|h_2(\mathbf{x}) - f(\mathbf{x})| = O(N^{-2/d})$. As we have assumed a bounded domain D, we get $\int_D |h_2(\mathbf{x}) - f(\mathbf{x})| \, d\mathbf{x} = O(N^{-2/d})$.
(i) For constant hats we analogously find $|h_1(\mathbf{x}) - f(\mathbf{x})| = O(r) = O(N^{-1/d})$. □

The performance gain of using Algorithm 4 compared to the Ahrens method (i.e., rejection from a piecewise constant hat) is twofold:

1. There is no need to estimate the maximum of f in each of the (many) rectangles, except those in the boundary region of D.
2. The rejection constant is reduced by some factor that is 1 in the worst case and 1/2 in the best.

Of course, the latter is of practical relevance only if the rejection constant is not too large (at least below 100). In particular, for a rejection constant close to 1 the benefits become obvious. However, the faster (asymptotic) convergence of the new method is not of great help here. Unfortunately, to achieve rejection constants below 10 requires many hyperrectangles even in a moderate number (6–10) of dimensions; see Karawatzki (2006) for some computational experiences with the Ahrens method.

Remark 2 The improvement by a factor of at most 2 is rather disappointing. Nevertheless, if we try to accomplish the same improvement by refining the partition of the Ahrens method instead of using linear non-constant hats in lieu of constant ones, we need about 2^d times more rectangles.

4.2 Computational Experience

We implemented and tested our algorithms. For generating from linear densities we were astonished to see that the speed-up was even higher than the reduction of the expected number of repetitions reached by the reflection principle. Depending on the value of f_m, Algorithm 1 was up to three times faster than Algorithm 2. In one dimension Algorithm 1 is also faster than the inversion method so we can call it the fastest method to generate from linear densities.

For concave densities the speed-up is less spectacular. Depending mainly on the reduction of the expected number of repetitions we observed speed-ups in the range of 10–20%.

4.3 Importance Sampling

It has been shown that in the computation of expectations of functions of random variates by Monte Carlo methods it is often more efficient to replace the rejection algorithm by importance sampling (IS) with the hat function as the importance density; see e.g., Hörmann and Leydold (2005). This is in particular the case when the evaluation of the density $f(x)$ is expensive compared to the integrand, as by using IS we keep all information about our density f. This is even more inevitable when the rejection constant is high (which becomes very likely when the dimension increases).

Some methods like VEGAS (Lepage 1978) approximate the integrand in question by a piecewise constant function. Again using tangents decreases the approximation error. The rejection constant can be seen as a rough measure for the deviation from the target distribution. Notice that in the case of IS the hat function need not be a majorizing function as for the rejection algorithm. Thus we can safely drop the concavity assumption.

5 Conclusion

We have introduced a simple and fast algorithm for efficient sampling from linear multivariate densities with point-symmetric domains. It was demonstrated that the new idea can be utilized to considerably simplify and speed up both the set-up and the sampling of table methods to generate from multivariate concave distributions. The new idea may also be applied to importance sampling and for generating non-uniform distributions on the sphere.

References

Ahrens, J. H. 1993. Sampling from general distributions by suboptimal division of domains. *Grazer Math. Berichte* 319.

Devroye, L. 1986. *Non-Uniform Random Variate Generation*. New York: Springer-Verlag.

Hörmann, W., and J. Leydold. 2005. Monte Carlo integration using importance sampling and Gibbs sampling. In *Proceedings of the International Conference on Computational Science and Engineering*, eds. H. Dag and Y. Deng, pp. 92–97.

Hörmann, W., J. Leydold, and G. Derflinger. 2004. *Automatic Nonuniform Random Variate Generation*. Berlin Heidelberg: Springer-Verlag.

Jadach, S. 2003. Foam: a general-purpose cellular Monte Carlo event generator. *Computer Physics Communications* 152:55–100.

Karawatzki, R. 2006. The multivariate Ahrens sampling method. Technical Report 30, Department of Statistics and Mathematics, WU Wien. `http://epub.wu-wien.ac.at/`.

Lepage, G. P. 1978. A new algorithm for adaptive multidimensional integration. *Journal of Computational Physics* 27(2):192–203.

Factor Screening in Simulation Experiments: Review of Sequential Bifurcation

Jack P. C. Kleijnen

Abstract Factor screening means searching for the most important factors (or inputs) among the many factors that may be varied in an experiment with a real or a simulated system. This chapter gives a review of Sequential Bifurcation (SB), which is a screening method for simulation experiments in which many factors may be varied. SB is most efficient and effective if its assumptions are satisfied. SB was originally studied back in 1990. This review first summarizes SB. Then it summarizes a recent case study, namely, a supply-chain simulation with 92 factors where SB identifies a shortlist with 10 factors after simulating only 19 combinations. The review also references recent research. It ends with a discussion of possible topics for future research.

1 Introduction

Factor screening (or briefly "screening") means that the analysts are searching for the most important factors among the many factors that can be varied in their experiment. In practice, however, experiments with real-world systems usually can vary only a few factors, whereas experiments with simulation models can indeed vary hundreds or more factors (also see Kleijnen et al. 2005). In general, scientists assume that effects are "sparse"; they do not wish to report that "everything depends on everything else." The scientists' clients do not want to be "confused by details." Furthermore, philosophy of science exploits the *parsimony* principle or *Occam's razor*, which implies that a simpler explanation is preferred to a more complex explanation—all other things being equal. The psychologist Miller (1956) claims that people cannot handle more than "seven plus or minus two" factors when processing information. Many simulation modelers assume that the *Pareto* principle or *20-80* rule holds, i.e., only a few factors are really important (or "active", as some authors say). Many authors on simulation modeling (be it deterministic or random simulation) mention the *curse of dimensionality*; see, e.g., the panel discussion

J.P.C. Kleijnen (✉)
Department of Information Management, Tilburg University, 5000 LE Tilburg, Netherlands
e-mail: kleijnen@uvt.nl

C. Alexopoulos et al. (eds.), *Advancing the Frontiers of Simulation: A Festschrift in Honor of George Samuel Fishman*, International Series in Operations Research & Management Science 133, DOI 10.1007/b110059_8, © Springer Science+Business Media, LLC 2009

reported by Simpson et al. (2004). Altogether, screening is necessary in realistic simulation modeling. Unfortunately, practitioners do not yet apply screening methods; instead, they experiment with a few intuitively selected factors only. Hopefully, this chapter contributes to a change of attitude.

A practical example may illustrate the need for screening.Bettonvil and Kleijnen (1996) summarize a case study on the CO_2 greenhouse effect, using a deterministic simulation model with 281 factors (that simulation is also discussed by Kleijnen et al. 1992). The politicians wanted to take measures to reduce the release of CO_2; they realized that they should start with legislation for a limited number of factors. Another example will be summarized in Section 3.3.

A most efficient and effective screening method may be *Sequential Bifurcation (SB)*. After a slow start, SB has gained momentum in recent years—as the following overview shows.

1. Bettonvil's (1990) Ph.D. dissertation introduced SB.
2. Bettonvil and Kleijnen (1996) provide a summary of that dissertation (after experiencing rather much delay, finding a publication outlet).
3. Recently, SB has attracted the attention of several researchers in the UK and USA:

   - Cheng (1997) further explores SB for random simulation; also see Cheng and Holland (1999).
   - Wan et al. (2006a) improve SB's control of the type-I (or type-α) and type-II (or type-β) error probabilities in discrete-event simulation. Next, Wan et al. (2006b) extend their method to account for interactions between the factors of the simulation model. Shen and Wan (2006) combine this approach with classic fractional factorial designs. Finally, Xu et al. (2007) extend SB to nonnormal distributions, namely, binary responses (outputs) of simulation models for software reliability studies.
   - Ankenman et al. (2006) develop a more efficient but also more complicated SB variant based on "polytopes," requiring repeated solution of a sequence of Linear Programming (LP) problems.
   - Kleijnen et al. (2006) summarize SB and apply SB to a practical discrete-event simulation of a supply chain centered around the Ericsson company in Sweden, involving 92 factors; they identify a shortlist with 10 factors after simulating only 19 combinations. (This chapter updates the summary in Kleijnen et al. 2006, and summarizes their case study.)

The rest of this chapter is organized as follows. Section 2 summarizes several screening methods that may compete with SB. Section 3 summarizes SB, including its assumptions. Section 3.1 gives an outline of the simplest type of SB. Section 3.2 covers some mathematical details of this simplest SB. Section 3.3 summarizes a case study, namely, a supply-chain simulation for Ericsson in Sweden. Section 3.4 extends SB, accounting for two-factor interactions. Section 4 presents conclusions and possible topics for future research. Many references are given to enable further study of screening methods.

2 SB's Competitors

There are several types of screening designs. All these designs treat the simulation model as a *black box*; i.e., the simulation model transforms observable inputs into observable outputs, whereas the values of internal variables and specific functions implied by the simulation's computer modules are unobservable.

The importance of factors depends on the *experimental domain*—also called the experimental area or experimental frame; see Zeigler et al. (2000). The users should supply information on this domain, including realistic ranges of the individual factors and limits on the admissible factor combinations (or "scenarios"); e.g., factor combinations are admissible only if they add up to 100% because these combinations represent chemical compositions (also called recipes). In practice, *user involvement* is therefore crucial for the application of screening methods.

This section summarizes several design types that may be considered to compete with SB (empirical comparisons of these various designs is beyond the scope of this chapter).

2.1 Classic Two-Level Factorial Designs

Classic two-level factorial designs are often considered to provide screening designs; these designs are detailed in many textbooks, e.g., Kleijnen (2008) and Myers and Montgomery (2002). In particular, so-called resolution-III designs are often called screening designs; see, e.g., Georgiou (2007) and Yu (2007). By definition, a *resolution-III* design gives unbiased estimators of all the main effects or first-order effects, provided a first-order polynomial is a "valid metamodel" ("adequate approximation") of the Input/Output (I/O) function that is implicitly determined by the underlying simulation model. These designs require the simulation of at least $n = k + 1$ factor combinations where k denotes the number of factors in the experiment. In such a design, each factor has two values or levels; these levels may denote quantitative or qualitative values. The case study reported in Section 3.3 has $k = 92$ factors, so at least 93 factor combinations would need to be simulated; simulating one combination takes 40 minutes (after modification of the simulation code, which originally took 3 hours per combination). Moreover, a random simulation model like this case study requires replication of each combination to obtain an estimate of the *signal/noise ratio*; i.e., if the *noise* (variance of the simulation output) is large compared with the *signal* (the simplest signal is the factor's main effect), then *replication* (simulation with non-overlapping PseudoRandom Numbers, PRNs) is unavoidable.

Note: A different type of resolution-III design changes only one factor at a time. In the supply-chain example, such a design still requires 93 combinations if not more than two values per factor are simulated (simulation practitioners often study three values per factor, when changing one factor at a time). Moreover, this approach is less efficient; i.e., the variances of the estimated main effects are larger than the variances resulting from the Design Of Experiments (DOE) literature.

Note: A full-factorial design is often used by practitioners when the number of factors is small. However, full factorials are impossible in screening studies; e.g., the supply-chain example with its 92 factors would require $2^{92} \approx 5 \times 10^{27}$ combinations.

Another class of designs called "conference designs" requires $n = 2k$ combinations; see Elster and Neumaier (1995). These designs are not practical screening designs if the simulation model is *expensive*; i.e., a single run with the model takes relatively much computer time.

2.2 Frequency Domain Experimentation (FDE)

Whereas classic designs keep the factor values constant during a simulation run, FDE oscillates these levels during a run. More precisely, each factor has its own oscillation frequency. FDE tries to find which input oscillations affect output oscillations. Originally, Schruben and Cogliano (1987) proposed this approach. Sanchez et al. (2006) apply FDE for second-order polynomial metamodels with an arbitrary number of factors; they give an example of a kanban simulation with 34 factors. Unfortunately, FDE requires rather complicated Fourier spectral analysis. Moreover, FDE has not yet been applied to simulation models with hundreds of factors.

2.3 Supersaturated Designs

By definition, supersaturated designs have fewer combinations than factors: $n < k$. These designs are not sequential, so they are relatively inefficient. Indeed, sequential statistical procedures are known to require fewer observations than fixed-sample (one-shot) procedures; see, e.g., Park et al. (2002). By definition, sequential designs imply that observations are analyzed—so the data generating process is better understood—before the next input combination is selected. This property implies that the design depends on the specific underlying process (simulation model); i.e., the design is customized (tailored or application-driven, not generic; also see Kleijnen and Van Beers 2004 and Van Beers and Kleijnen 2008). Moreover, computer experiments (unlike many real-world experiments) proceed sequentially. Nevertheless, sequential procedures may lose some efficiency; e.g., switching between the SB procedure and the simulation model may be awkward. Recent discussions of supersaturated designs are presented in Allen and Bernshteyn (2003), Gilmour (2006), Li and Li (2005), Wu and Hamada (2000), Yamada et al. (2006), and Zhang et al. (2007). Note that Tu and Jones (2003) also give a supersaturated design, but they use Moving Least Squares instead of classic linear regression analysis.

2.4 Group-Screening Designs

Group-screening designs aggregate (or confound) individual factors into groups so that k factors may be evaluated in $n < k$ combinations. Consequently, these designs are also supersaturated—but they are executed in two or more steps (stages).

There are several types of screening designs. Examples are One-factor-At-a-Time (OAT), Morris's OAT, Cotter's design, Andres's Iterated Fractional Factorial Design (IFFD), multi-stage group screening, and SB; see Andres (1997), Andres and Hajas (1993), Campolongo et al. (2007), Campolongo et al. (2000), De Vos et al. (2006), Morris (2006), Saltelli et al. (2004, 2005), and Schonlau and Welch (2006). Note that Chipman (2006) gives a Bayesian analysis of screening experiments, but Bayesian approaches are not further considered in this chapter. The following web address gives access to a package (written in the R statistical software tool) that implements Morris's OAT.

```
http://cran.r-project.org/src/contrib/Descriptions/.
sensitivity.html
```

Different group-screening designs are based on different mathematical assumptions concerning the *characteristics* of their metamodels; e.g., their *smoothness* and *monotonicity*. Reviewing the assumptions and procedures of all these designs is beyond the scope of this chapter. This chapter focuses on SB because SB is a very efficient and effective method if its assumptions are satisfied.

3 Sequential Bifurcation

SB uses the following metamodel assumptions, which will be detailed in the next subsections; Assumption 1(b) may replace Assumption 1(a).

Assumption 1(a): a first-order polynomial is a valid metamodel.
Assumption 1(b): a first-order polynomial augmented with two-factor interactions is a valid metamodel.
Assumption 2: all main effects have known signs and are nonnegative.
Assumption 3: there is "strong heredity" if Assumption 1(b) holds.

3.1 Outline of Simplest SB

As its name suggests, SB is a *sequential* procedure. Its first step aggregates all factors into a single group, and tests whether or not that group of factors has an important effect (this statistical test will be presented in Section 3.2). If that group indeed has an important effect (which is most likely), then the second step *splits* the group into two subgroups—*bifurcates*—and tests each of these subgroups for importance. The next steps continue in a similar way; i.e., SB splits important subgroups into smaller subgroups, and discards unimportant subgroups. In the final steps, all individual factors that are not in subgroups identified as unimportant are estimated and tested—which terminates the procedure.

The simplest type of SB is based on Assumptions 1(a) and 2, which are now detailed.

Assumption 1(a) a valid metamodel is a *first-order polynomial* plus noise:

$$y = \beta_0 + \beta_1 x_1 + \cdots + \beta_k x_k + e. \tag{1}$$

In this equation, the factors x_j $(j = 1, \ldots, k)$ are standardized such that they are either -1 or $+1$ (scaling in DOE is further discussed by Kleijnen 2008). This scaling implies that the factors may be ranked (sorted) by $|\beta_j|$; i.e., the most important factor is the one with the largest absolute value of its main effect; the least important factor is the one with the effect closest to zero. Note that the larger the range of an original (untransformed) factor is, the larger the response difference and hence the main effect of that factor is; also see the "unit cost" effects in Cheng and Holland (1999). The noise e in (1) arises from both approximation error and the use of PRNs. If the metamodel is *valid*, then this noise has zero mean: $E(e) = 0$.

To estimate the parameters in (1), it is most efficient to experiment with only two levels per factor. In practice, it is important that these levels are realistic *extreme* values; so the users of the underlying simulation model should provide these values. Readers are also referred to the discussion on scaling in Wan et al. (2006a) and the discussion on the experimental domain above, in Section 2.

Assumption 2 all main effects have *known signs* and are non-negative:

$$\beta_j \geq 0 \, (j = 1, \ldots, k).$$

Without Assumption 2, main effects within a (sub)group might cancel each other. However, if Assumption 2 holds, then the analysts can define the two levels of an individual factor such that changing the level from the standardized value -1 to $+1$ does not decrease the expected simulation output (i.e., that change either increases the output or does not change it at all). An example is the M/M/1 model: if the arrival rate increases, then the expected steady-state waiting time also increases; if the queuing discipline changes from First-In-First-Out (FIFO) to Shortest-Processing-Time-first (SPT), then the expected waiting time decreases; consequently, the level -1 should correspond to SPT and the level $+1$ to FIFO.

Assumption 2 is related to the *monotonicity* of the I/O function. By definition, a function $w = f(x_1, \ldots, x_k)$ is monotonically increasing if $\partial w / \partial x_j > 0$ for all j, for all values of $x_{j'}$ $(j, j' = 1, \ldots, k;\ j \neq j')$. Experience shows that Assumption 2 may be easily satisfied in practice; i.e., it is straightforward to define the upper and lower level of each factor such that changing a factor from its lower to its upper level does not decrease the expected response. For example, in the supply-chain case study, some factors refer to transportation speeds: the higher these speeds, the lower the Work In Process (WIP) and hence the lower the cost—which is the output of interest in the screening experiment. More examples are given by other authors; e.g., Lewis and Dean (2001) and Lim and Glynn (2006).

In unconstrained optimization, the function to be maximized or minimized is usually assumed not to be monotonically increasing (otherwise, the maximum or minimum would lie at the limits of the experimental area). This assumption may still be compatible with the known signs assumption; i.e., switching the standardized factor values from -1 to $+1$ may increase the output, so this factor will be found to have an important effect. However, a counterexample is an I/O function that is not monotonic and happens to give roughly the same output values at the two observed input levels -1 and $+1$; in this example, the factor effect seems to be zero and SB eliminates this factor; also see Kleijnen (2008, pp. 162–163).

Nevertheless, if a particular case study does not satisfy Assumption 2 for a few specific factors, then these factors should be treated *individually*; i.e., none of these factors should be grouped with other factors in SB. For example, De Vos et al. (2006) create some subgroups of size one in a multi-stage group-screening design; their design is less efficient than SB, but it also uses aggregation. In general, treating such factors individually is safer than assuming that the probability of cancellation within a subgroup is negligible.

The *efficiency* of SB (measured by the number of simulated factor combinations and hence simulation time) may be improved in the following ways.

- The individual factors are labeled such that factors are placed in *increasing order of importance*; see Bettonvil (1990, p. 44). This labelling makes the important factors clustered. To realize this labelling, it is crucial to utilize prior knowledge of users and analysts about the real system being simulated. For example, if the analysts conjecture that environmental factors are most important, then they should place these factors at the end of their list of factors. Indeed, in the supply-chain case study, Kleijnen et al. (2006) place the environmental factor "demand" at the very end of the list with 92 individual factors; Section 3.3 returns to this labelling.
- *Similar factors* are placed within the same subgroup. In the supply-chain study, all "test yield" factors are grouped together; the conjecture is that if one yield factor is unimportant, then all yield factors are likely to be unimportant too.
- Subgroups are split such that the number of factors for the first new subgroup is a *power of two*; e.g., split the first 48 factors into a subgroup of 32 ($= 2^5$) factors and a subgroup with the 16 remaining factors (so the important factors are placed into the smallest subgroup, assuming the factors are sorted from unimportant to most important). This splitting, however, is not recommended if it implies splitting up a group of related factors. In any case, splitting a subgroup into subgroups of *equal* size (as some authors do) does not need to be optimal. Further discussion is found in Bettonvil (1990, pp. 40–43).

The way SB proceeds may be interpreted through the following *metaphor*. Imagine a lake that is controlled by a dam. The goal of the experiment is to identify the highest (most important) rocks (actually, SB not only identifies, but also measures the height of these "rocks"). The dam is controlled in such a way that the level of the murky water slowly drops. Obviously, the highest rock first emerges from the water. The most-important-but-one rock turns up next, etc. SB stops when the simulation analysts feel that all the "important" factors are identified; once SB stops, the analysts know that all remaining (unidentified) factors have smaller effects than the effects of the factors that have been identified. This property of SB seems quite important for its use in practice.

Some reflection shows that the aggregated effect of a given subgroup is an *upper limit* for the value of any individual main effect within that subgroup. Examples will be given in the supply-chain study in Section 3.3. If the analysts must terminate SB *prematurely* (e.g., because their computer breaks down or their clients get impatient), then SB still allows identification of the factors with main effects larger than the current upper limit.

SB is extended by Wan et al. (2006a), improving the control over the type-I error rates ("false positives"), using either a two-stage approach or a fully sequential approach. Theoretically, this control does not satisfy the classic statistical requirements concerning a prespecified experimentwise error rate and a prespecified power for the *final* results—after *all* stages have been executed. Nevertheless, the numerical results look very promising.

SB is also extended to the so-called *polytope* method by Ankenman et al. (2006). Their method is more efficient (requiring fewer combinations), but also more complicated (requiring the solution of a LP problem after each additional observation). Moreover they assume main effects only (no interactions). Note that the LP problem arises because this method computes the Ordinary Least Squares (OLS) estimate under the constraint stipulating that all regression coefficients be non-negative (see Assumption 2 above).

3.2 Mathematical Details of Simplest SB

To explain some mathematical details of SB, the following additional notation is used.

$w_{(j);r}$: observed simulation output with the factors 1 through j set to their *high* levels and the remaining factors set to their low levels, in replication r (with $j = 1, \ldots, k$ and $r = 1, \ldots, m$ with $m > 1$);

$\beta_{j'-j}$: *sum* of main effects of factors j' through j; i.e.,

$$\beta_{j'-j} = \sum_{h=j'}^{j} \beta_h. \tag{2}$$

A simple estimate (a complicated estimate is given by Ankenman et al. 2006) of this group effect based on replication r is

$$\widehat{\beta_{j'-j;r}} = \frac{w_{(j);r} - w_{(j'-1);r}}{2}. \tag{3}$$

Section 3.1 mentioned that SB starts with simulating the two most extreme scenarios; i.e., scenario 1 implies that all k factors are at their low levels, so $x_j = -1$; scenario 2 implies that all these factors are high, so $x_j = 1$. If the metamodel in (1) is valid, then

$$E(w_{(0)}) = \beta_0 - \beta_1 - \cdots - \beta_k \tag{4}$$

and

$$E(w_{(k)}) = \beta_0 + \beta_1 + \cdots + \beta_k, \tag{5}$$

so

$$E(w_{(k)}) - E(w_{(0)}) = 2(\beta_1 + \cdots + \beta_k), \tag{6}$$

which shows that the group effect estimator defined in (3) is *unbiased*.

Likewise it follows that the individual main effect β_j is estimated through the analogue of (3):

$$\widehat{\beta_{j;r}} = \frac{w_{(j)r} - w_{(j-1);r}}{2}. \tag{7}$$

The replicates enable the estimation of the mean and the variance for each (aggregated or individual) estimated effect; e.g., (7) gives

$$\overline{\widehat{\beta_j}} = \frac{\sum_{r=1}^{m} \widehat{\beta_{j;r}}}{m} \text{ and } s(\overline{\widehat{\beta_j}}) = \sqrt{\frac{\sum_{r=1}^{m} (\widehat{\beta_{j;r}} - \overline{\widehat{\beta_j}})^2}{m(m-1)}}. \tag{8}$$

This variance estimator allows variance heterogeneity of the simulation outputs, as well as Common Random Numbers (CRN); also see the discussion on variance heterogeneity and CRN in Kleijnen (2008).

To *test* the importance of the estimated (either aggregated or individual) main effects statistically, SB uses a classic t statistic. Different scenarios probably produce observations with different variances, and may use CRN. SB applies a one-sided test because SB assumes that all individual main effects are nonnegative. SB uses a prespecified type-I error rate (e.g., $\alpha = 0.05$) per test; i.e., SB does not adjust for multiple testing. (Response Surface Methodology or RSM is also a sequential procedure that does not control the type-I and type-II error rates over the whole procedure, but is much applied; see Kleijnen 2008 and Myers and Montgomery 2002.) However, Wan et al. (2006a) do use multiple testing procedures in SB.

To *verify* (or validate) the shortlist resulting from SB, the effects of the "unimportant" factors may be tested through the following two scenarios, each simulated m times:

i. Set all factors that SB declared to be unimportant at their low levels, while keeping the important factors fixed (e.g., at their base levels).
ii. Switch all these unimportant factors to their high levels, still keeping the important factors fixed.

Obviously, these two scenarios are not used in SB if verification fixes the important factors at base values (coded as 0) that are not extreme values (coded as either -1 or 1, which are used in SB). The difference between the outputs of these two scenarios may be tested through a t statistic; this difference is expected not to differ significantly from zero if the factors are actually unimportant.

How SB proceeds sequentially is illustrated in the following case study. A formal computer procedure for the SB steps is given by Wan et al. (2006a).

3.3 Case Study: Ericsson's Supply-Chain Simulation

An example of the application of SB to a simulation with many factors is the following case study. Originally, Persson and Olhager (2002) developed a supply-chain simulation for the Ericsson company in Sweden, and simulated only nine factor combinations. Kleijnen et al. (2006), however, revisit this simulation model and distinguish $k = 92$ factors! (Moreover, they study two other variants of this supply chain with fewer factors, which are not reported in this chapter.)

They replicate each combination $m = 5$ times. The first extreme scenario with all 92 factors at their low levels gives the average output $\overline{w_{(0)}} = 3{,}981{,}627$. The other extreme scenario with all factors at their high levels gives the average output $\overline{w_{(92)}} = 34{,}013{,}832$. So, the estimated group effect of all 92 factors is obtained from (2), (6), and (8), and is $\widehat{\beta_{1-92}} = (34{,}013{,}832 - 3{,}983{,}627)/2 = 15{,}016{,}102$. The standard error of this estimated group effect follows from (8), and turns out to be $s(\widehat{\beta_{1-92}}) = 42{,}051.18$. So this effect is very significant; and in hindsight, fewer replicates might have been simulated at this early stage; e.g., only $m = 2$ replicates would have shown that one or more factors among the 92 factors must be important.

Next, SB divides the current group of 92 factors into two subgroups. The first subgroup consists of all the 79 "decision" (or "controllable") factors (labeled from 1 through 79); the other subgroup contains all 13 "environmental" factors. Simulation of this scenario must give an expected output between the expected outputs of the preceding extreme scenarios. Comparison of the simulation observations $\overline{w_{(79)}}$ and $\overline{w_{(0)}}$ gives the estimated (sub)group effect $\widehat{\beta_{1-79}}$. Similarly, comparison of $\overline{w_{(92)}}$ and $\overline{w_{(79)}}$ gives $\widehat{\beta_{80-92}}$. Thus, this step splits the total estimated effect $\widehat{\beta_{1-92}}$ into its two additive components. This step decreases the upper limit for any individual effect in the first subgroup and the second subgroup respectively.

Kleijnen et al. (2006) give details on the successive SB steps for this case study. SB does not split a subgroup any further when its estimated aggregated main effect is not significantly positive; e.g., the estimated aggregated main effect of factors 50 through 79 turns out to be a small negative value.

In this case study, it turns out that SB stops after only 19 replicated observations (combinations). The upper limit for the main effect of any remaining individual factor is then reduced to 87,759 (whereas $\widehat{\beta_{1-92}} = 15{,}016{,}102$). SB produces a shortlist with only 10 factors; its most important factor is factor 92. Section 3.1 mentioned that the SB efficiency improves when factors are labeled from least important to most important; indeed, factor 92 turns out to be the most important factor and no factor labelled smaller than 43 is declared to be important. The most important individual factor (namely, factor 92) has already been identified and estimated after only six replicated observations.

3.4 SB with Two-Factor Interactions

This section summarizes SB for situations in which Assumption 1(a) is replaced by Assumptions 1(b) and 3, which are detailed now.

Assumption 1(b) a valid metamodel is a first-order polynomial augmented with two-factor interactions $\beta_{j';j}$ ($j' < j$; $j' = 1, \ldots, k-1$; $j = 2, \ldots, k$) and noise:

$$y = \beta_0 + \beta_1 x_1 + \cdots + \beta_k x_k + \beta_{1;2} x_1 x_2 + \cdots + \beta_{k-1;k} x_{k-1} x_k + e. \qquad (9)$$

The signs of these interactions are irrelevant (see below).

Assumption 3 if a factor has no important main effect, then this factor does not interact with any other factor.

Assumption 3 is called the *strong heredity* assumption; see Wu and Hamada (2000) and also Saltelli et al. (2005). Strong heredity is related to *functional marginality*, which is discussed by Tsai et al. (2007).

Let $w_{-(j)}$ denote the *mirror* observation of $w_{(j)}$; i.e., $w_{-(j)}$ is the simulation output with the factors 1 through j set to their *low* levels and the remaining factors set to their high levels. For example, the analogues of (4) and (5) are for $j = 48$:

$$\begin{aligned} E(w_{-(49)}) = \beta_0 &+ (-\beta_1 - \cdots - \beta_{49}) + (\beta_{50} + \cdots + \beta_{92}) \\ &+ (\beta_{1;2} + \cdots + \beta_{48;49}) + (-\beta_{1;50} - \cdots - \beta_{49;92}) \\ &+ (\beta_{50;51} + \cdots + \beta_{91;92}) \end{aligned}$$

and

$$\begin{aligned} E(w_{(49)}) = \beta_0 &+ (\beta_1 + \cdots + \beta_{49}) + (-\beta_{50} - \cdots - \beta_{92}) \\ &+ (\beta_{1;2} + \cdots + \beta_{48;49}) + (-\beta_{1;50} - \cdots - \beta_{49;92}) \\ &+ (\beta_{50;51} + \cdots + \beta_{91;92}), \end{aligned}$$

so subtracting these two equations *cancels all interactions*. The analogue of (3) gives the unbiased group estimator

$$\widehat{\beta_{j'-j;r}} = \frac{(w_{(j);r} - w_{-(j);r}) - (w_{(j'-1);r} - w_{-(j'-1);r})}{4}. \qquad (10)$$

The analogue of (7) gives the unbiased individual estimator

$$\widehat{\beta_{j;r}} = \frac{(w_{(j);r} - w_{-(j);r}) - (w_{(j-1);r} - w_{-(j-1);r})}{4}. \qquad (11)$$

In other words, SB enables the estimation of first-order effects unbiased by two-factor interactions provided SB simulates the mirror combinations besides the original combinations. Hence, the number of simulated combinations *doubles*. Wan et al. (2006b) point out that—in the case of mirror observations—fewer replications per combination may be needed. They further state that the SB efficiency may increase when applying CRN separately to all positive levels and negative levels respectively.

SB with mirror scenarios may still give misleading results if (say) two factors have unimportant main effects but their interaction is important. Therefore SB assumes *strong heredity* (Assumption 3). If the analysts suspect that this assumption is violated for a specific factor, then they should investigate that factor after the screening phase.

SB with mirror observations does not enable estimation of *individual* interactions, but it does show whether interactions are important—as follows. Estimate the main effects from the original scenarios—ignoring the mirror scenarios. If the analyses of the mirror observations and of the original observations give the same conclusions, then interactions are unimportant. This happened, e.g., in the ecological simulation reported in Bettonvil (1990) and Bettonvil and Kleijnen (1996). In that study, the factor values change relatively little (larger changes give unrealistic simulation outputs), so a first-order polynomial is adequate. In the supply-chain study, however, interactions turn out to be important; see Kleijnen et al. (2006). (In a follow-up experiment with the factors declared to be important in SB, the sizes of the individual interactions are estimated from a resolution-V design, which by definition enables the unbiased estimation of all the individual two-factor interactions; details on resolution-V designs are given by Kleijnen 2008.) Note that the mirror observations and the original observations may give different SB paths through the list of individual factors.

4 Conclusions and Future Research

This chapter may be summarized as follows. There are different screening designs, including resolution-III, supersaturated, and group-screening designs. This chapter, however, focused on SB, and stated the various assumptions of SB. These assumptions may not be too restrictive in practice, as the Ericsson case-study illustrated. If its assumptions are satisfied, then SB is a most efficient and effective screening method.

There is a need for more research:

- It is a challenge to derive the number of replicates that control the *overall* probability of correctly classifying the individual factors as important or unimportant. So far, SB applies a statistical test to each subgroup individually. (Furthermore, SB may terminate "prematurely," and yet estimate the most important factors—instead of classifying all factors with effects that exceed a prespecified threshold.)

- It might be that the simulation of *mirror* factor combinations can be stopped as soon as it seems that no interactions are important.
- After SB stops, the resulting shortlist of important factors should be *validated.* (A procedure was proposed above.)
- *Software* needs to be developed that implements sequential screening of simulation experiments. This software should generate an input file, once a particular design type (e.g., SB) has been chosen. Such a file can then be executed sequentially and efficiently in batch mode; i.e., no human intervention is required while the computer executes the sequential design. Furthermore, good computer programming avoids fixing the inputs at specific numerical values within the code; instead, the computer reads input values so that the program can be run for many combinations of these values. (Of course, the computer should check whether these values are admissible; i.e., do these combinations fall within the experimental domain?) Such a practice can automatically provide a long list of potential factors.
- A contest may be organized that challenges the experts in the different screening methods to estimate the most important factors in a set of simulation models. Such "testbeds" are popular in Mathematical Programming. Note that nobody is expert in *all* screening methods.
- *Multivariate* output may consist of univariate outputs that require different SB paths. This problem has not yet been touched in the literature!

Acknowledgments I thank Bert Bettonvil and Wim van Beers (both at Tilburg University) for their comments on an earlier version.

References

Allen, T. T., and M. Bernshteyn. 2003. Supersaturated designs that maximize the probability of finding the active factors. *Technometrics* 45(1):1–8.

Andres, T. H. 1997. Sampling methods and sensitivity analysis for large parameter sets. *Journal of Statistical Computation and Simulation* 57(1/4):77–110.

Andres, T. H., and W. C. Hajas. 1993. Using iterated fractional factorial design to screen parameters in sensitivity analysis of a probabilistic risk assessment model. In *Proceedings of the Joint International Conference on Mathematical Methods and Supercomputing in Nuclear Applications*, Karlsruhe, Germany, April 19–23, 2:328–337.

Ankenman, B. E., R. C. H. Cheng, and S. M. Lewis. 2006. A polytope method for estimating factor main effects efficiently. In *Proceedings of the 2006 Winter Simulation Conference*, eds. L. F. Perrone, F. P. Wieland, J. Liu, B. G. Lawson, D. M. Nicol, and R. M. Fujimoto, pp. 369–375. Piscataway, New Jersey: Institute of Electrical and Electronics Engineers.

Bettonvil, B. 1990. *Detection of important factors by sequential bifurcation.* Ph.D. dissertation, Tilburg University Press, Tilburg.

Bettonvil, B., and J. P. C. Kleijnen. 1996. Searching for important factors in simulation models with many factors: Sequential bifurcation. *European Journal of Operational Research* 96:180–194.

Campolongo, F., J. Cariboni, and A. Saltelli. 2007. An effective screening design for sensitivity analysis of large models. *Environmental Modelling & Software* 22(10):1509–1518.

Campolongo, F., J. P. C. Kleijnen, and T. Andres. 2000. Screening methods. In *Sensitivity Analysis*, eds. A. Saltelli, K. Chan, and E. M. Scott, pp. 65–89. Chichester, England: Wiley.

Cheng, R. C. H. 1997. Searching for important factors: Sequential bifurcation under uncertainty. In *Proceedings of the 1997 Winter Simulation Conference*, eds. S. Andradóttir, K. J. Healy, D. H. Withers, and B. L. Nelson, pp. 275–280. Piscataway, New Jersey: Institute of Electrical and Electronics Engineers.

Cheng, R. C. H., and W. Holland. 1999. Stochastic sequential bifurcation: Practical issues. In *Proceedings of UK Simulation Society Conference*, eds. D. Al-Dabass and R. C. H. Cheng, pp. 74–80. Nottingham: UKSIM.

Chipman, H. 2006. Prior distributions for Bayesian analysis of screening experiments. In *Screening: Methods for Experimentation in Industry, Drug Discovery, and Genetics*, eds. A. Dean and S. Lewis, pp. 235–267. New York: Springer-Verlag.

De Vos, C., H. W. Saatkamp, M. Nielen, and R. B. M. Huirne. 2006. Sensitivity analysis to evaluate the impact of uncertain factors in a scenario tree model for classical swine fever introduction. *Risk Analysis* 26(5):1311–1322.

Elster, C., and A. Neumaier. 1995. Screening by conference designs. *Biometrika* 82(3):589–602.

Georgiou, S. D. 2007. New two-variable full orthogonal designs and related experiments with linear regression models. *Statistics & Probability Letters* 77(1):25–31.

Gilmour, S. G. 2006. Factor screening via supersaturated designs. In *Screening: Methods for Experimentation in Industry, Drug Discovery, and Genetics*, eds. A. Dean and S. Lewis, pp. 169–190. New York: Springer-Verlag.

Kleijnen, J. P. C. 2008. *Design and Analysis of Simulation Experiments*. New York: Springer Science + Business Media.

Kleijnen, J. P. C., B. Bettonvil, and F. Persson. 2006. Screening for the important factors in large discrete-event simulation: sequential bifurcation and its applications. In: *Screening Methods for Experimentation in Industry, Drug Discovery, and Genetics*, eds. A. Dean and S. Lewis, pp. 287–307. New York: Springer-Verlag.

Kleijnen, J. P. C., S. M. Sanchez, T. W. Lucas, and T. M. Cioppa. 2005. State-of-the-art review: A user's guide to the brave new world of designing simulation experiments. *INFORMS Journal on Computing* 17(3):263–289.

Kleijnen, J. P. C., and W. C. M. van Beers. 2004. Application-driven sequential designs for simulation experiments: Kriging metamodeling. *Journal of the Operational Research Society* 55(9):876–883.

Kleijnen, J. P. C., G. van Ham, and J. Rotmans. 1992. Techniques for sensitivity analysis of simulation models: A case study of the CO_2 greenhouse effect. *Simulation* 58(6):410–417.

Lewis, S. M., and A. M. Dean. 2001. Detection of interactions in experiments on large numbers of factors (including discussion). *Journal Royal Statistical Society B* 63:633–672.

Li, L., and W. Li. 2005. Tabu search and perturbation methods in the construction of supersaturated designs. *American Journal of Mathematical and Management Sciences* 25(1&2):189–205.

Lim, E., and P. W. Glynn. 2006. Simulation-based response surface computation in the presence of monotonicity. In *Proceedings of the 2006 Winter Simulation Conference*, eds. L. F. Perrone, F. P. Wieland, J. Liu, B. G. Lawson, D. M. Nicol, and R. M. Fujimoto, pp. 264–271. Piscataway, New Jersey: Institute of Electrical and Electronics Engineers.

Miller, G. A. 1956. The magical number seven plus or minus two: Some limits on our capacity for processing information. *The Psychological Review* 63:81–97.

Morris, M. D. 2006. An overview of group factor screening. In *Screening: Methods for Experimentation in Industry, Drug Discovery, and Genetics*, eds. A. Dean and S. Lewis, pp. 191–206. New York: Springer-Verlag.

Myers, R. H., and D. C. Montgomery. 2002. *Response Surface methodology: Process and Product Optimization Using Designed Experiments*, 2nd edition. New York: Wiley.

Park, S., J. W. Fowler, G. T. Mackulak, J. B. Keats, and W. M. Carlyle. 2002. D-optimal sequential experiments for generating a simulation-based cycle time-throughput curve. *Operations Research* 50(6):981–990.

Persson, J. F., and J. Olhager. 2002. Performance simulation of supply chain designs. *International Journal of Production Economics* 77:231-245.

Saltelli, A., M. Ratto, S. Tarantola, and F. Campolongo. 2005. Sensitivity analysis of chemical models. *Chemical Reviews* 105(7):2811–2827.

Saltelli, A., S. Tarantola, F. Campolongo, and M. Ratto. 2004. *Sensitivity Analysis in Practice: A Guide to Assessing Scientific Models.*Chichester, England: Wiley.

Sanchez, S. M., F. Moeeni, and P. J. Sanchez. 2006. So many factors, so little time... Simulation experiments in the frequency domain. *International Journal of Production Economics* 103(1):149–165.

Schonlau, M., and W. J. Welch. 2006. Screening the input factors to a computer code via Analysis of Variance and visualization. In *Screening: Methods for Experimentation in Industry, Drug Discovery, and Genetics*, eds. A. Dean and S. Lewis, pp. 308–327. New York: Springer-Verlag.

Schruben, L. W., and V. J. Cogliano. 1987. An experimental procedure for simulation response surface model identification. *Communications of the ACM* 30(8):716–730.

Shen, H., and H. Wan. 2006. A hybrid method for simulation factor screening. In *Proceedings of the 2006 Winter Simulation Conference*, eds. L. F. Perrone, F. P. Wieland, J. Liu, B. G. Lawson, D. M. Nicol, and R. M. Fujimoto, pp. 382–389. Piscataway, New Jersey: Institute of Electrical and Electronics Engineers.

Simpson, T. W., A. J. Booker, D. Ghosh, A. A. Giunta, P. N. Koch, and R.-J. Yang. 2004. Approximation methods in multidisciplinary analysis and optimization: a panel discussion. *Structural and Multidisciplinary Optimization* 27(5):302–313.

Tsai, P.-W., S. G. Gilmour, and R. Mead. 2007. Three-level main-effects designs exploiting prior information about model uncertainty. *Journal of Statistical Planning and Inference* 137(2):619–627.

Tu, J., and D. R. Jones. 2003. Variable screening in metamodel design by cross-validated Moving Least Squares method. In *Proceedings of the 44th AIAA / ASME / ASCE / AHS / ASC Structures, Structural Dynamics, and Materials Conference*, AIAA-2003-1669, Norfolk, Virginia, April 7–10, 2003.

Van Beers, W. C. M., and J. P. C. Kleijnen. 2008. Customized sequential designs for random simulation experiments: Kriging metamodeling and bootstrapping. *European Journal of Operational Research* 186(3):1099–1113.

Wan, H., B. E. Ankenman, and B. L. Nelson. 2006a. Controlled sequential bifurcation: A new factor-screening method for discrete-event simulation. *Operations Research* 54(4):743–755.

Wan, H., B. E. Ankenman, and B. L. Nelson. 2006b. Simulation factor screening with controlled sequential bifurcation in the presence of interactions. Working Paper, Purdue University, West Lafayette, Indiana.

Wu, C. F. J., and M. Hamada. 2000. *Experiments: Planning, Analysis, and Parameter Design Optimization.* New York: Wiley.

Xu, J., F. Yang, and H. Wan. 2007. Controlled sequential bifurcation for software reliability study. In *Proceedings of the 2007 Winter Simulation Conference*, eds. S. G. Henderson, B. Biller, M.-H. Hsieh, J. Shortle, J. D. Tew, and R. R. Barton, pp. 281–288. Piscataway, New Jersey: Institute of Electrical and Electronics Engineers.

Yamada, S., M. Matsui, T. Matsui, D. K. J. Lin, and T. Takahashi. 2006. A general construction method for mixed-level supersaturated design *Computational Statistics & Data Analysis* 50(1):254–265.

Yu, H.-F. 2007. Designing a screening experiment with a reciprocal Weibull degradation rate. *Computers & Industrial Engineering* 52(2):175–191.

Zeigler, B. P., H. Praehofer, and T. G. Kim. 2000. *Theory of Modeling and Simulation*, 2nd edition. San Diego: Academic Press.

Zhang, Q.-Z., R.-C. Zhang, and M.-Q. Liu. 2007. A method for screening active effects in supersaturated designs. *Journal of Statistical Planning and Inference* 137(6):2068–2079.

$\mathbb{F}_2$-Linear Random Number Generators

Pierre L'Ecuyer and François Panneton

Abstract Random number generators based on linear recurrences modulo 2 are among the fastest long-period generators currently available. The uniformity and independence of the points they produce, by taking vectors of successive output values from all possible initial states, can be measured by theoretical figures of merit that can be computed quickly, and the generators having good values for these figures of merit are statistically reliable in general. Some of these generators can also provide disjoint streams and substreams efficiently. In this paper, we review the most interesting construction methods for these generators, examine their theoretical and empirical properties, describe the relevant computational tools and algorithms, and make comparisons.

1 Introduction

Given that computers work in binary arithmetic, it seems natural to construct random number generators (RNGs) defined via recurrences in arithmetic modulo 2, so that these RNGs can be implemented efficiently via elementary operations on bit strings, such as shifts, rotations, exclusive-or's (xor's), and bit masks. Very fast RNGs whose output sequences have huge periods can be constructed in this way. Among them, we find the Tausworthe or linear feedback shift register (LFSR), generalized feedback shift register (GFSR), twisted GFSR (TGFSR), Mersenne twister, the WELL, and xorshift generators (Fishman 1996; L'Ecuyer 1996, 2006; L'Ecuyer and Panneton 2002; Matsumoto and Nishimura 1998; Panneton 2004; Panneton and L'Ecuyer 2005; Panneton et al. 2006; Tezuka 1995). A common characterization of all these generators is that they are special cases of a general class of generators whose state evolves according to a (matrix) linear recurrence modulo 2. The bits that form their output are also determined by a linear transformation modulo 2 applied

P. L'Ecuyer (✉)
Département d' Informatique et de Recherche Opérationnelle (DIRO),
Université de Montréal, C.P. 6128, Succ. Centre-Ville,
Montréal, H3C 3J7, Canada
e-mail: lecuyer@iro.umontreal.ca

C. Alexopoulos et al. (eds.), *Advancing the Frontiers of Simulation: A Festschrift in Honor of George Samuel Fishman*, International Series in Operations Research & Management Science 133, DOI 10.1007/b110059_9,

to the state. Since doing arithmetic modulo 2 can be interpreted as working in $\mathbb{F}_2$, the finite field of cardinality 2 with elements $\{0, 1\}$, we shall refer to this general class as $\mathbb{F}_2$-*linear generators*.

It must be underlined right away that some widely-used RNGs of this form are *not* statistically reliable and should be discarded. But other well-designed instances are good, reliable, and fast. Which ones? What defects do the others hide? What mathematical tools can be used to analyze and practically assess their quality from a theoretical viewpoint? Is it easy to jump ahead quickly in their sequence in order to split it into multiple streams and substreams? In the remainder of this paper, we address these questions and provide a state-of-the-art overview of $\mathbb{F}_2$-linear RNGs.

In the next section, we define a general framework that covers all $\mathbb{F}_2$-linear generators. We provide some basic properties of these RNGs, such as maximal-period conditions, a simple way to jump ahead, and a simple combination method of $\mathbb{F}_2$-linear generators (via a bitwise xor) to construct larger (and often better-behaved) $\mathbb{F}_2$-linear generators. We describe efficient algorithms to compute the characteristic polynomial of an RNG and to check if it has maximal period. In Section 3, we discuss the theoretical measures of uniformity and independence that are typically used in practice as figures of merit to assess its quality. The $\mathbb{F}_2$-linear RNGs turn out to have a lattice structure in spaces of polynomials and formal series over $\mathbb{F}_2$. There are counterparts in those spaces of the spectral test and other lattice-based tests and properties that have been developed for linear congruential generators. Interestingly, these tests are strongly linked with computing the measures of uniformity of $\mathbb{F}_2$-linear generators. Section 4 outlines this theory. We explain how to construct and analyze the polynomial lattices and how to use them for computing the uniformity measures of interest. In Section 5, we describe specific families of $\mathbb{F}_2$-linear generators proposed over the years, show how they fit the general framework, and summarize what we know about their strengths and weaknesses. In Section 6, we compare specific implementations in terms of their speed and (theoretical) figures of merit, and discuss their behavior in empirical statistical tests. Compared with the most widely used RNG that offers multiple streams and substreams in simulation software, the best $\mathbb{F}_2$-linear RNGs are faster by a factor of 1.5–3, depending on the computing platform. Section 7 concludes the paper.

2 $\mathbb{F}_2$-Linear Generators

2.1 General Framework

We consider an RNG defined by a matrix linear recurrence over the finite field $\mathbb{F}_2$, as follows:

$$\mathbf{x}_n = \mathbf{A}\mathbf{x}_{n-1}, \tag{1}$$

$$\mathbf{y}_n = \mathbf{B}\mathbf{x}_n, \tag{2}$$

$$u_n = \sum_{\ell=1}^{w} y_{n,\ell-1} 2^{-\ell} = .y_{n,0}\ y_{n,1}\ y_{n,2}\ \cdots, \tag{3}$$

where $\mathbf{x}_n = (x_{n,0}, \ldots, x_{n,k-1})^{\mathsf{t}} \in \mathbb{F}_2^k$ is the k-bit *state vector* at step n (^{t} means "transposed"), $\mathbf{y}_n = (y_{n,0}, \ldots, y_{n,w-1})^{\mathsf{t}} \in \mathbb{F}_2^w$ is the w-bit *output vector* at step n, k and w are positive integers, $\mathbf{A}$ is a $k \times k$ *transition matrix* with elements in $\mathbb{F}_2$, and $\mathbf{B}$ is a $w \times k$ *output transformation matrix* with elements in $\mathbb{F}_2$. The real number $u_n \in [0, 1)$ is the *output* at step n. All operations in (1) and (2) are performed in $\mathbb{F}_2$, i.e., modulo 2. This setting is from L'Ecuyer and Panneton (2002). Several popular classes of RNGs fit this framework as special cases, by appropriate choices of the matrices $\mathbf{A}$ and $\mathbf{B}$. Many will be described in Section 5.

The period of this RNG cannot exceed $2^k - 1$, because there are only $2^k - 1$ possible nonzero values for $\mathbf{x}_n$. When this maximum is reached, we say that the RNG has *maximal period*. To discuss the periodicity and see how we can construct maximal-period $\mathbb{F}_2$-linear RNGs, we use the following basic definitions and properties from linear algebra and finite fields. Let $\mathbb{F}_2[z]$ denote the ring of polynomials with coefficients in $\mathbb{F}_2$. The *characteristic polynomial* of the matrix $\mathbf{A}$ is

$$P(z) = \det(z\mathbf{I} - \mathbf{A}) = z^k - \alpha_1 z^{k-1} - \cdots - \alpha_{k-1} z - \alpha_k,$$

where $\mathbf{I}$ is the identity matrix and each α_j is in $\mathbb{F}_2$. This $P(z)$ is also the characteristic polynomial of the linear recurrence (in $\mathbb{F}_2$)

$$x_n = \alpha_1 x_{n-1} + \cdots + \alpha_k x_{n-k}. \tag{4}$$

We shall assume that $\alpha_k = 1$. Usually, we know a priori that this is true by construction of the matrix $\mathbf{A}$. In that case, the recurrence (4) has *order* k and it is purely periodic, i.e., there is some integer $\rho > 0$ such that $(x_\rho, \ldots, x_{\rho+k-1}) = (x_0, \ldots, x_{k-1})$; this ρ is called the *period* of the recurrence. The *minimal polynomial* of $\mathbf{A}$ is the polynomial $Q(z) \in \mathbb{F}_2[z]$ of smallest degree for which $Q(\mathbf{A}) = 0$. Every other polynomial $R(z) \in \mathbb{F}_2[z]$ for which $R(\mathbf{A}) = 0$ must be a multiple of the minimal polynomial. This implies in particular that $P(z)$ is a multiple of $Q(z)$. In the context of RNG construction, $Q(z)$ and $P(z)$ are almost always identical, at least for good constructions.

The fact that the sequence $\{\mathbf{x}_n, n \geq 0\}$ obeys (1) implies that it satisfies the recurrence that corresponds to the minimal polynomial of $\mathbf{A}$ (or any other polynomial that is a multiple of $Q(z)$):

$$\mathbf{x}_n = (\alpha_1 \mathbf{x}_{n-1} + \cdots + \alpha_k \mathbf{x}_{n-k}) \qquad (\text{in } \mathbb{F}_2). \tag{5}$$

This means that the sequence $\{x_{n,j}, n \geq 0\}$ obeys (4) for each j, $0 \leq j < k$. The sequence $\{y_{n,j}, n \geq 0\}$, for $0 \leq j < w$, also obeys that same recurrence. However, these sequences may also follow recurrences of order smaller than k. For any periodic sequence in $\mathbb{F}_2$, there is a linear recurrence of *minimal order* obeyed by this sequence, and the characteristic polynomial of that recurrence is called the *minimal polynomial* of the sequence. This minimal polynomial can be computed by the Berlekamp-Massey algorithm (Massey 1969). The sequences $\{x_{n,j},\ n \geq 0\}$ may have different minimal polynomials for different values of j, and also different minimal polynomials than the sequences $\{y_{n,j},\ n \geq 0\}$. But all these minimal polynomials must necessarily divide $P(z)$. If $P(z)$ is *irreducible* (i.e., it has no divisor other than 1 and itself), then $P(z)$ must be the minimal polynomial of all these sequences. Reducible polynomials $P(z)$ do occur when we combine generators (Section 2.3); in that case, $P(z)$ is typically the minimal polynomial of the output bit sequences $\{y_{n,j},\ n \geq 0\}$ as well, but the sequences $\{x_{n,j},\ n \geq 0\}$ often have much smaller minimal polynomials (divisors of $P(z)$).

It is well-known that the recurrences (4) and (5) have maximal period if and only if $P(z)$ is a primitive polynomial over $\mathbb{F}_2$ (Niederreiter 1992, Knuth 1998). Primitivity is a stronger property than irreducibility: $P(z)$ is *primitive* if and only if it is irreducible and for all prime divisors p_i of $r = 2^k - 1$, $z^{r/p_i} \not\equiv 1 \bmod P(z)$. A good way to verify if a polynomial is primitive is to verify irreducibility first, and then check the second condition. Note that when r is prime (this type of prime is called a *Mersenne prime*), the second condition is automatically satisfied.

In the context of RNG construction, we are interested essentially only in maximal-period recurrences. The RNG is constructed either from a single maximal-period recurrence, or from a combination of maximal-period recurrences, as we shall explain later. Assuming that we are interested only in primitive polynomials $P(z)$, we can compute $P(z)$ and check its primitivity as follows.

We first run the generator for k steps from some arbitrary non-zero initial state $\mathbf{x}_0$, and we compute the minimal polynomial $Q_0(z)$ of $\{x_{n,0},\ n \geq 0\}$ with the Berlekamp-Massey algorithm. If $Q_0(z)$ has degree less than k, then $P(z)$ is necessarily reducible and we reject this generator; otherwise $P(z) = Q_0(z)$ and it remains to verify its primitivity. For this, we can use the following algorithm from Rieke et al. (1998) and Panneton (2004); it verifies the set of necessary and sufficient conditions stated in Knuth (1998, p. 30), but it also specifies in what order to perform the polynomial exponentiations:

Algorithm P

{ Given $P(z)$ of degree k, returns TRUE iff $P(z)$ is primitive }
Factorize $r = 2^k - 1 = p_1^{e_1} \cdots p_b^{e_b}$ where $p_1, \ldots, p_b$ are distinct primes;
Compute $q := r/(p_1 \cdots p_b)$ and $q_b(z) := z^q \bmod P(z)$;
For $i = b, \ldots, 1$, let $q_{i-1}(z) := q_i(z)^{p_i} \bmod P(z)$;
If $q_0(z) \neq 1$ or $q_1(z) = 1$, return FALSE;
For $i = b, \ldots, 2$, {
    Compute $t_i(z) := q_i(z)^{p_{i-1} \cdots p_1} \bmod P(z)$;
    If $t_i(z) = 1$, return FALSE; }
Return TRUE.

When k is large, it is worthwhile to first apply an irreducibility test that can detect reducibility faster than this primitivity test. Note that $P(z)$ is reducible if and only if it has an irreducible factor of degree $\leq \lfloor k/2 \rfloor$, where $\lfloor \cdot \rfloor$ denotes the well-known floor function, which truncates its argument to an integer. A key theorem in finite fields theory states that for any integer $n \geq 1$, the product of all irreducible polynomials whose degree d divides n is equal to $z^{2^n} + z$. This means that $P(z)$ is irreducible if and only if $\gcd(z^{2^n} + z,\ P(z)) = 1$ for all $n \leq \lfloor k/2 \rfloor$ (gcd means "greatest common divisor"). This gives the following algorithm:

Algorithm I
{ Given $P(z)$ of degree k, returns TRUE iff $P(z)$ is irreducible }
For $n = 1, \ldots, \lfloor k/2 \rfloor$: if $\gcd(z^{2^n} + z,\ P(z)) \neq 1$, return FALSE;
Return TRUE.

When searching for primitive polynomials for RNG construction, we typically select k and impose a special form on the matrix $\mathbf{A}$, so that a fast implementation is available (see Section 5). Then we search (often at random), in the space of matrices $\mathbf{A}$ that satisfy these constraints, for instances having a primitive characteristic polynomial. The following old result (see, e.g., Lidl and Niederreiter 1986) may give a rough idea of our chances of success. It gives the probability that a random polynomial, generated uniformly over the set of all polynomials of degree k, is primitive. It is important to underline, however, that when generating $\mathbf{A}$ randomly from a special class, the polynomial $P(z)$ does not necessarily have the uniform distribution over the set of polynomials, so the probability that it is primitive might differ from the formula given in the theorem.

Theorem 1 *Among the 2^k polynomials of degree k in $\mathbb{F}_2[z]$, the proportion of primitive polynomials is exactly*

$$\frac{1}{k} \prod_{i=1}^{b} \frac{p_i - 1}{p_i}$$

where $p_1, \ldots, p_b$ are the distinct prime factors of $r = 2^k - 1$.

This result suggests that to improve our chances, it is better to avoid values of r having several small factors. If r is a Mersenne prime, the proportion is exactly $1/k$.

2.2 Jumping Ahead

A key requirement of modern stochastic simulation software is the availability of random number generators with multiple disjoint streams and substreams. These streams and substreams can provide parallel RNGs and are also important to support the use of variance reduction techniques (Fishman 1996, Law and Kelton 2000, L'Ecuyer et al. 2002). They are usually implemented by partitioning the output sequence of a long-period generator into long disjoint subsequences and subsubsequences whose starting points are found by making large jumps in the original sequence.

Jumping ahead directly from $\mathbf{x}_n$ to $\mathbf{x}_{n+\nu}$ for a very large integer ν is easy in principle with this type of generator. It suffices to precompute the matrix $\mathbf{A}^\nu \bmod 2$ (this can be done in $O(k^3 \log \nu)$ operations by a standard method) and then multiply $\mathbf{x}_n$ by this binary matrix, modulo 2. The latter step requires $O(k^2)$ operations and $O(k^2)$ words of memory to store the matrix. This approach works fine for relatively small values of k (e.g., up to 100 or so), but becomes rather slow when k is large. For example, the Mersenne twister of Matsumoto and Nishimura (1998) has $k = 19937$ and the above method is impractical in that case.

A more efficient method is proposed by Haramoto et al. (2008). For a given step size ν, the method represents the state $\mathbf{x}_{n+\nu}$ as $g_\nu(\mathbf{A})\mathbf{x}_n$, where $g_\nu(z) = \sum_{j=0}^{k-1} d_j z^j$ is a polynomial of degree less than k in $\mathbb{F}_2[z]$. The product

$$g_\nu(\mathbf{A})\mathbf{x}_n = \sum_{j=0}^{k-1} d_j \mathbf{A}^j \mathbf{x}_n = \sum_{j=0}^{k-1} d_j \mathbf{x}_{n+j}$$

can be computed simply by running the generator for $k - 1$ steps to obtain $\mathbf{x}_{n+1}, \ldots, \mathbf{x}_{n+k-1}$ and adding (modulo 2) the $\mathbf{x}_{n+j}$'s for which $d_j = 1$. For large k, the cost is dominated by these additions. Their number can be reduced (e.g., by a factor of about 4 when $k = 19937$) by using a sliding window technique, as explained in Haramoto et al. (2008). This method still requires $O(k^2)$ operations, but with a smaller hidden constant and (most importantly) much less memory than the standard matrix multiplication. Yet jumping ahead for $\mathbb{F}_2$-linear generators of large order k (such as the Mersenne twister) remains slow with this method. One way to make the jumping-ahead more efficient is to adopt a combined generator, as discussed in Section 2.3 below, and do the ν-step jumping-ahead separately for each component.

2.3 Combined $\mathbb{F}_2$-Linear Generators

A simple way of combining $\mathbb{F}_2$-linear generators is as follows. For some integer $C > 1$, consider C distinct recurrences of the form (1)–(2), where the cth recurrence has parameters $(k, w, \mathbf{A}, \mathbf{B}) = (k_c, w, \mathbf{A}_c, \mathbf{B}_c)$ and state $\mathbf{x}_{c,n}$ at step n, for $c = 1, \ldots, C$. The output of the combined generator at step n is defined by

$$\mathbf{y}_n = \mathbf{B}_1 \mathbf{x}_{1,n} \oplus \cdots \oplus \mathbf{B}_C \mathbf{x}_{C,n},$$

$$u_n = \sum_{\ell=1}^{w} y_{n,\ell-1} 2^{-\ell},$$

where $\oplus$ denotes the bitwise exclusive-or (xor) operation. One can show (Tezuka 1995, Tezuka and L'Ecuyer 1991) that the period ρ of this combined generator is the least common multiple of the periods ρ_c of its components. This combined generator is equivalent to the generator (1)–(3) with $k = k_1 + \cdots + k_C$, $\mathbf{A} = \operatorname{diag}(\mathbf{A}_1, \ldots, \mathbf{A}_C)$, and $\mathbf{B} = (\mathbf{B}_1, \ldots, \mathbf{B}_C)$. If $P_c(z)$ is the characteristic polynomial of $\mathbf{A}_c$ for each c, then the characteristic polynomial of $\mathbf{A}$ is $P(z) = P_1(z) \cdots P_C(z)$. This polynomial is obviously reducible, so the combined RNG cannot have maximal period $2^k - 1$. However, if we select the parameters carefully so that each component has maximal period $\rho_c = 2^{k_c} - 1$ and if the ρ_c are pairwise relatively prime (the $P_c(z)$ must be distinct irreducible polynomials), then the period of the combined generator is the product of the periods of the components: $\rho = \prod_{c=1}^{C}(2^{k_c} - 1)$. In fact, within one cycle, all combinations of nonzero states for the C components are visited exactly once. When the k_c's are reasonably large, this ρ is not far from $2^k - 1$; the difference is that instead of discarding a single k-bit zero state, we must discard the zero state for each component (i.e., all k-bit states in which at least one of the components is in the zero state). Concrete constructions of this form are given in Tezuka and L'Ecuyer (1991), Wang and Compagner (1993), L'Ecuyer (1996) and Tezuka (1995).

Why would we want to combine generators like this? We already gave one good reason in the previous subsection: efficient jumping-ahead is easier for a combined generator of order k having several components of smaller order than for a non-combined generator with the same k. Another important reason is that matrices $\mathbf{A}$ that give very fast implementations typically lead (unfortunately) to poor quality RNGs from the statistical viewpoint, because of a too simplistic structure. Combined generators provide a way out of this dilemma: select simple components that allow very fast implementations and such that the corresponding combined generator has a more complicated structure, good figures of merit from the theoretical viewpoint, and good statistical properties. Many of the best $\mathbb{F}_2$-linear generators are defined via such combinations. As an illustration, one may have four components of periods $2^{63} - 1, 2^{58} - 1, 2^{55} - 1, 2^{47} - 1$, so the state of each component fits a 64-bit integer and the overall period is near 2^{223}.

There could be situations where instead of combining explicitly known $\mathbb{F}_2$-linear components, we would go the other way around; we may want to generate matrices $\mathbf{A}$ randomly from a given class, then find the decomposition of their (reducible) characteristic polynomials, analyze their periodicity and figures of merit, and so on. This approach is used by Brent and Zimmermann (2003), for example. In that case, we can decompose $P(z) = P_1(z) \cdots P_C(z)$, where the $P_c(z)$ are irreducible, and also decompose the matrix $\mathbf{A}$ in its *Jordan normal form*: $\mathbf{A} = \mathbf{P}\tilde{\mathbf{A}}\mathbf{P}^{-1}$, where $\mathbf{P}$ is an invertible matrix and $\tilde{\mathbf{A}} = diag(\tilde{\mathbf{A}}_1, \ldots, \tilde{\mathbf{A}}_C)$ is a block-diagonal matrix for which each block $\tilde{\mathbf{A}}_c$ has irreducible characteristic polynomial $P_c(z)$ (Golub and Van Loan 1996, Strang 1988). Once we have this decomposition, we know that the generator is equivalent to a combined RNG with transition matrix $\tilde{\mathbf{A}}$ and output transformation matrix $\tilde{\mathbf{B}} = \mathbf{BP}$, and we can analyze it in the same way as if we had first selected its components and then combined them. It is important to note that the purpose of the decomposition in this case is not to provide an efficient implementation for the combined generator, nor an efficient algorithm to jump ahead, but only to analyze the periodicity and other theoretical properties of the generator.

3 Quality Criteria

In general, good RNGs must have a long period ρ (say, $\rho \approx 2^{200}$ or more), must run fast, should not waste memory (the state should be represented in no more than roughly $\log_2 \rho$ bits of memory), must be repeatable and portable (able to reproduce exactly the same sequence in different software/hardware environments), and must allow efficient jumping-ahead in order to obtain multiple streams and substreams. But these required properties do not suffice to imitate independent random numbers.

Recall that a sequence of random variables $U_0, U_1, U_2, \ldots$ are i.i.d. $U[0, 1)$ if and only if for all integers $i \geq 0$ and $t > 0$, the vector $(U_i, \ldots, U_{i+t-1})$ is uniformly distributed over the t-dimensional unit hypercube $[0, 1)^t$. Of course, this cannot hold for algorithmic RNGs that have a finite period. For RNGs that fit our $\mathbb{F}_2$-linear framework, any vector of t successive output values of the generator belongs to the finite set

$$\Psi_t = \{(u_0, \ldots, u_{t-1}) : \mathbf{x}_0 \in \mathbb{F}_2^k\},$$

i.e., the set of output points obtained when the initial state runs over all possible k-bit vectors. This set Ψ_t always has cardinality 2^k when viewed as a multiset (i.e., if the points are counted as many times as they appear).

If $\mathbf{x}_0$ is drawn at random from the set of k-bit vectors $\mathbb{F}_2^k$, with probability 2^{-k} for each vector, then $(u_0, \ldots, u_{t-1})$ is a random vector having the uniform distribution over Ψ_t. Thus, to approximate well the uniform distribution over $[0, 1)^t$, Ψ_t must cover the hypercube $[0, 1)^t$ very uniformly (L'Ecuyer 1994, 2006). More generally, we may also want to measure the uniformity of sets of the form

$$\Psi_I = \{(u_{i_1}, \ldots, u_{i_t}) \mid \mathbf{x}_0 \in \mathbb{F}_2^k\},$$

where $I = \{i_1, \ldots, i_t\}$ is a fixed ordered set of non-negative integers such that $0 \leq i_1 < \cdots < i_t$. For $I = \{0, \ldots, t-1\}$, we recover $\Psi_t = \Psi_I$.

The uniformity of Ψ_I is usually assessed by measures of the *discrepancy* between the empirical distribution of its points and the uniform distribution over $[0, 1)^t$ (Hellekalek and Larcher 1998, L'Ecuyer and Lemieux 2002, Niederreiter 1992). These measures can be defined in many ways and they are in fact equivalent to goodness-of-fit tests for the multivariate uniform distribution. They must be computable *without enumerating the points*, because the cardinality of Ψ_t makes the enumeration practically infeasible when the period is large enough. For this reason, the uniformity measures are usually tailored to the general structure of the RNG. The selected discrepancy measure can be computed for each set I in some predefined class $\mathcal{J}$; then these values can be weighted or normalized by factors that may depend on I, and the worst-case (or average) over $\mathcal{J}$ can be adopted as a *figure of merit* used to rank RNGs. The choices of $\mathcal{J}$ and of the weights are arbitrary. They are a question of compromise and practicality. Typically, $\mathcal{J}$ would contain sets I such that t and $i_t - i_1$ are rather small. We generally try to optimize this figure of merit when searching (by computer) for concrete RNG parameters, within a given class of constructions.

For $\mathbb{F}_2$-linear generators, the uniformity of the point sets Ψ_I is typically assessed by measures of equidistribution defined as follows (L'Ecuyer 2004, 1996; L'Ecuyer and Panneton 2002; Tezuka 1995). For an arbitrary vector $\mathbf{q} = (q_1, \ldots, q_t)$ of non-negative integers, partition the unit hypercube $[0, 1)^t$ into 2^{q_j} intervals of the same length along axis j, for each j. This determines a partition of $[0, 1)^t$ into $2^{q_1+\cdots+q_t}$ rectangular boxes of the same size and shape. If a given set Ψ_I has exactly 2^q points in each box of this partition, where the integer q must satisfy $k-q = q_1+\cdots+q_t$, we say that Ψ_I is $\mathbf{q}$-*equidistributed*. This means that among the 2^k points $(u_{i_1}, \ldots, u_{i_t})$ of Ψ_I, if we consider all $(k-q)$-bit vectors formed by the q_j most significant bits of u_{i_j} for $j = 1, \ldots, t$, each of the 2^{k-q} possibilities occurs exactly the same number of times. Of course, this is possible only if $q_1+\cdots+q_t \le k$. When $q_1+\cdots+q_t \ge k$, i.e., when the number of boxes is larger or equal to the number of points, we say that Ψ_I is $\mathbf{q}$-*collision-free* (CF) if no box contains more than one point (L'Ecuyer 1996).

If Ψ_I is $(\ell, \ldots, \ell)$-equidistributed for some $\ell \ge 1$, it is called t-*distributed with* ℓ *bits of accuracy*, or (t, ℓ)-*equidistributed* (L'Ecuyer 1996). (We will avoid this last notation because it conflicts with that for $(q_1, \ldots, q_t)$-equidistribution.) The largest value of ℓ for which this holds is called the *resolution* of the set Ψ_I and is denoted by ℓ_I. It cannot exceed $\ell_t^* = \min(\lfloor k/t \rfloor, w)$. We define the *resolution gap* of Ψ_I as $\delta_I = \ell_t^* - \ell_I$. Potential figures of merit can then be defined by

$$\Delta_{\mathcal{J},\infty} = \max_{I \in \mathcal{J}} \omega_I \delta_I \quad \text{and} \quad \Delta_{\mathcal{J},1} = \sum_{I \in \mathcal{J}} \omega_I \delta_I$$

for some non-negative weights ω_I, where $\mathcal{J}$ is a preselected class of index sets I. The weights are often taken all equal to 1.

We also denote by t_ℓ the largest dimension t for which Ψ_t is t-distributed with ℓ bits of accuracy, and we define the *dimension gap* for ℓ bits of accuracy as

$$\tilde{\delta}_\ell = t_\ell^* - t_\ell,$$

where $t_\ell^* = \lfloor k/\ell \rfloor$ is an upper bound on t_ℓ. We may then consider the *worst-case weighted dimension gap* and the *weighted sum of dimension gaps*, defined as

$$\tilde{\Delta}_\infty = \max_{1 \le \ell \le w} \omega_\ell \tilde{\delta}_\ell \quad \text{and} \quad \tilde{\Delta}_1 = \sum_{\ell=1}^{w} \omega_\ell \tilde{\delta}_\ell$$

for some non-negative weights ω_ℓ, as alternative figures of merit for our generators. Often, the weights are all 1 and the word "weighted" is removed from these definitions.

When $\tilde{\Delta}_\infty = \tilde{\Delta}_1 = 0$, the RNG is said to be *maximally equidistributed* (ME) or *asymptotically random* for the word size w (L'Ecuyer 1996, Tezuka 1995, Tootill et al. 1973). This property ensures perfect equidistribution of all sets Ψ_t, for any partition of the unit hypercube into subcubes of equal sizes, as long as $\ell \le w$ and the number of subcubes does not exceed the number of points in Ψ_t. As an additional requirement, we may ask that Ψ_t be $(\ell, \ldots, \ell)$-collision-free whenever $t\ell \ge k$. Then we say that the RNG is *collision-free* (CF) (L'Ecuyer 1999b). Large-period ME (or almost ME) and ME-CF generators can be found in L'Ecuyer (1999b), L'Ecuyer and Panneton (2002), Panneton and L'Ecuyer (2004), and Panneton et al. (2006), for example.

The $(k-q)$-bit vectors involved in assessing the $\mathbf{q}$-equidistribution of Ψ_I can be expressed as a linear function of the k-bit initial state $\mathbf{x}_0$, that is, as $\mathbf{z}_0 = \mathbf{M_q x}_0$ for some $(k-q) \times k$ binary matrix $\mathbf{M_q}$. Clearly, Ψ_I is $\mathbf{q}$-equidistributed if and only if $\mathbf{M_q}$ has full rank. Thus, $\mathbf{q}$-equidistribution can easily be verified by constructing this matrix $\mathbf{M_q}$ and checking its rank via (binary) Gaussian elimination (Fushimi 1983, L'Ecuyer 1996, Tezuka 1995). This is a major motivation for adopting this measure of uniformity.

To construct the matrix $\mathbf{M_q}$ that corresponds to Ψ_I, one can proceed as follows. For $j \in \{1, \ldots, k\}$, start the generator in initial state $\mathbf{x}_0 = \mathbf{e}_j$, where $\mathbf{e}_j$ is the unit vector with a 1 in position j and zeros elsewhere, and run the generator for i_t steps. Record the q_1 most significant bits of the output at step i_1, the q_2 most significant bits of the output at step $i_2, \ldots,$ and the q_t most significant bits of the output at step i_t. These bits form the jth column of the matrix $\mathbf{M_q}$.

In the case of a combined generator as in Section 2.3, the matrix $\mathbf{M_q}$ can be constructed by first constructing the corresponding matrices $\mathbf{M}_\mathbf{q}^{(c)}$ for the individual components, and simply juxtaposing these matrices, as suggested in L'Ecuyer (1999b). To describe how this is done, let us denote by $\Psi_I^{(c)}$ the point set that corresponds to component c alone, and let $\mathbf{x}_0^{\mathrm{t}} = ((\mathbf{x}_0^{(1)})^{\mathrm{t}}, \ldots, (\mathbf{x}_0^{(C)})^{\mathrm{t}})$ where $\mathbf{x}_0^{(c)}$ is the initial state for component c. If $\mathbf{z}_0^{(c)}$ is the $(k-q)$-bit vector relevant for the $\mathbf{q}$-equidistribution of $\Psi_I^{(c)}$, then we have $\mathbf{z}_0^{(c)} = \mathbf{M}_\mathbf{q}^{(c)} \mathbf{x}_0^{(c)}$ for some $(k-q) \times k_c$ binary matrix $\mathbf{M}_\mathbf{q}^{(c)}$ that can be constructed as explained earlier. Note that the point set Ψ_I can be written as the direct sum

$$\Psi_I = \Psi_I^{(1)} \oplus \cdots \oplus \Psi_I^{(C)} = \{\mathbf{u} = \mathbf{u}^{(1)} \oplus \cdots \oplus \mathbf{u}^{(C)} \mid \mathbf{u}^{(c)} \in \Phi_I^{(c)} \text{ for each } c\},$$

coordinate by coordinate, and observe that

$$\mathbf{z}_0 = \mathbf{z}_0^{(1)} \oplus \cdots \oplus \mathbf{z}_0^{(C)} = \mathbf{M}_\mathbf{q}^{(1)} \mathbf{x}_0^{(1)} \oplus \cdots \oplus \mathbf{M}_\mathbf{q}^{(C)} \mathbf{x}_0^{(C)}.$$

This means that $\mathbf{M_q}$ is just the juxtaposition $\mathbf{M_q} = \mathbf{M}_\mathbf{q}^{(1)} \cdots \mathbf{M}_\mathbf{q}^{(C)}$. That is, $\mathbf{M}_\mathbf{q}^{(1)}$ gives the first k_1 columns of $\mathbf{M_q}$, $\mathbf{M}_\mathbf{q}^{(2)}$ gives the next k_2 columns, and so on.

For very large values of k, the matrix $\mathbf{M_q}$ is expensive to construct and reduce, but a more efficient method based on the computation of the shortest nonzero vector in a lattice of formal series, studied in Couture and L'Ecuyer (2000), can be used in that case to verify $(\ell, \ldots, \ell)$-equidistribution; see Section 4.

The figures of merit defined above look at the *most significant bits* of the output values, but give little importance to the least significant bits. We could of course extend them so that they also measure the equidistribution of the least significant bits, simply by using different bits to construct the output values and computing the corresponding **q**-equidistributions. But this becomes quite cumbersome and expensive to compute in general because there are too many ways of selecting which bits are to be considered. Certain classes of $\mathbb{F}_2$-linear generators (the Tausworthe/LFSR RNGs defined in Section 5.1) have the interesting property that if all output values are multiplied by a given power of two, modulo 1, all equidistribution properties remain unchanged. In other words, they enjoy the nice property that their least significant bits have the same equidistribution as the most significant ones. We call such generators *resolution-stationary* (Panneton and L'Ecuyer 2007).

Aside from excellent equidistribution, good $\mathbb{F}_2$-linear generators are also required to have characteristic polynomials $P(z)$ whose number N_1 of nonzero coefficients is not too far from half the degree, i.e., near $k/2$ (Compagner 1991, Wang and Compagner 1993). Intuitively, if N_1 is very small and if the state $\mathbf{x}_n$ happens to contain many 0's and only a few 1's, then there is a high likelihood that the $N_1 - 1$ bits used to determine any given new bit of the next state are all zero, in which case this new bit will also be zero. In other words, it may happen frequently in that case that only a small percentage of the bits of $\mathbf{x}_n$ are modified from one step to the next, so the state can contain many more 0's than 1's for a large number of steps. Then, in the terminology of cryptologists, the recurrence has *low diffusion capacity*. An illustration of this with the Mersenne twister can be found in Panneton et al. (2006). In particular, generators for which $P(z)$ is a trinomial or a pentanomial, which have often been used in the past, should be avoided. They fail rather simple statistical tests (Lindholm 1968, Matsumoto and Kurita 1996). The fraction N_1/k of nonzero coefficients in $P(z)$ can be used as a secondary figure of merit for an RNG.

Other measures of uniformity are popular in the context where k is small and the entire point set Ψ_t is used for quasi-Monte Carlo integration (Hellekalek and Larcher 1998, L'Ecuyer and Lemieux 2002, Niederreiter 1992); for example, the smallest q for which Ψ_t is a (q, k, t)-net (commonly known as a (t, m, s)-net, using a different notation), the P_α measure and its weighted versions, the diaphony, etc. However, no one knows how to compute these measures efficiently when $k > 50$ (say), which is always the case for good $\mathbb{F}_2$-linear RNGs.

4 Lattice Structure in a Space of Formal Series

The lattice structure of linear congruential generators (LCGs) is well-known in the simulation community (Knuth 1998, Law and Kelton 2000). $\mathbb{F}_2$-linear RNGs do not have a lattice structure in the real space, but they do have a similar form of lattice structure in a space of formal series (Couture and L'Ecuyer 2000, L'Ecuyer 2004, Lemieux and L'Ecuyer 2003, Tezuka 1995), which we now outline. In comparison with the lattices of LCGs, the real space $\mathbb{R}$ is replaced by the space $\mathbb{L}_2$ of formal power series with coefficients in $\mathbb{F}_2$, of the form $\sum_{\ell=\omega}^{\infty} x_\ell z^{-\ell}$ for some integer ω, and the integers are replaced by polynomials over $\mathbb{F}_2$.

Some $\mathbb{F}_2$-linear RNGs (e.g., the LFSR generators) have a *dimension-wise* lattice structure where the lattice contains vectors of t-dimensional formal series, whose coordinate j is the generating function for the binary expansion of the jth output value, for a given initial state (L'Ecuyer 1994, Lemieux and L'Ecuyer 2003, Tezuka 1995, Tezuka and L'Ecuyer 1991). This dimension-wise lattice can be used to study equidistribution, but it only applies to a subclass of $\mathbb{F}_2$-linear RNGs. For this reason, we will not discuss it any further here. We will concentrate instead on the *resolution-wise* lattice introduced by Tezuka (1995), which applies to all $\mathbb{F}_2$-linear generators.

The sequence of values taken by the jth bit of the *output*, from a given initial state $\mathbf{x}_0$, has *generating function*

$$G_j(z) = \sum_{n=1}^{\infty} y_{n-1,j} z^{-n}$$

(which depends on $\mathbf{x}_0$). When multiplying this formal series by $P(z)$, we obtain the polynomial $g_j(z) = G_j(z)P(z)$ in $\mathbb{F}_2[z]/P(z)$ (the space of polynomials of degree less than k, with coefficients in $\mathbb{F}_2$), because the successive terms of the series satisfy a recurrence with this characteristic polynomial. For $\ell = 1, \ldots, w$, let $\mathbf{G}^{(\ell)}(z) = (G_0(z), \ldots, G_{\ell-1}(z))$.

We first consider the case where $P(z)$ is an irreducible polynomial. In that case, if $G_0(z) \neq 0$, then $g_0(z)$ has an inverse modulo $P(z)$ and there is a unique initial state of the RNG that corresponds to the vector

$$\begin{aligned}\bar{\mathbf{G}}^{(\ell)}(z) &= g_0^{\ 1}(z)\mathbf{G}^{(\ell)}(z)\\ &= (1, g_0^{-1}(z)g_1(z), \ldots, g_0^{-1}(z)g_{\ell-1}(z))/P(z)\end{aligned}$$

(Panneton 2004, Lemma 3.2). Thus, if we rename momentarily $g_0^{-1}(z)g_j(z)$ as $g_j(z)$, we see that it is always possible to select the initial state of the RNG so that $g_0(z) = 1$, i.e,

$$\bar{\mathbf{G}}^{(\ell)}(z) = (1, g_1(z), \ldots, g_{\ell-1}(z))/P(z).$$

When $P(z)$ is irreducible, any given bit of the output follows the same recurrence, with minimal polynomial $P(z)$, but with a lag between the recurrences for the different bits, i.e., they have different starting points. The vector $\bar{\mathbf{G}}^{(\ell)}(z)$ tells us about these lags. More specifically, if $g_i(z) \equiv g_0(z)z^{t_i} \bmod P(z)$, then the lag between the recurrences for bit 0 and bit i is t_i.

Let $\mathbb{L}_2 = \mathbb{F}_2((z^{-1}))$ be the space of formal series of the form $\sum_{n=i}^{\infty} d_{n-1} z^{-n}$ where $i \in \mathbb{Z}$ and $d_{n-1} \in \mathbb{F}_2$ for each n. Let $\mathbb{L}_{2,0}$ be those series for which $i \geq 1$. Suppose that the first ℓ rows of the matrix $\mathbf{B}$ are linearly independent. Then the vectors $\mathbf{v}_1(z) = \bar{\mathbf{G}}^{(\ell)}(z)$, $\mathbf{v}_2(z) = \mathbf{e}_2(z), \ldots, \mathbf{v}_\ell(z) = \mathbf{e}_\ell(z)$ form a basis of a *lattice* $\mathcal{L}_\ell$ in $\mathbb{L}_2$, defined by

$$\mathcal{L}_\ell = \left\{ \mathbf{v}(z) = \sum_{j=1}^{\ell} h_j(z)\mathbf{v}_j(z) \text{ such that } h_j(z) \in \mathbb{F}_2[z] \right\}.$$

This lattice is called the ℓ-bit *resolution-wise lattice* associated with the RNG. The matrix $\mathbf{V}$ whose rows are the $\mathbf{v}_j$'s has an inverse $\mathbf{W} = \mathbf{V}^{-1}$ whose columns

$$\begin{aligned} \mathbf{w}_1(z) &= (P(z), 0, \ldots, 0)^{\mathrm{t}}, \\ \mathbf{w}_2(z) &= (-g_1(z), 1, \ldots, 0)^{\mathrm{t}}, \\ &\vdots \\ \mathbf{w}_\ell(z) &= (-g_{\ell-1}(z), 0, \ldots, 1)^{\mathrm{t}} \end{aligned}$$

form a basis of the *dual lattice*

$$\mathcal{L}_\ell^* = \{\mathbf{h}(z) \in \mathbb{L}_2^\ell : \mathbf{h}(z) \cdot \mathbf{v}(z) \in \mathbb{F}_2[z] \text{ for all } \mathbf{v}(z) \in \mathcal{L}_\ell\},$$

where $\mathbf{h}(z) \cdot \mathbf{v}(z) = \sum_{j=1}^{\ell} h_j(z) v_j(z)$ (the scalar product). This resolution-wise lattice fully describes all the possible output sequences of the RNG via the following theorem. It says that the set of all vectors of generating functions that we can get, from all possible initial states $\mathbf{x}_0$, is exactly the set of lattice points that belong to $\mathbb{L}_{2,0}$. (Here we do not assume that $g_0(z) = 1$.)

Theorem 2 (Couture and L'Ecuyer 2000) *We have*

$$\mathcal{L}_\ell \cap \mathbb{L}_{2,0} = \{(g_0(z), \ldots, g_{\ell-1}(z))/P(z) : \mathbf{x}_0 \in \mathbb{F}_2^k\}.$$

For any $\mathbf{h}(z) = (h_1(z), \ldots, h_\ell(z)) \in (\mathbb{F}_2[z])^\ell$, *we define the* length *of* $\mathbf{h}(z)$ *by* $\|\mathbf{0}\| = 0$ *and*

$$\log_2 \|\mathbf{h}(z)\| = \max_{1 \le j \le \ell} \deg h_j(z) \quad \text{for} \mathbf{h}(z) \neq \mathbf{0}.$$

Theorem 3 (Couture and L'Ecuyer 2000, Tezuka 1995) Ψ_t *is* t*-distributed with* ℓ *bits of accuracy if and only if*

$$\min_{\mathbf{0} \neq \mathbf{h}(z) \in \mathcal{L}_\ell^*} \log_2 \|\mathbf{h}(z)\| > \ell.$$

This theorem shows that checking equidistribution amounts to computing a shortest nonzero vector in the dual lattice $\mathcal{L}_\ell^*$, just like the spectral test commonly applied to LCGs but with a different lattice. As it turns out, very similar algorithms can be used to compute the shortest vector in both cases (Couture and L'Ecuyer 2000).

The algorithm of Lenstra (1985) computes a reduced lattice basis in the sense of Minkowski for a polynomial lattice; the first (shortest) vector of that reduced basis is a shortest nonzero vector in the lattice.

This approach is more efficient than applying Gaussian elimination to the matrix $\mathbf{M_q}$ (see Section 3) when t is large. However, it applies only to the point set Ψ_t formed by t successive output values, and not to the more general point sets Ψ_I.

To construct a basis of the dual lattice for all $\ell \leq w$, we only need the polynomials $g_0(z), g_1(z), \ldots, g_{w-1}(z)$. These polynomials can be computed as follows. Start the generator in some arbitrary nonzero initial state $\mathbf{x}_0$, run it for $k-1$ steps, and observe the corresponding output bits $\mathbf{y}_n = (y_{n,0}, \ldots, y_{n,w-1})$, for $n = 0, \ldots, k-1$. This gives the first k coefficients of $G_j(z)$ for $j = 0, \ldots, w-1$. The coefficients of each $g_j(z) = \sum_{i=1}^{k} c_{j,i} z_{k-i}$ can then be obtained via (Lemieux and L'Ecuyer 2003, Proposition 3.6):

$$\begin{pmatrix} c_{j,1} \\ c_{j,2} \\ \vdots \\ c_{j,k} \end{pmatrix} = \begin{pmatrix} 1 & 0 & \ldots & 0 \\ \alpha_1 & 1 & \ldots & 0 \\ \vdots & \ddots & \ddots & \vdots \\ \alpha_{k-1} & \ldots & \alpha_1 & 1 \end{pmatrix} \begin{pmatrix} y_{0,j} \\ y_{1,j} \\ \vdots \\ y_{k-1,j} \end{pmatrix}.$$

Then, to obtain $g_0(z) = 1$, it suffices to compute the inverse of $g_0(z)$ modulo $P(z)$ and to multiply each $g_j(z)$ by this inverse.

When $P(z)$ is reducible, we can no longer use the argument that $g_0(z)$ has an inverse, but everything else still applies. Suppose $P(z) = P_1(z) \cdots P_C(z)$, where the $P_c(z)$'s are *distinct* irreducible polynomials; all interesting RNGs should satisfy this assumption, usually with a small value of C. In that case, the RNG can then be interpreted as a combined $\mathbb{F}_2$-linear generator that fits the framework of Section 2.3 and a basis of the dual lattice can be constructed by decomposition, as we now explain. If $\mathcal{L}_\ell^{(c)}$ denotes the resolution-wise lattice associated with component c alone and $\mathcal{L}_\ell^{(c)*}$ its dual, it can be seen easily that

$$\mathcal{L}_\ell = \mathcal{L}_\ell^{(1)} \oplus \cdots \oplus \mathcal{L}_\ell^{(C)}$$

(the direct sum of lattices) and

$$\mathcal{L}_\ell^* = \mathcal{L}_\ell^{(1)*} \cap \cdots \cap \mathcal{L}_\ell^{(C)*}.$$

To find a basis of this dual lattice, we can first compute a basis of the dual lattice $\mathcal{L}_\ell^{(c)*}$ for each component c, as described earlier. Let $-g_1^{(c)}(z), \ldots, -g_{\ell-1}^{(c)}(z)$ be the polynomials found in the first coordinates of these dual basis vectors (we assume that $g_0^{(c)}(z) = 1$). For each c, compute $Q_c(z) = (P(z)/P_c(z))^{-1} \bmod P_c(z)$; then for $j = 1, \ldots, \ell-1$, compute

$$g_j(z) = \sum_{c=1}^{C} \left(g_j^{(c)}(z) Q_c(z) P(z)/P_j(z) \right) \bmod P(z),$$

so that $g_j(z) \equiv g_j^{(c)}(z) \bmod P(z)$ for each j. Then define $\mathbf{v}_1 = (1, g_2(z), \ldots, g_{\ell-1}(z)) / P(z)$, $\mathbf{v}_j(z) = \mathbf{e}_j(z)$ for $j \geq 2$, $\mathbf{w}_1(z) = (P(z), 0, \ldots, 0)^{\mathsf{t}}$, and $\mathbf{w}_j(z) = \mathbf{e}_j - g_j(z)\mathbf{e}_1$ for $j \geq 2$, as before. Under the assumption that the $P_c(z)$'s are pairwise relatively prime, the proof of Proposition 4.13 of Lemieux and L'Ecuyer (2002), which is an expanded version of Lemieux and L'Ecuyer (2003), implies the following result:

Proposition 1 *The vectors* $\mathbf{v}_1, \ldots, \mathbf{v}_\ell$ *form a basis of* $\mathcal{L}_\ell$ *and* $\mathbf{w}_1, \ldots, \mathbf{w}_\ell$ *are a basis of the dual lattice* $\mathcal{L}_\ell^*$.

This way of doing most of the computations for the components separately before putting the results together is more efficient than working directly with the combined generator, especially if the components are much smaller than the combination.

5 Specific Classes of Generators

5.1 The LFSR Generator

The *Tausworthe* or *linear feedback shift register* (LFSR) generator (L'Ecuyer 1996, Tausworthe 1965, Tezuka 1995) is defined by a linear recurrence modulo 2, from which a block of w bits is taken every s steps, for some positive integers w and s:

$$x_n = a_1 x_{n-1} + \cdots + a_k x_{n-k}, \tag{6}$$

$$u_n = \sum_{\ell=1}^{w} x_{ns+\ell-1} 2^{-\ell}. \tag{7}$$

where $a_1, \ldots, a_k$ are in $\mathbb{F}_2$ and $a_k = 1$. This fits our framework by taking $\mathbf{A} = (\mathbf{A}_0)^s$ (in $\mathbb{F}_2$) where

$$\mathbf{A}_0 = \begin{pmatrix} & 1 & & \\ & & \ddots & \\ & & & 1 \\ a_k & a_{k-1} & \ldots & a_1 \end{pmatrix}, \tag{8}$$

and blank entries in this matrix are zeros (we use that convention in this paper). If $w \leq k$, the matrix $\mathbf{B}$ would contain the first w rows of the $k \times k$ identity matrix. However, we may also have $w > k$, in particular when implementing an LFSR used as a component of a combined generator. In that case, it is convenient to expand $\mathbf{A}$ into a $w \times w$ matrix with the same minimal polynomial (of degree k), as follows: For $j = 1, \ldots, w - k$, add the row $(a_1^{(j)}, \ldots, a_k^{(j)})$, where the coefficients $a_i^{(j)}$ are such that $x_{n+j} = a_1^{(j)} x_{n-1} + \cdots + a_k^{(j)} x_{n-k}$. This can be done in the same way as when we build the matrix $\mathbf{M_q}$ in Section 3. Then, we add $w-k$ columns of zeros.

Note that $P(z)$ is the characteristic polynomial of the matrix $\mathbf{A} = (\mathbf{A}_0)^s$, not that of the recurrence (6), and the choice of s is important for determining the quality of this generator. A frequently encountered case is when a single a_j is nonzero in addition to a_k; then, the characteristic polynomial of $\mathbf{A}_0$ is a trinomial and we have a *trinomial-based* LFSR generator. Typically, s is small to make the implementation efficient. These trinomial-based generators are known to have important statistical weaknesses (Matsumoto and Kurita 1996, Tezuka 1995) but they can be used as components of combined RNGs (L'Ecuyer 1996, Tezuka and L'Ecuyer 1991, Wang and Compagner 1993). They also enjoy the important properties of being resolution-stationary (Panneton and L'Ecuyer 2007). Tables of specific parameters for maximally equidistributed combined LFSR generators, together with concrete implementations for 32-bit and 64-bit computers, can be found in L'Ecuyer (1999b). These generators are amongst the fastest ones currently available.

To show how an LFSR generator can be implemented efficiently, we outline an algorithm for the following situation. Suppose that $a_j = 1$ for $j \in \{j_1, \ldots, j_d\}$ and $a_j = 0$ otherwise, with $k/2 \le j_1 < \cdots < j_d = k \le w$ and $0 < s \le j_1$. We work directly with the w-bit vectors $\mathbf{y}_n = (x_{ns}, \ldots, x_{ns+w-1})$, assuming that w is the computer's word length. Under these conditions, a left shift of $\mathbf{y}_n$ by $k - j_i$ bits, denoted $\mathbf{y}_n \ll (k - j_i)$, gives a vector that contains the first $w - k + j_i$ bits of $\mathbf{y}_{n+k-j_i}$ followed by $k - j_i$ zeros (for $i = d$, $j_i = k$ so there is no shift). Adding these d shifted vectors by a bitwise xor, for $j = 1, \ldots, d$, gives a vector $\tilde{\mathbf{y}}$ that contains the first $w - k + j_1$ bits of $\mathbf{y}_{n+k} = \mathbf{y}_{n+k-j_1} \oplus \cdots \oplus \mathbf{y}_{n+k-j_d}$ followed by $k - j_1$ other bits (which do not matter). Now we shift $\tilde{\mathbf{y}}$ by $k - s$ positions to the right, denoted $\tilde{\mathbf{y}} \gg (k - s)$; this gives $k - s$ zeros followed by the last $w - k + s$ bits of $\mathbf{y}_{n+s}$ (the $k - j_1$ bits that do not matter have disappeared, because $s \ge j_1$). Zeroing the last $w - k$ bits of $\mathbf{y}_n$ and then shifting it to the left by s bits gives the first $k - s$ bits of $\mathbf{y}_{n+s}$. Adding this to $\tilde{\mathbf{y}}$ then gives $\mathbf{y}_{n+s}$. This is summarized by the following algorithm, in which & denotes a bitwise "and" and `mask` contains k 1's followed by $w - k$ 0's.

Algorithm L

{ One step of a simple LFSR generator }

$\tilde{\mathbf{y}} = \mathbf{y}_n$;

For $i = 2, \ldots, d$, $\tilde{\mathbf{y}} = \tilde{\mathbf{y}} \oplus (\mathbf{y}_n \ll (k - j_i))$;

$\mathbf{y}_{n+s} = (\tilde{\mathbf{y}} \gg (k - s)) \oplus ((\mathbf{y}_n \& \texttt{mask}) \ll s)$;

For this to work properly, we must make sure that $\mathbf{y}_0$ is initialized to a valid state, i.e., that the values $x_k, \ldots, x_{w-1}$ satisfy the recurrence $x_j = a_1 x_{j-1} + \cdots + a_k x_{j-k}$ for $j = k, \ldots, w-1$. We can take $(x_0, \ldots, x_{k-1})$ as an arbitrary nonzero vector, and then simply compute $x_k, \ldots, x_{w-1}$ from the recurrence. L'Ecuyer (1996) explains how to implement this.

5.2 The GFSR, Twisted GFSR, and Mersenne Twister

Here we suppose that $\mathbf{A}$ is a $pq \times pq$ matrix with the general form

$$\mathbf{A} = \begin{pmatrix} \mathbf{S}_1 & \mathbf{S}_2 & & \mathbf{S}_{q-1} & \mathbf{S}_q \\ \mathbf{I}_p & & & & \\ & \mathbf{I}_p & & & \\ & & \ddots & & \\ & & & \mathbf{I}_p & \end{pmatrix}$$

for some positive integers p and q, where $\mathbf{I}_p$ is the $p \times p$ identity matrix, and each $\mathbf{S}_j$ is a $p \times p$ matrix. Often, $w = p$ and $\mathbf{B}$ contains the first w rows of the $pq \times pq$ identity matrix. If $\mathbf{S}_r = \mathbf{S}_q = \mathbf{I}_p$ for some r and all the other $\mathbf{S}_j$'s are zero, this generator is the trinomial-based *generalized feedback shift register* (GFSR), for which $\mathbf{x}_n$ is obtained by a bitwise exclusive-or of $\mathbf{x}_{n-r}$ and $\mathbf{x}_{n-q}$ and where $\mathbf{x}_n$ gives the w bits of u_n (Lewis and Payne 1973). This provides an extremely fast RNG. However, its period cannot exceed $2^q - 1$, because each bit of $\mathbf{x}_n$ follows the same binary recurrence of order $k = q$, with characteristic polynomial $P(z) = z^q - z^{q-r} - 1$.

More generally, we can define $\mathbf{x}_n$ as the bitwise exclusive-or of $\mathbf{x}_{n-r_1}$, $\mathbf{x}_{n-r_2}, \ldots, \mathbf{x}_{n-r_d}$ where $r_d = q$, so that each bit of $\mathbf{x}_n$ follows a recurrence in $\mathbb{F}_2$ whose characteristic polynomial $P(z)$ has $d+1$ nonzero terms. This corresponds to taking $\mathbf{S}_j = \mathbf{I}_p$ for $j \in \{r_1, \ldots, r_d\}$ and $\mathbf{S}_j = \mathbf{0}$ otherwise. However, the period is still bounded by $2^q - 1$, whereas considering the pq-bit state, we should expect a period close to 2^{pq}. This was the main motivation for the *twisted GFSR* (TGFSR) generator. In the original version introduced by Matsumoto and Kurita (1992), $w = p$, $\mathbf{S}_q$ is defined as the transpose of $\mathbf{A}_0$ in (8) with k replaced by p, $\mathbf{S}_r = \mathbf{I}_p$, and all the other $\mathbf{S}_j$'s are zero. The characteristic polynomial of $\mathbf{A}$ is then $P(z) = P_S(z^q + z^{q-r})$, where $P_S(\zeta) = \zeta^p - a_p\zeta^{p-1} - \cdots - a_1$ is the characteristic polynomial of $\mathbf{S}_q$, and its degree is $k = pq$. If the parameters are selected so that $P(z)$ is primitive over $\mathbb{F}_2$, then the TGFSR has period $2^k - 1$. Matsumoto and Kurita (1994) pointed out important weaknesses of the original TGFSR, for which $\mathbf{B}$ contains the first rows of the identity matrix, and introduced an improved version that uses a well-chosen matrix $\mathbf{B}$ whose rows differ from those of the identity. The operations implemented by this matrix are called *tempering* and their purpose is to improve the uniformity of the points produced by the RNG. To our knowledge, this was the first version of an $\mathbb{F}_2$-linear RNG with a $\mathbf{B}$ that differs from the truncated identity.

The *Mersenne twister* (Matsumoto and Nishimura 1998, Nishimura 2000) (MT) is a variant of the TGFSR where k is slightly less than pq and can be a prime number. It uses a pq-bit vector to store the k-bit state, where $k = pq - r$ is selected so that $r < p$ and $2^k - 1$ is a Mersenne prime. The matrix $\mathbf{A}$ is a $(pq - r) \times (pq - r)$ matrix similar to that of the TGFSR and the implementation is also quite similar. The main reason for using a k of that form is to simplify the search for primitive characteristic polynomials (see Algorithm P). If we take $k = pq$, then we know that we cannot have a Mersenne prime because $2^{pq} - 1$ is divisible by $2^p - 1$ and $2^q - 1$. A specific instance proposed by Matsumoto and Nishimura (1998), and named MT19937, has become quite popular; it is fast, and has the huge period of $2^{19937} - 1$.

A weakness of this RNG is underlined and illustrated in Panneton et al. (2006): if the generator starts in (or reaches) a state that has very few ones, it may take up to several hundred thousand steps before the ratio of ones in the output and/or the average output value are approximately 1/2. For example, for MT19937, if we average the output values at steps $n + 1$ to $n + 100$ (a moving average) and average this over all 19937 initial states $\mathbf{x}_0$ that have a single bit at one, then we need at least $n > 700{,}000$ before the average gets close to 1/2, as it should (this is graphically illustrated in Panneton et al. 2006). Likewise, if two states differ by a single bit, or by only a few bits, a very large number of steps are required on average before the states or the outputs differ by about half of their bits. The source of the problem is that this RNG has a (huge) 19937-bit state and very few of these bits are modified from one step to the next, as explained near the end of Section 3; it has only $N_1 = 135$ nonzero coefficients out of 19938 in its characteristic polynomial. Moreover, the figure of merit $\tilde{\Delta}_1$ takes the large value 6750 for this generator.

It has been proved that the TGFSR and Mersenne twister construction methods used in Matsumoto and Kurita (1994) and Matsumoto and Nishimura (1998) *cannot* provide ME generators in general. They typically have large equidistribution gaps. But combining them via a bitwise xor *can* yield generators with the ME property. Concrete examples of ME combined TGFSR generators with periods around 2^{466} and 2^{1250} are given in L'Ecuyer and Panneton (2002). These generators have the additional property that the resolution gaps δ_I are also zero for a class of index sets I of small cardinality and whose elements are not too far apart. Of course, they are somewhat slower than their original (uncombined) counterparts.

5.3 The WELL RNGs

These RNGs were developed by Panneton (2004) and are described by Panneton et al. (2006). The idea was to "sprinkle" a small number of very simple operations on w-bit words (where w is taken as the size of the computer word), such as xor, shift, bit mask, etc., into the matrix $\mathbf{A}$ in a way that the resulting RNG satisfied the following requirements: (1) it has maximal period, (2) it runs about as fast as the Mersenne twister, and (3) it also has the best possible equidistribution properties, and a characteristic polynomial with around 50% nonzero coefficients.

The state $\mathbf{x}_n = (\mathbf{v}_{n,0}^{\mathsf{t}}, \ldots, \mathbf{v}_{n,r-1}^{\mathsf{t}})^{\mathsf{t}}$ is comprised of r blocks of $w = 32$ bits $\mathbf{v}_{n,j}$, and the recurrence is defined by a set of linear transformations that apply to these blocks, as described in Panneton et al. (2006). Essentially, the transformations modify $\mathbf{v}_{n,0}$ and $\mathbf{v}_{n,1}$ by using several of the other blocks. They are selected so that $P(z)$, a polynomial of degree $k = rw - p$, is primitive over $\mathbb{F}_2$. The output is defined by $\mathbf{y}_n = \mathbf{v}_{n,0}$.

The authors list specific parameters for WELL generators with periods ranging from $2^{512}-1$ to $2^{44497}-1$. Many of them are ME and the others are nearly ME. Their characteristic polynomials have nearly 50% coefficients equal to 1. These RNGs have much better diffusion capacity than the Mersenne twister and have comparable speed.

5.4 *Xorshift Generators*

Marsaglia (2003) proposed a class of very fast RNGs whose recurrence can be implemented by a small number of xorshift operations only, where a *xorshift operation* consists of replacing a w-bit block in the state by a (left or right) shifted version of itself (by a positions, where $0 < a < w$) xored with the original block. The constant w is the computer's word size (usually 32 or 64). The specific generators he proposed in his paper use three xorshift operations at each step. As it turns out, xorshifts are linear operations so these generators fit our $\mathbb{F}_2$-linear setting.

Panneton and L'Ecuyer (2005) analyzed the theoretical properties of a general class of xorshift generators that contains those proposed by Marsaglia. They studied maximal-period conditions, limits on the equidistribution, and submitted xorshift generators to empirical statistical testing. They concluded that three-xorshift generators are unsafe and came up with generators based on 7 and 13 xorshifts, whose speed is only 20% slower than those with three xorshifts to generate $U(0, 1)$ numbers. Aside from the tests that detect $\mathbb{F}_2$-linearity, these RNGs pass other standard statistical tests.

Brent (2004) proposed a family of generators that combine a xorshift RNG with a Weyl generator. The resulting generator is no longer $\mathbb{F}_2$-linear and it behaves well empirically (L'Ecuyer and Simard 2007).

5.5 *Linear Recurrences in $\mathbb{F}_{2^w}$*

Fix a positive integer w (e.g., $w = 32$) and let $q = 2^w$. Panneton (2004) and Panneton and L'Ecuyer (2004) consider fast RNGs based on recurrences in the finite field $\mathbb{F}_q$, which can be written as

$$m_n = b_1 m_{n-1} + \cdots + b_r m_{n-r}$$

for some integer r, where the arithmetic is performed in $\mathbb{F}_q$. The maximal period $\rho = 2^{rw} - 1$ is reached if and only if $\tilde{P}(z) = z^r - b_1 z^{r-1} - \cdots - b_{r-1}z - b_r$ is a primitive polynomial over $\mathbb{F}_q$.

To implement this recurrence, these authors select an algebraic element ζ of $\mathbb{F}_q$, take $\{1, \zeta, \ldots, \zeta^{r-1}\}$ as a basis of $\mathbb{F}_q$ over $\mathbb{F}_2$, and represent the elements $m_n = v_{n,0} + v_{n,1}\zeta + \cdots + v_{n,w-1}\zeta^{w-1}$ of $\mathbb{F}_q$ by the bit vectors $\mathbf{v}_n = (v_{n,0}, v_{n,1}, \ldots, v_{n,w-1})^{\mathsf{t}}$. The state of the RNG is thus represented by a rw-bit vector and the output is constructed as in (3), from the bits of $\mathbf{v}_n$. (More generally, one could define the output by taking $\mathbf{y}_n = (\mathbf{v}_n, \mathbf{v}_{n-1}, \ldots, \mathbf{v}_{n-r+1})$ for some $r \geq 1$.) This construction fits our $\mathbb{F}_2$-linear framework (1)–(3) and generalizes the TGFSR generators. Panneton and L'Ecuyer (2004) call them *LFSR generators in* $\mathbb{F}_{2^w}$.

The same authors also propose a slightly different construction called *polynomial LCG in* $\mathbb{F}_{2^w}$, based on the recurrence

$$q_n(z) = zq_{n-1}(z) \mod \tilde{P}(z)$$

in $\mathbb{F}_q[z]$ (the ring of polynomials with coefficients in $\mathbb{F}_q$), where $\tilde{P}(z) \in \mathbb{F}_q[z]$ is a primitive polynomial. To implement this, each coefficient of $q_n(z)$ is represented by a w-bit vector just as for m_n and the output is defined in a similar way. Again, this fits the $\mathbb{F}_2$-linear framework (1)–(3).

Panneton (2004) (see also Panneton and L'Ecuyer 2005) goes further by proving certain properties of the equidistribution of these RNGs. For instance, he shows that if $\tilde{P}(z)$ is irreducible over $\mathbb{F}_q$ and can be written as

$$\tilde{P}(z) = p_0(z) + \zeta p_1(z) + \cdots + \zeta^\gamma p_\gamma(z)$$

where each $p_i(z)$ is in $\mathbb{F}_2[z]$, then the RNG cannot be t-distributed with ℓ bits of accuracy if $t > r$ and $\ell > \gamma$. As a corollary, since the TGFSR has $\tilde{P}(z) = p_0(z) + \zeta p_1(z)$, it cannot be t-distributed with more than a single bit of accuracy in any dimension $t > r$. He also shows that if $\tilde{P}(z)$ is irreducible over $\mathbb{F}_q$ and has at least three nonzero coefficients, then among the $2^{rw} - 1$ two-dimensional point sets $\Psi_{\{0,j\}}$ where $1 \leq j < 2^{kw}$, exactly $2^w - 1$ are not 2-distributed with w bits of accuracy. For example, if $w = 32$ and $r = 25$ (so $k = 800$), only one two-dimensional projection out of 2^{768} is not equidistributed!

Panneton (2004) and Panneton and L'Ecuyer (2004) propose tables of good parameters for LFSRs and polynomial LCGs in $\mathbb{F}_q$. These parameters were found by computer searches based on the figure of merit $\tilde{\Delta}_1$. They also provide concrete implementations in the C language. These implementations are fast, comparable to the Mersenne twister for instance, but one drawback is that they use precomputed multiplication tables that require a non-negligible amount of memory. (In the case of multiple streams, a single copy of the tables is shared by all the streams.) The output transformation by a non-trivial matrix $\mathbf{B}$ is integrated into these multiplication tables to improve the efficiency.

6 Speed and Performance in Statistical Tests

6.1 Speed Comparisons

Table 1 reports the speed of some RNGs available in the Java-based SSJ simulation package (L'Ecuyer and Buist 2005). The timings are for the SSJ implementation (with SUN's JDK 1.6) and a C implementations, both on a 2.4 GHz 64-bit

Table 1 CPU time (sec) to generate 10^9 random numbers, and CPU time to jump ahead 10^6 times, with some RNGs available in SSJ

RNG	$\rho \approx$	CPU time in SSJ (Java)			CPU time in C	
		gen. 64	gen. 32	jump	gen. 64	gen. 32
LFSR113	2^{113}	20	70	0.1	10	39
LFSR258	2^{258}	22	105	0.2	12	58
WELL512	2^{512}	24	57	234	12	38
WELL1024	2^{1024}	30	55	917	11	37
MT19937	2^{19937}	33	51	—	16	42
MRG31k3p	2^{185}	48	60	0.9	21	71
MRG32k3a	2^{191}	65	93	1.1	21	99

AMD-Athlon computer and on a 2.8 GHz 32-bit Intel processor. The first and second columns of the table give the generator's name and its approximate period. All these generators are implemented for a 32-bit computer, although the C implementation of the two MRG generators (last two lines) used on the 64-bit computer was different; it exploits the 64-bit arithmetic, which explains the large speed gains. The SSJ implementations of all generators have more overhead because they support multiple streams, can generate either integers or real numbers, etc. We estimate this overhead at about 10–20% in general, but there are cases where it is higher than that. The jumping ahead in SSJ is implemented via a multiplication by $\mathbf{A}^{\nu}$ as explained in Section 2.2. For the combined LFSR generators, the linear recurrence that corresponds to the matrix $\mathbf{A}^{\nu}$ is implemented directly using the algorithm of Section 5.1, for each component of the combination. It is much faster for this reason. Columns 3 and 4 of the table give the CPU times (sec) to generate 10^9 random numbers and add them up, on the 64-bit (gen. 64) and 32-bit (gen. 32) computers, respectively. Column 5 gives the CPU time needed to jump ahead 10^6 times by a very large number of steps (to get a new stream), in SSJ, on the 64-bit computer. For comparison, columns 6 and 7 give the times to generate 10^9 numbers with the C implementation available in TestU01 (L'Ecuyer and Simard 2007), also on the 64-bit and 32-bit computers. The difference in speed between Java and C depends on the performance of the Java interpreter or just-in-time compiler; we have observed a significant difference between JDK 1.5 and 1.6, for example.

The first five RNGs are $\mathbb{F}_2$-linear and the last two are combined multiple recursive generators (MRGs). The first two are combined LFSRs proposed by L'Ecuyer (1999b) for 32-bit and 64-bit computers, with four and five components, respectively. The two WELL RNGs are proposed in Panneton et al. (2006). Other WELL generators with much longer periods (up to nearly 2^{44497}) proposed in that paper have approximately the same speed as those given here to generate random numbers, but are much slower than WELL1024 for jumping ahead because of their larger value of k. For the Mersenne twister MT19937, proposed by Matsumoto and Nishimura (1998), jumping ahead is also too slow and is not implemented in SSJ. All these $\mathbb{F}_2$-linear RNGs have roughly the same speed for generating random numbers. Other ones with about the same speed are also proposed by Matsumoto and Kurita (1994) and Panneton and L'Ecuyer (2004), e.g., with periods near 2^{800}. It is interesting to note that in additional experiments in Java without the streams and substreams, on the 32-bit computer, the LFSR113 took 39 seconds, the same as in C. It took 17 seconds on the 64-bit computer, compared with 10 seconds in C.

The timings of the two MRGs in the table are reported for comparison. The first one (MRG31k3p) was proposed by L'Ecuyer and Touzin (2000) while the second one (MRG32k3a) was proposed by L'Ecuyer (1999a) and is used in several simulation packages to provide multiple streams and substreams. This latter RNG has been heavily tested over the years and is very robust. On the other hand, the $\mathbb{F}_2$-linear generators are faster.

6.2 Statistical Testing

All the RNGs in Table 1 have been submitted to empirical statistical testing using the batteries Smallcrush, Crush, and Bigcrush of the TestU01 package (L'Ecuyer and Simard 2007). They passed all the tests in these batteries with the following notable exceptions: All $\mathbb{F}_2$-linear generators fail the tests that look for linear relationships in the sequences of bits they produce, namely, the matrix-rank test (Marsaglia 1985) for huge binary matrices and the linear complexity tests (Erdmann 1992). The reason for this general failure is obvious: We know from their definitions that these generators produce bit sequences that obey linear recurrences, so they cannot have the linear complexity of a truly random sequence. This is definitely a limitation of these RNGs. But whenever the bit sequences are transformed nonlinearly by the application (e.g., to generate real-valued random numbers from non-uniform distributions), the linear relationships between the bits usually disappear, and the linearity is then very unlikely to cause a problem. For situations where simulation results can be noticeably affected by the linear dependencies among the bits, to make these RNGs safer without slowing them down too much, we could either combine them with a generator from another class (such as an MRG, for instance), or combine them with a small nonlinear RNG implemented via precomputed tables as suggested by L'Ecuyer and Granger-Piché (2003), or add a nonlinear output transformation that is fast to compute.

7 Conclusion

$\mathbb{F}_2$-linear RNGs are convenient for simulation because they are fast and the high-dimensional uniformity of their point sets can be measured by theoretical figures of merit that can be computed efficiently. Combined $\mathbb{F}_2$-linear generators with relatively small components have the important advantage of faster jumping-ahead, because the (smaller) components can be dealt with separately. Some $\mathbb{F}_2$-linear generators proposed in the literature have huge periods, but it is not always true that larger is better. A huge state has the disadvantage of using more memory (this is important when there is a large number of streams in a simulation). It also makes jumping ahead much slower, and it requires more operations to modify a large fraction of the bits in the state. Of course, very long bit sequences produced by $\mathbb{F}_2$-linear generators will always fail statistical tests that measure their linear complexity. This can be viewed as a weak limitation, which could be overcome by adding a nonlinear output transformation or combining the $\mathbb{F}_2$-linear RNG with a generator from another class.

Acknowledgments This work has been supported by NSERC-Canada grant No. ODGP0110050 and a Canada Research Chair to the first author. This paper was written while the first author was a visiting scientist at the University of Salzburg, Austria, in 2005, and at IRISA, Rennes, in 2006. A short draft of it appeared in the *Proceedings of the 2005 Winter Simulation Conference*. Richard Simard helped doing the speed tests.

References

Brent, R. P. 2004. Note on Marsaglia's xorshift random number generators. *Journal of Statistical Software* 11(5):1–4.

Brent, R. P., and P. Zimmermann. 2003. Random number generators with period divisible by a Mersenne prime. In *Computational Science and its Applications—ICCSA 2003*, eds. V. Kumar, M. L. Gavrilova, C. J. K. Tan, and P. L'Ecuyer, Volume 2667 of *Lecture Notes in Computer Science*, pp. 1–10. Berlin: Springer-Verlag.

Compagner, A. 1991. The hierarchy of correlations in random binary sequences. *Journal of Statistical Physics* 63:883–896.

Couture, R., and P. L'Ecuyer. 2000. Lattice computations for random numbers. *Mathematics of Computation* 69(230):757–765.

Erdmann, E. D. 1992. Empirical tests of binary keystreams. Master's thesis, Department of Mathematics, Royal Holloway and Bedford New College, University of London.

Fishman, G. S. 1996. *Monte Carlo: Concepts, Algorithms, and Applications*. New York: Springer-Verlag.

Fushimi, M. 1983. Increasing the orders of equidistribution of the leading bits of the Tausworthe sequence. *Information Processing Letters* 16:189–192.

Golub, G. H., and C. F. Van Loan. 1996. *Matrix Computations*, 3rd edition. Baltimore: John Hopkins University Press.

Haramoto, H., M. Matsumoto, T. Nishimura, F. Panneton, and P. L'Ecuyer. 2008. Efficient jump ahead for $\mathbf{F}_2$-linear random number generators. *INFORMS Journal on Computing* 20(3):385–390.

Hellekalek, P., and G. Larcher (eds.) 1998. *Random and Quasi-Random Point Sets*, Volume 138 of *Lecture Notes in Statistics*. New York: Springer-Verlag.

Knuth, D. E. 1998. *The Art of Computer Programming, Volume 2: Seminumerical Algorithms*, 3rd edition. Reading, MA: Addison-Wesley.

Law, A. M., and W. D. Kelton. 2000. *Simulation Modeling and Analysis*, 3rd edition. New York: McGraw-Hill.

L'Ecuyer, P. 1994. Uniform random number generation. *Annals of Operations Research* 53: 77–120.

L'Ecuyer, P. 1996. Maximally equidistributed combined Tausworthe generators. *Mathematics of Computation* 65(213):203–213.

L'Ecuyer, P. 1999a. Good parameters and implementations for combined multiple recursive random number generators. *Operations Research* 47(1):159–164.

L'Ecuyer, P. 1999b. Tables of maximally equidistributed combined LFSR generators. *Mathematics of Computation* 68(225):261–269.

L'Ecuyer, P. 2004. Polynomial integration lattices. In *Monte Carlo and Quasi-Monte Carlo Methods 2002*, ed. H. Niederreiter, pp. 73–98. Berlin: Springer-Verlag.

L'Ecuyer, P. 2006. Uniform random number generation. In *Simulation*, eds. S. G. Henderson and B. L. Nelson, Handbooks in Operations Research and Management Science, pp. 55–81. Amsterdam, The Netherlands: Elsevier.

L'Ecuyer, P., and E. Buist. 2005. Simulation in Java with SSJ. In *Proceedings of the 2005 Winter Simulation Conference*, eds. M. E. Kuhl, N. M. Steiger, F. B. Armstrong, and J. A. Joines, pp. 611–620. Pistacaway, NJ: IEEE Press.

L'Ecuyer, P., and J. Granger-Piché. 2003. Combined generators with components from different families. *Mathematics and Computers in Simulation* 62:395–404.

L'Ecuyer, P., and C. Lemieux. 2002. Recent advances in randomized quasi-Monte Carlo methods. In *Modeling Uncertainty: An Examination of Stochastic Theory, Methods, and Applications*, eds. M. Dror, P. L'Ecuyer, and F. Szidarovszky, pp. 419–474. Boston: Kluwer Academic.

L'Ecuyer, P., and F. Panneton. 2002. Construction of equidistributed generators based on linear recurrences modulo 2. In *Monte Carlo and Quasi-Monte Carlo Methods 2000*, eds. K.-T. Fang, F. J. Hickernell, and H. Niederreiter, pp. 318–330. Berlin: Springer-Verlag.

L'Ecuyer, P., and R. Simard. 2007, August. TestU01: A C library for empirical testing of random number generators. *ACM Transactions on Mathematical Software* 33(4):Article 22.

L'Ecuyer, P., R. Simard, E. J. Chen, and W. D. Kelton. 2002. An object-oriented random-number package with many long streams and substreams. *Operations Research* 50(6):1073–1075.

L'Ecuyer, P., and R. Touzin. 2000. Fast combined multiple recursive generators with multipliers of the form $a = \pm 2^q \pm 2^r$. In *Proceedings of the 2000 Winter Simulation Conference*, eds. J. A. Joines, R. R. Barton, K. Kang, and P. A. Fishwick, pp. 683–689. Pistacaway, NJ: IEEE Press.

Lemieux, C., and P. L'Ecuyer. 2002. Randomized polynomial lattice rules for multivariate integration and simulation. Extended version available online at http://www.iro.umontreal.ca/ lecuyer

Lemieux, C., and P. L'Ecuyer. 2003. Randomized polynomial lattice rules for multivariate integration and simulation. *SIAM Journal on Scientific Computing* 24(5):1768–1789.

Lenstra, A. K. 1985. Factoring multivariate polynomials over finite fields. *Journal of Computer and System Sciences* 30:235–248.

Lewis, T. G., and W. H. Payne. 1973. Generalized feedback shift register pseudorandom number algorithm. *Journal of the ACM* 20(3):456–468.

Lidl, R., and H. Niederreiter. 1986. *Introduction to Finite Fields and Their Applications*. Cambridge: Cambridge University Press.

Lindholm, J. H. 1968. An analysis of the pseudo-randomness properties of subsequences of long m-sequences. *IEEE Transactions on Information Theory* IT-14(4):569–576.

Marsaglia, G. 1985. A current view of random number generators. In *Computer Science and Statistics, Sixteenth Symposium on the Interface*, pp. 3–10. North-Holland, Amsterdam: Elsevier Science Publishers.

Marsaglia, G. 2003. Xorshift RNGs. *Journal of Statistical Software* 8(14):1–6.

Massey, J. L. 1969. Shift-register synthesis and BCH decoding. *IEEE Transactions on Information Theory* IT-15:122–127.

Matsumoto, M., and Y. Kurita. 1992. Twisted GFSR generators. *ACM Transactions on Modeling and Computer Simulation* 2(3):179–194.

Matsumoto, M., and Y. Kurita. 1994. Twisted GFSR generators II. *ACM Transactions on Modeling and Computer Simulation* 4(3):254–266.

Matsumoto, M., and Y. Kurita. 1996. Strong deviations from randomness in m-sequences based on trinomials. *ACM Transactions on Modeling and Computer Simulation* 6(2):99–106.

Matsumoto, M., and T. Nishimura. 1998. Mersenne twister: a 623-dimensionally equidistributed uniform pseudo-random number generator. *ACM Transactions on Modeling and Computer Simulation* 8(1):3–30.

Niederreiter, H. 1992. *Random Number Generation and Quasi-Monte Carlo Methods*, Volume 63 of *SIAM CBMS-NSF Regional Conference Series in Applied Mathematics*. Philadelphia, PA: SIAM.

Nishimura, T. 2000. Tables of 64-bit Mersenne twisters. *ACM Transactions on Modeling and Computer Simulation* 10(4):348–357.

Panneton, F. 2004, August. *Construction d'ensembles de points basée sur des récurrences linéaires dans un corps fini de caractéristique 2 pour la simulation Monte Carlo et l'intégration quasi-Monte Carlo*. Ph.D. thesis, Département d'informatique et de recherche opérationnelle, Université de Montréal, Canada.

Panneton, F., and P. L'Ecuyer. 2004. Random number generators based on linear recurrences in F_{2^w}. In *Monte Carlo and Quasi-Monte Carlo Methods 2002*, ed. H. Niederreiter, pp. 367–378. Berlin: Springer-Verlag.

Panneton, F., and P. L'Ecuyer. 2005. On the xorshift random number generators. *ACM Transactions on Modeling and Computer Simulation* 15(4): 346–361.

Panneton, F., and P. L'Ecuyer. 2007. Resolution-stationary random number generators. *Mathematics and Computers in Simulation*, to appear.

Panneton, F., P. L'Ecuyer, and M. Matsumoto. 2006. Improved long-period generators based on linear recurrences modulo 2. *ACM Transactions on Mathematical Software* 32(1):1–16.

Rieke, A., A.-R. Sadeghi, and W. Poguntke. 1998, August. On primitivity tests for polynomials. In *Proceedings of the 1998 IEEE International Symposium on Information Theory*. Cambridge, MA.

Strang, G. 1988. *Linear Algebra and its Applications*, 3rd edition. Philadelphia, PA: Saunders.

Tausworthe, R. C. 1965. Random numbers generated by linear recurrence modulo two. *Mathematics of Computation* 19:201–209.

Tezuka, S. 1995. *Uniform Random Numbers: Theory and Practice*. Norwell, MA: Kluwer Academic Publishers.

Tezuka, S., and P. L'Ecuyer. 1991. Efficient and portable combined Tausworthe random number generators. *ACM Transactions on Modeling and Computer Simulation* 1(2):99–112.

Tootill, J. P. R., W. D. Robinson, and D. J. Eagle. 1973. An asymptotically random Tausworthe sequence. *Journal of the ACM* 20:469–481.

Wang, D., and A. Compagner. 1993. On the use of reducible polynomials as random number generators. *Mathematics of Computation* 60:363–374.

Opportunities and Challenges in Health Care Simulation

Andrew F. Seila and Sally Brailsford

Abstract The delivery of health care is a complex and expensive process that presently is failing to meet the expectations and standards of patients, physicians, administrators and government entities. Simulation tools, both discrete-event simulation and system dynamics, enable managers to better understand the behavior of complex systems and predict their response to changes. After reviewing some successful applications of simulation in health care, this article examines how well-suited simulation is to analyze health care systems, and explores the reasons why simulation has not been adopted as a routine part of health care systems analysis. Some ideas are presented regarding what can be done to encourage greater use of simulation in health care.

1 Introduction

Throughout the world, in both high-income and low-income countries, health care systems deliver care to people. Most governments consider the delivery of health care to be one of their primary duties, along with defending their citizens, providing education and managing criminal and civil justice. Many governments consider health care to be a right, and a service that they are obligated to provide.

1.1 Similarities among Health Care Systems

Much is written about the differences in the world's health care systems, but they have much more in common. First and foremost, the performance of these systems is literally a life-and-death matter. Measures used to evaluate performance include average lifespan and infant mortality. Delays in emergency room access and treatment, and delays in cancer diagnosis, for example, translate directly to reductions in lifespan and increases in mortality.

A.F. Seila (✉)
University of Georgia, 1240 Saint Andrews Drive, Bogart, Georgia 30602
e-mail: aseila@gmail.com

C. Alexopoulos et al. (eds.), *Advancing the Frontiers of Simulation: A Festschrift in Honor of George Samuel Fishman*, International Series in Operations Research & Management Science 133, DOI 10.1007/b110059_10, © Springer Science+Business Media, LLC 2009

Whether in Britain or Bangladesh, America or Angola, the health care system is large and complex. It consists of an interconnected web of hospitals, clinics, diagnostic resources, physicians, nurses, dentists, patients, information systems, pharmacies, and alternative medicine practitioners. In low income countries, the resources are more primitive, but the same basic elements nevertheless exist.

The health care system is people-centered. Patients are people, but people also deliver health care, and payroll costs are a substantial portion of the health care budget. Decision making by health care professionals is a core component of the delivery process, as well as patient behavior. Patients may interact with or otherwise be subject to decisions by numerous people during an episode of care, including doctors, clinical officers, nurse practitioners, nurses, pharmacists, clinical and hospital administrators, and others. The human element is central to the delivery of care, and will be for the foreseeable future. In all regions, the health care system depends upon an educational infrastructure, consisting of institutions such as medical and nursing schools, which may exist outside the region, to provide the professionals that deliver health care.

The health care system consumes vast amounts of money. In the United States, the health care system is estimated to have cost upwards of $2 trillion in 2005, which is $6,400 per capita or 15.3% of GDP (OECD 2007b). In the UK, per capita health care spending for 2005 was $2,724 or about 8.3% of GDP (OECD 2007a). Costs are much lower in developing countries. In most countries, especially the US, it is not only the cost of the current system that causes alarm, but also the rate of increase of that cost which is much higher than the rate of inflation. Funding adequate health care is considered a serious problem in virtually all countries.

The delivery of health care is, by its nature, stochastic. The environment and operation of the health care system is unpredictable, and the system must be able to respond to changes. The patterns of disease in the community may change unpredictably. Moreover, individual patients vary in their behavior and response to treatment, adding another stochastic element to processes. Within hospitals and clinics, service and response times can vary greatly, and clinical outcomes are usually uncertain.

The health care system has conflicting goals due to the variety of stakeholders. Patients want a complete set of services delivered quickly and inexpensively, but providers want to maximize profit or minimize cost. Government entities have to balance efficacy and cost while being concerned about patient—and citizen—satisfaction.

All health care systems involve the government in an intrinsic way as a regulator, payer and provider. In many countries, including the United States, federal and state governments regulate spending on health care facilities through such programs as Certificate of Need. In Britain and Canada, among many others, the government is the main payer for health care and sets the price for various services. Reforms within the UK National Health Service in recent years have led to a more entrepreneurial structure, in which local health care organizations have greater financial autonomy and the private sector plays an increasing role. In the US, the government pays approximately 45% of all health care dollars through various federal programs such as Medicare, Medicaid, and the Veterans Administration (OECD 2007b). Medicare also exerts heavy influence on private insurers through their procedures and payment schedules.

Technology is an important component of the health care system, both by providing sophisticated tools for diagnosis and treatment, and through information technology for medical record keeping and managing the health care enterprise. Technology adds more expense to the health care system, both through acquisition and use, but, if applied well, it can greatly increase the efficiency of the system. Significant questions arise about how to utilize technology most efficiently. For example, how many new devices such as MRI machines should be purchased and how much value does new technology actually add to the system? Additionally, questions arise about how the existing clinical processes should be modified to accommodate the new technology most effectively.

Finally, all health care systems have an administrative hierarchy that is responsible for the strategic and day-to-day system management. Administrators must decide, within the constraints they face, what resources will be utilized and how they will be utilized, to improve and maintain the health of the population. The structure and operations of these systems have considerable inertia, and are often very difficult to change. Persons and organizations with financial and political interests are reluctant to allow changes that will reduce their income or influence, even if this would mean more efficient delivery of health care. Administrators are also reluctant to experiment with the system because it is expensive and risky. Lives can be disrupted and lost if changes reduce the quality of care.

1.2 Dissimilarities among Health Care Systems

With all of their similarities, health care systems throughout the world are also very dissimilar. They differ in structural and operational details. A rural clinic in a low-income country like Bangladesh might have a single doctor or clinical officer and a small number of nurses serving 10,000 patients, coping with shortages of medicine, and lacking readily available radiology or laboratory services. No ambulance service is available in rural areas, and hospital inpatient care is reserved for patients with the most severe and acute illnesses. In contrast, in a developed country such as the UK, hospitals, ambulance services, basic primary care, and advanced medical procedures are available to all citizens. Health care systems also differ in their immediate goals and decision alternatives available to achieve those goals. A less developed country, where rural areas have almost no modern health care available, might consider paramount the use of mobile clinics that emphasize education, vaccinations and basic primary care. Decision alternatives might focus on the number of rural clinics and policies concerning training midwives for safe at-home deliveries. In Canada and the UK, the current problems involve waiting times for elective procedures, and decision alternatives concern policies and procedures to reduce waiting lines for these procedures.

Moreover, health care models are not usually built for national public health systems, but rather for subsystems such as an emergency department, a neighborhood clinic, or a hospital's inpatient operations. These subsystems involve equipment choices and processes that are specified locally, so they are quite different from place to place, even in the same region or city. So, while two emergency departments may have the same certification and deliver the same set of services to their populations, the processes and system designs they employ can differ greatly. As a result, it is challenging to design a model that can represent most or all health care subsystems in a particular category such as ED, ICU or a network of neighborhood clinics. A model must be built for each individual system.

All of these characteristics suggest that health care systems are ideal candidates for modeling and simulation. Modeling and simulation is an excellent tool for evaluating decisions, and designing and improving processes in systems where experimentation is either impossible or expensive and risky.

In this article, we will consider the following three questions:

1. How can simulation be used as a tool to solve problems in health care, especially health care delivery?
2. Why has simulation not been employed as much as it could have been? What are the impediments to the effective use of simulation to improve the efficiency and effectiveness of health care systems?
3. What can be done to encourage routine use of simulation as a tool in health care management?

In Section 2, we will examine two simulation world views and discuss their fit to health care systems. We will also present a categorization of health care models that has proved useful. In Section 3, we will address the first question by examining some successful applications of simulation in health care and exploring some problems that appear to be effectively analyzed using simulation. We will discuss the special merits of simulation as an analysis tool in health care systems in Section 4. Section 5 deals with the second question. Here, we present our view of the impediments, or challenges, to the effective use of simulation as an aid in the design and management of health care systems. Our ideas concerning the third question are the subject of Section 6. Section 7 contains concluding remarks.

2 Simulation in Health Care Systems

Operations research has been applied in the domain of health care for more than 40 years. The UK OR Society and the UK National Health Service (NHS) held a joint colloquium on hospital appointment systems as far back as 1962 (Jackson 1964). Since the 1960s OR models have been successfully used to assist clinical decision making, facility location and planning, resource allocation, evaluation of treatments, and organizational redesign. Simulation is one of the most commonly used OR approaches, and is regarded by many as the technique of choice in health care (Davies and Davies 1994).

2.1 Discrete-Event Simulation

There are two main simulation world views that have been applied to health care: discrete-event simulation (DES) and system dynamics (SD). In either approach, the system is described by a collection of state variables that change over time. A DES model identifies events which are discrete points in time when the state of the system may change. The times between events as well as the new values of state variables may be, and usually are, stochastic. A more detailed explanation of DES is provided in Chapters 1 and 2 of Fishman (2001) and Chapter 5 of Seila et al. (2003). Many DES models have more or less the structure of a queueing network, in which individual entities flow around a network of stations and queue for services. This approach appears to be especially useful for hospital systems and other health care constructs in which patients join waiting lists for appointments, examinations and treatments. In DES, entities have properties which determine their pathway through the network, in exactly the same way that patients have individual characteristics which determine their pathway through the hospital or clinical system. This pleasing analogy contributes greatly to the enduring popularity of DES as a modeling approach. Psychologically, DES is appealing because it enables the modeler to give the entities all of the necessary human characteristics of age, gender, diagnosis, blood group, disease status, sexual preference, hair color, or whatever.

DES has many advantages from a mathematical perspective, too. Markov models must account for past medical history only through the current state of the system and must represent times between state changes with exponential distributions. In contrast, in a DES model, future medical events, service time distributions and other actions within events can be dependent on the entire previous history and environmental circumstances as well as individual patient characteristics, and any parametric or empirical distribution can be chosen to model activity durations. Complex logical rules can be used to determine patients' routing through the system, or the outcome of a treatment. The model can include any amount of randomness, as long as enough simulation runs are performed to obtain statistically significant results. Indeed, depending upon the flexibility of the software chosen and the modeler's imagination, virtually any system can be modeled using DES.

Another advantage of DES is the wealth of software packages available and the maturity of the modeling and statistical methodology. The DES modeling paradigm dates back to the 1950s and is now a mature and well accepted methodology. See Nance (1993) for a history of DES software. Today, many software packages are available, at a wide range of prices, including freely available open source packages, for implementing and running discrete-event simulations. Many commercial packages have a graphical interface which enables the user to visualize the model and even build the model using this interface. This can be invaluable as a communication aid with health care professionals.

Often unrecognized, but of equal importance, is the maturity of the statistical and computational methodology which underlies DES. Due to the dependent nature of the series of observations output by discrete-event simulations, special statistical methods are required to compute reliable estimates of system performance measures. This methodology, which was pioneered by George Fishman (Fishman 1973, 1978, 2001) and others, is now accepted and trusted to compute reliable estimates. Modelers can build models, run the simulations, and analyze data with the confidence that the results obtained are correct for the model implemented.

2.2 System Dynamics

System dynamics (SD) (Sterman 2000), on the other hand, lacks many of these desirable features, yet has distinctive virtues of its own. In SD the individual entities are not modeled, but are represented as a continuous mass which flows around the model like water in a central heating system, accumulating in "stocks" corresponding to tanks or radiators, with inflows and outflows governed by valves or "rates." Psychologically, this is definitely less attractive, especially to health care professionals who by training are people-focused and do not like the idea of reducing a group of human beings to a continuous mass. There is no longer a stochastic element—SD models are deterministic. All output from an SD model will be identical on every simulation run using the same input. Moreover, in general, SD software does not have the appealing graphics of some DES software.

However, SD has many key features which DES lacks. The fundamental principle of SD is that "structure determines behavior," so that if we understand the structural relationships between elements in a system, we will understand the emergent behavior of that system as a whole. In a DES model, we often cannot see the forest for the trees because we are inclined to focus on a very detailed model representation. In SD, we cannot model this level of detail, but we can gain understanding of the dynamic complexity of the system. SD is concerned with feedback and unanticipated effects. Unlike DES, SD models are intended to be used at a more speculative, strategic level for larger populations and longer time horizons. A key advantage of SD is that the models generally run very fast (and of course do not require multiple iterations), so they can be run interactively in real time with decision makers. SD is also useful for large-scale systems that would be too complex to model at a level of detail needed for DES and would require too much time and computational resources to run.

For excellent surveys of applications of DES and SD in health care, see Jun et al. (1999), Dangerfield (1999), Royston et al. (1999), and Fone et al. (2003). Of course, simulation has been widely and successfully used for many years in many fields, including defense, manufacturing industry, services industries, communication systems, transportation systems, finance and for training purposes. It is interesting to speculate whether health care is in any way different from these areas. The entities in health care systems are vulnerable human beings, with feelings, emotions and human behaviors, often in life-threatening situations. Health care systems are safety-critical and often highly technological. There are different cultures, too. The clinical hierarchy has its own professional boundaries. Managers can be seen by clinicians as interfering outsiders concerned only with costs and targets. Patients are obviously not inanimate widgets in a production process. The entities in a model of an airport terminal or a call center are also human beings, so we have to take account of human behavior and reactions in many other system models. Many of the characteristics of health care systems occur in other organizations, but we would argue that the particular characteristics and the way they interact is unique in health care.

2.3 A Taxonomy of Health Care Models

For the purposes of this article, health care models are classified into three groups. The first, termed Level 1 models, are models of the human body (see Fig. 1). These are frequently called "disease models" but they can also represent biological processes in healthy individuals. Models can be at the body system or organ level, or even at the cellular or microbiological level. They are often used for studying the clinical effectiveness or cost-effectiveness of some intervention. For example, by simulating the progression of breast cancer in the female population, it is possible to compare the effects of different screening policies for early detection.

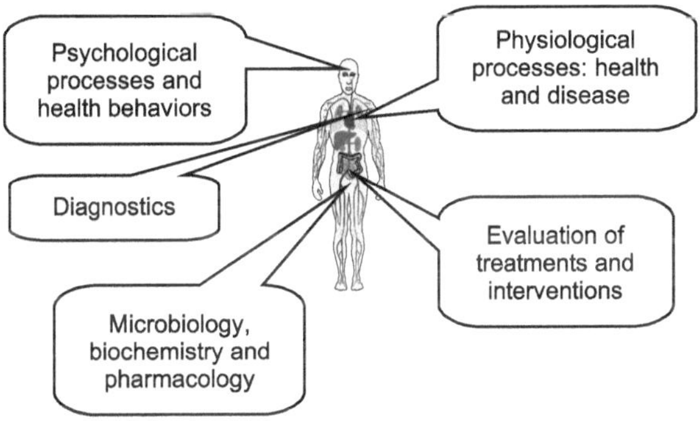

Fig. 1 Disease models

We can also model people's health behaviors, for example the effect of a smoking cessation campaign on the prevalence of coronary heart disease, or the spread of infectious diseases such as chlamydia or HIV/AIDS. Simulation has also been used to model the biochemical effects of drugs, a study know as pharmacodynamics. Examples of these are given in Section 3.1. Level 1 models are arguably the most interesting, as they often require creative modeling approaches or theoretical innovations.

The next group, Level 2 (Fig. 2), denotes operational or tactical models at the health care unit level. By this we mean a clinic, a ward or hospital department such as the operating suite or emergency room. These usually, although not always, still model the behavior and movement of individual patients, but are not concerned with modeling the physiological or clinical processes going on inside them. Rather, they are concerned with modeling the movement of patients within the system, and

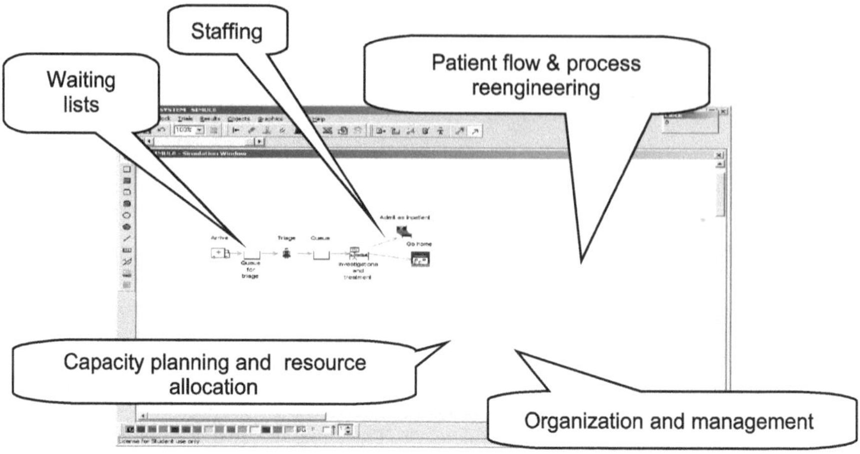

Fig. 2 Operational models

identifying and eliminating bottlenecks. These models are used for capacity planning, resource allocation and process redesign. For example, how many ICU beds are needed to be 95% sure there will always be a free bed when required? How many nurses and doctors are needed to ensure that no patient spends more than 4 hours in the emergency department? Should inpatients or outpatients take priority for diagnostic imaging services? These are classical areas for the application of operations research in general and discrete-event simulation modeling in particular, and the academic literature contains hundreds, if not thousands, of Level 2 models. Some examples are given in Section 3.2.

Level 3, or strategic models (Fig. 3) are system-wide models which often do not model individual patients at all. They are also comparatively few in number in the literature, possibly reflecting the relatively low use of operations research for strategic planning. Unlike Levels 1 and 2, where DES is usually, but not always, the chosen approach, Level 3 models almost always use SD, as it lends itself to answering more long-term, broad-brush questions using a relatively low level of detail. For example, how should a large city configure its services for emergency and unscheduled care? How might the electronic health record impact upon the health of the next generation of children? How many doctors will the US need in 2030? How can we improve the way health professionals work alongside Social Services? Some examples of Level 3 models are given in Section 3.3.

3 Some Illustrative Health Care Simulation Models

In this section, we provide a more or less representative selection of successful simulation models of health care systems that have been developed over the past ten to fifteen years. These examples will (a) provide concrete examples to further define the taxonomy in the previous section, (b) show the types of systems that have

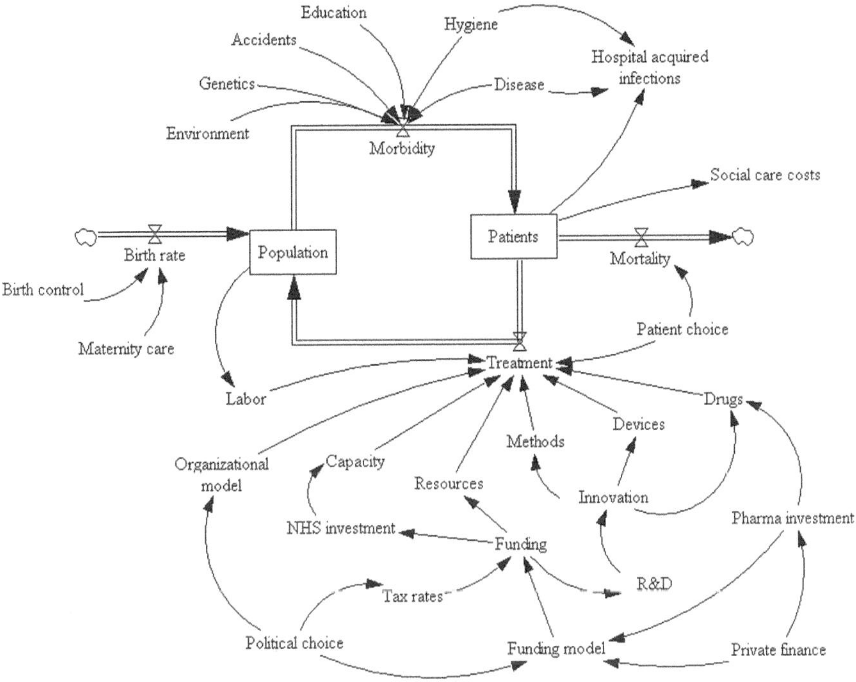

Fig. 3 System-wide models

been successfully modeled and problems that have been solved, (c) provide some specific examples of difficulties that have been encountered and overcome in health care simulation modeling, and (d) put the process of modeling health care systems in a managerial and political context. Readers familiar with some simulation models of health care systems may skip this section without loss of continuity.

3.1 Level 1: Disease Models

We describe three Level 1 models: a DES model for screening diabetic patients for eye complications, an SD model for the sexually transmitted infection chlamydia, and a DES model for policy evaluation for organ transplantation. We provide this sample of models and those in Sections 3.2 and 3.3 in order to show the characteristics of these models at a more detailed level.

3.1.1 Screening for Diabetic Retinopathy

Diabetic patients are at risk of a condition called diabetic retinopathy (DR), which can lead to blindness if untreated. Early signs of DR can be detected before the patient is even aware of any problems, and the condition can be successfully treated using a laser, preventing future loss of sight. Screening can be carried out using various methods and in a variety of settings—in hospital, at an optometrist or ophthalmologist or in a primary care facility—and while it is clearly beneficial, since the tests are relatively cheap and easy, there is no consensus about the best setting, method or frequency of screening. Davies et al. (2000) and Davies and Brailsford (2004) developed a DES model which was used to address this problem. In fact, two models were developed, one for Type 1 diabetes and the other for Type 2 diabetes. The two types of diabetes cause different types of retinopathy.

A key feature of this model is that patients must participate in multiple activities simultaneously. This modeling approach, which was developed by Davies et al. (1993) and termed Patient Oriented Simulation Technique (POST), is not needed by most manufacturing, transportation, and communication models. In this model, disease progression is modeled as a virtual queueing system by treating the disease state dwelling times as activity durations where the activities are assumed unconstrained by resources (i.e., an infinite-server queueing system). At the same time, patient entities may also be taking part in a genuine resource-constrained queueing system, for example, waiting for hospital treatment, medical treatment or screening test. Thus, in a realistic model of a health care system, patient entities may be participating in several concurrent activities or queues. Moreover, these activities and queues are interdependent: if a patient changes disease state, his/her treatment may need to be changed, he/she may no longer require a screening test, and hospital appointments may need to be rescheduled.

The models were populated with data about the natural history (the untreated progression) of DR from one of the world's largest studies, the Wisconsin epidemiologic study of diabetic retinopathy (Klein et al. 1985). Various screening and treatment policies in current use in the UK were tested, with current cost data (NHS 2001) applied. Interestingly, among the policies tested, no striking differences were found, although there was a preference in terms of cost effectiveness for community-based rather than hospital-based screening.

3.1.2 Screening for Chlamydia

Chlamydia was the most common sexually transmitted infection in the US in 2001 (Centers for Disease Control and Prevention 2004) and the UK in 2003 (Health Protection Agency 2004), and is a major public health problem. Most cases of chlamydia infection are asymptomatic and easily treated with antibiotics, but if untreated the infection can have serious long term consequences (called sequelae). Screening programs have been shown to be effective, but there are concerns that blanket screening of the whole population at risk will add an extra financial and operational burden to the already over-stretched health care system. Unlike diabetic retinopathy, where the population at risk is relatively small and well-defined, in this case all sexually active people are at risk. In 2003 the UK Department of Health introduced screening of all people between the ages of 16 and 25 in 10 centers, with the view to extend this program to the rest of the country within the next few years as part of the National Chlamydia Screening Programme (UK Department of Health 2005).

Evenden et al. (2005) developed an SD model which demonstrated that certain high-risk subgroups within the general population are critical in the infection dynamics, and improved targeting of these high-risk populations achieves greater cost-effectiveness. Figure 4 shows the model structure. The model used data from an opportunistic screening trial held in the Portsmouth, UK area from October 1999 to September 2000 (Harindra et al. 2002).

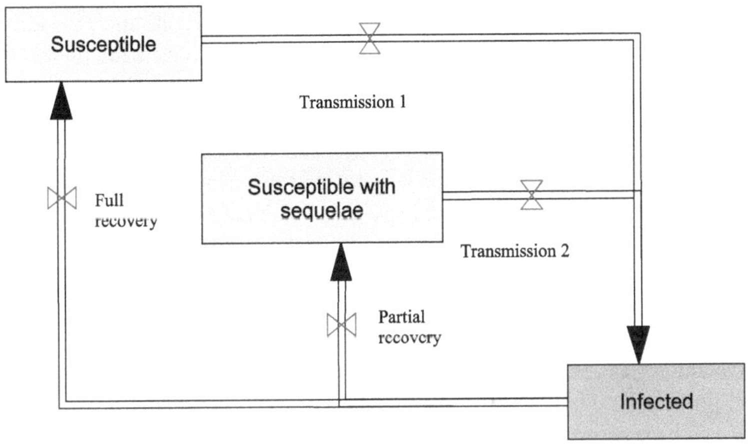

Fig. 4 SD model for chlamydia screening

The SD approach is ideally suited to modeling infection rates and population movements between infected and susceptible states. A particularly relevant aspect was that the repeated reinfection mechanism was captured, along with the increased risk of sequelae this creates.

These models were replicated for two risk groups of patients. Based on the findings of Townshend and Turner (2000), 2.5% of the population were assumed to be high-risk. Various assumptions were made about the rates of sexual contact within and between these groups, about prevalence of infection (5%, 8%, 10%), and about the screening rates in each risk group. The low-risk screening rate was varied from a minimum value equal to the low-risk rate to a maximum of 120% of the low-risk rate. The model results showed that the more targeted screening became, the more cost-effective it was. Therefore, rather than increase the general population screening rate by 1%, which would be hugely expensive, it was far more advantageous to identify a few more high-risk people and screen them.

In a further paper, Evenden et al. (2007) describe the use of statistical risk group clustering techniques with the Portsmouth data to identify indicators that are strong predictors in determining high-risk status. The results are combined with geomapping techniques which visually display prevalence geographically across the region, thus identifying high prevalence postal code clusters and informing public health planners where to target intervention and screening strategies. These findings are then combined with the results from the simulation model to provide a unique holistic view of the problem.

3.1.3 Evaluating Organ Transplantation Policies

Organ transplantation is now done routinely in many large medical centers. However, organs for transplantation are a scarce resource. Organ allocation policies are used to specify which waiting patient will receive an organ graft as it becomes available. The policies must balance medical consequences, economic considerations and political issues. For example, patients who are older or have been on the waiting list longer tend to be sicker and will tend to survive fewer years than those who are younger and have been on the list a shorter time. Economic considerations concern the revenues received by the transplant centers as well as transportation costs and costs to patients and health plans for surgery and days in hospital. Political issues involve perceptions of fairness of the policy. For example, many people believe available organs should be provided to patients who have been waiting longest, the first-come-first-served rule.

The Organ Procurement and Transplantation Network (OPTN) was established by the United States Congress in 1984 to manage the procurement and allocation of transplant organs in the US. The United Network for Organ Sharing (UNOS) is a non-profit organization that has managed OPTN under contract since 1986 and continues to be responsible for proposing, evaluating and adopting policies for organ allocation. In 1995, UNOS contracted with Pritsker Corporation to develop a model to evaluate policies for allocating livers for transplantation. This model is described in Pritsker et al. (1995).

Pritsker Corporation and other partners developed the UNOS Liver Allocation Model (ULAM) for comparing proposed alternative allocation policies (Pritsker 1998; Pritsker et al. 1995, 1996). The model was intended to be used to evaluate policies currently under consideration at the time, but more importantly, it was intended to be updated regularly and used in an ongoing manner to evaluate new policies as they were proposed. ULAM is a DES model that follows the process of transplant patients entering the waiting list, progressing to other medical status levels, possibly being a match for a liver graft, being offered the graft, accepting or declining the graft, and so forth. The model categorizes patients' medical status in 7 categories, according to medical urgency. It also includes other patient attributes such as age, weight, sex and blood type. When a graft becomes available at a particular geographic location, an allocation policy is applied to determine to whom the graft will be offered first. Policies employ a point ranking scheme that incorporates waiting time on the list, medical status, location of the patient relative to the graft, compatibility, and possibly other criteria. The model applies the policy to offer the graft to successive patients until it has been accepted or all eligible patients have been offered the graft. It also incorporates processes of mortality and graft rejection after transplantation. Patients can decide to not relist after rejection.

The model presented several technical challenges. Data for the patient arrival process did not fit any of the well-known arrival processes, so an exponential-polynomial-trigonometric rate function was developed for a nonhomogeneous Poisson process to model the arrivals. Secondly, application of allocation policies required that the waiting list be recreated and patients on the list be reranked upon every graft arrival. This presented a huge computational burden to the model and required that data structures and computational algorithms be modified in order for the model to execute in a reasonable amount of time.

The model was built from a number of modules that implemented various subprocesses in the system such as the arriving patient stream, allocation policies and patient medical status. It underwent an extensive verification and validation process. ULAM was validated by comparing the model output for the current policy with observed results nationally. It was important that members of the transplant community have confidence in the model. To achieve this goal, ULAM was demonstrated to many committees and other groups within the transplant community. An animation was developed to show the movement of grafts and patients. The software was also provided with a user-friendly interface so it could be used independently by UNOS staff.

Model outputs were numerous and included the following:

1. Number of days in each medical status.
2. Number of repeated and non-repeated transplants.
3. Number of pediatric transplants.
4. Number of post-transplant deaths within 12 months.
5. Percent survival greater than 12 months/24 months.
6. Percent of transplants in local/regional/national location.
7. Average distance to transplantation site.
8. Blood type exact match percentage.
9. Pre-transplant deaths.
10. Post-transplant deaths.
11. Mean and median days from arrival on waiting list to transplant.

A more complete list is given in Pritsker (1998).

ULAM was used to evaluate five allocation policies. Details of the model and simulation runs can be found in Pritsker et al. (1995). For death measures, the results of the simulation runs reinforced intuitive expectations, with the current policy leading to more pre-transplant deaths and other policies providing somewhat longer average times on the waiting list. Based on the results of these runs, the decision was made to continue with the current policy. In its initial use, the model did not lead to adoption of a new policy but it did provide support for the current policy. Later applications did support modifications of the allocation policy; see Pritsker (1998) for death measures.

3.2 Level 2: Operational Models of Health Care Units

In this section, we present three DES models developed to address organizational issues of resource allocation and capacity planning. Discrete-event simulation has been a workhorse simulation methodology for models of operational systems in health care, including units such as the emergency department, intensive care unit, operating suites, laboratories, patient wards and others.

3.2.1 Hospital Bed Capacity Modeling

Harper and Shahani (2002) describe a simulation model developed for an 800-bed hospital in Reading, UK. This model has a generic structure in which patients flow through "care units" which can be a ward, a group of wards, a specialty bed pool or even a whole hospital. The model structure is highly flexible and very detailed, and requires a great deal of data input, for example user-defined patient groups, arrival rates (hourly, daily, monthly), length-of-stay (LOS) distributions, bed numbers, admission rules, deferral rules when no bed is available, and priority listings. Because of the need for so much detailed input data, the model was designed to interface with the routine patient management and reporting systems used by the hospital. The fitting of arrival processes and LOS distributions was fully automated through a system called Apollo, a classification and regression tree (CART) method which allowed the construction of homogeneous and clinically meaningful patient groups, thus reducing the need for input from hospital staff. Arrival and LOS distributions were fitted for the groups thus constructed.

The model was used in three areas in this hospital. For example, the model was used in adult medicine for estimating the future seasonal bed requirements. A key outcome of the model use was the realization by hospital planners that occupancy rates and their corresponding refusal rates are linked in a highly complex way, and are a function of the case mix, the number of beds available and the variability of LOS. Previously the hospital had used a rule of thumb of average LOS multiplied by average demand—assumed fixed—and their views changed greatly after this modeling exercise.

3.2.2 Intensive Care Units

Another popular area for the use of capacity planning models is the ICU. This is a very expensive hospital resource both in terms of highly skilled staff and costly specialist equipment. Therefore it is vital to provide the optimal number of ICU beds and staff. Determining staff and beds are equivalent problems since most ICU patients require at least 1:1 nursing ratio. Having too many beds can have serious financial consequences, but having too few results in low quality of care.

Griffiths et al. (2005) developed a model for a large teaching hospital which used a more flexible approach to beds. Officially, the ICU had 14 beds, but in times of high demand extra beds would be made available elsewhere in the hospital. The problem addressed by this model was not how many ICU beds were needed, but instead how many nurses. A fixed number of nurses, permanently based in the ICU and known as establishment staff, are rostered to work in the ICU for each 8-hour shift. In busy periods supplementary nurses may be required, and these can be either bank staff (establishment staff doing overtime) or agency staff. The latter are very costly and moreover may be unfamiliar with the unit, resulting in lower quality of care. Thus for both cost and patient safety reasons, the hospital would prefer to keep the number of supplementary nurses to a minimum.

A DES model was developed and used detailed data for nearly 1,100 patients admitted to the ICU in 2000. The model was highly complex. Patients were categorized by referral source and, for surgical patients, into elective or unplanned categories. Moreover, distinct day-of-week and time-of-day patterns were observed and a separate arrival rate was required for each of the 168 hours in the week. Length of stay distributions were fitted for all patient types. Because it is nurses rather than beds which are the constraint in this model, an estimated number of 30 beds were available, although the maximum number of beds used in 2000 was only 19. The rostered number of nurses per shift was fixed for each experiment, but the number of supplementary nurses available was assumed to be unbounded. The number of supplementary nurses used was a performance measure for the model.

In 2000, the actual number of nurses per shift was 14. The model showed that had 16 nurses been rostered, the potential savings would have been equivalent to 690 times the cost of employing one establishment nurse for one shift. The model was extended to consider increased future demand and also a pioneering method in use at the hospital for early detection of patients in general wards who might need ICU treatment at some future point. The physicians believed that by offering early ICU care, the eventual LOS in ICU could be reduced by between 10 and 20%. The model was able to evaluate the effect of this, concluding that a 20% reduction in LOS would result in one less nurse being rostered each shift, with further consequent savings.

3.2.3 Moving to a Replacement Hospital

The process of moving a patient population from an existing hospital facility or group of facilities to a replacement facility is something that does not happen often, but when it is necessary, a considerable amount of planning is required in order to accomplish the move with minimal disruption in patient care and minimal opportunity for adverse events for patients, especially those in intensive care. Ashby et al. (2007) developed a discrete-event simulation model to explore alternative strategies for making such a move. In this project, a large county hospital was operating from two facilities, but needed to move patient populations from both facilities to a single facility located nearby. During the move, it was necessary to keep emergency services operating, along with their supporting departments to serve arriving emergency patients.

The move was complicated by a number of factors: There was a large number of patients to transfer. Each patient had to be "packed" prior to the move and "unpacked" after arrival. The specific tasks in packing and unpacking differed from patient to patient and resulted in highly variable packing and unpacking times. The process of moving patients for each unit had to be done over a period of time in order to avoid the confusion that would result from many simultaneous moves. As a result, the move was estimated to require two days. There were other challenges in the move. The old units in the two existing facilities did not map one-to-one to the units in the new facility. The result was that each patient needed a unique routing that specified to which unit the patient would be transferred in the new facility. Other constraints included the speed of the elevators and the narrow corridors leading to the loading site in the old facility. In the new facility, the move was constrained by the number of locations for unloading and the number of elevators that could accommodate a gurney.

The simulation sought to answer a number of questions: Which of the constraints would present the most severe bottlenecks to the move, and would those constraints keep the move from being completed in two days? How many gurneys, wheel chairs, ambulances and staff would be needed to make the move smoothly? A second question concerned how to manage the existing resources to facilitate the move maximally. How many patients in the old facility should be packed and put in queue at any point in time to avoid delays waiting for patients to be packed or elevators to be available? How many elevators at the new facility should be reserved for patient transport? More generally, what protocol should be used to allocate elevators to competing demands? What strategies for managing patient teams, ambulances, gurneys and other resources used in the transportation of patients would be most effective?

Performance measures relating to patient care that were estimated by the simulation include time when a patient's access to care was limited (e.g., when in an ambulance) and time patients were "packed" awaiting transport. A second set of metrics measured the cost of the move. Cost was determined mainly by the amount of resources required, including personnel. Finally, various wait times for resources such as elevators and ambulances were computed, as well as resource utilization and the total time of the move. The model demonstrated to managers that the best protocol for utilizing ambulances, gurneys, and teams to move patients was to keep each team at one location, load the patient and gurney on the ambulance, unload the patient with the gurney at the receiving hospital, and return the ambulance to the sending hospital with another empty gurney. This protocol greatly improved the efficiency of the move and reduced patient waiting times as well as cost. The simulation also provided specific numbers for the teams, ambulances, wheel chairs and other resources that need to be available, as well as protocols for utilizing elevators and loading areas. This model is not at all generic and cannot be adapted easily to the needs of another hospital. However, the value of this model lies in the large cost, both in money and quality of care, that the move could incur and the ability of the simulation to find ways to reduce the cost of the move while not compromising patient care.

3.3 Level 3: Strategic Whole-System Models

Strategic whole-system models models are becoming increasingly popular in the UK. In this section we describe two applications of system dynamics, which is probably better suited to this area than DES. The first is perhaps one of the best-known examples of SD modeling in the mainstream UK health OR literature. The second describes the entire emergency and unscheduled health care system in the city of Nottingham, a large industrial city in the center of England.

3.3.1 SD Modeling of Emergency Admissions

Lane et al. (2000) developed a model designed to explore the relationships between waiting times in the ER and hospital bed closures. At the time a major London hospital (denoted pseudonymously "St. Danes") was planning to cut costs by closing beds, and intended to measure the impact of these bed closures by tracking waiting times in the ER. The hospital's argument was that bed closures led to cancelled elective admissions and this led to more people presenting in the ER, partly as a direct result of the deterioration of their health and partly as a behavioral response by primary care doctors wishing to get their patients admitted "by the back door." It was therefore expected that as beds were closed, the first sign of pressure would be that waiting times in the ER would rise because of this increase in demand.

The model was run for a range of different assumptions about the number of beds available, for various permanent increases in demand, and for a "crisis day" when there was a sudden increase of 13% in demand. The key finding was that the immediate impact of bed shortages was not observed in the ER at all, but was evident first in cancelled elective admissions, so that using ER waiting times to measure the effect of bed shortages was misleading. In fact, the model showed that it did not make sense to look at any single measure in isolation, but that this was indeed a complex system where changes in one area would have unforeseen knock-on effects elsewhere, and so a holistic view was required.

3.3.2 Emergency Care in Nottingham

Brailsford et al. (2004) developed an SD model as part of a research project which was itself part of a larger, ongoing project in Nottingham, known as the Emergency Care-on Demand (ECOD) project. In Nottingham, emergency hospital admissions had risen dramatically. The aim of the ECOD project was to look at the whole health care system to determine why demand was so high and to investigate what could be done to alleviate this pressure. The ECOD project had a Steering Group which contained representatives of all the providers of emergency and unscheduled health care in Nottingham.

The Nottingham study concerned a population of over 600,000 potential patients. Furthermore, although the specific pathways followed by individual patients were of interest, they were of less importance than understanding the major flows of people through the "front doors" to the system, and gaining insight into the general structure of the system and the relationships between its component parts. The problems experienced in the ER, for example, were not principally believed to be due to high variability in case mix or staffing levels, but more to the sheer volume of demand and consequent pressure on resources. Thus, SD was selected as the modeling tool.

The study involved both qualitative and quantitative modeling. A system map and a set of influence diagrams were developed through a series of about 30 interviews with different stakeholders and providers. Participants were asked to amend the map and describe the historical, organizational or political factors which influenced flows of patients in their own area. The resulting final map of the system was used as the basis of a quantitative SD model which was populated with activity data for 2000–01 from all the providers. After the model was validated for the current year, a range of scenarios were run (suggested by the ECOD Steering Group). These looked at the effects of various interventions, including early discharge, increasing the proportion of patients who used the Walk-in Center rather than the ER, reducing admissions for specific groups of patients (e.g., over 65) and providing additional diagnostic services in the community, since it was thought that some primary care physicians used the ER to bypass the normal referral process for diagnostics, so that patients ended up being admitted unnecessarily (the same effect as observed earlier by Lane et al. 2000).

The key findings of the model were that if current trends continued, both hospitals would be forced to cancel several hundred admissions for elective surgery every month within a couple of years. However, relatively small changes in one part of the system had significant impact elsewhere. For example, it was much better to prevent admissions through the use of a community diagnostic facility than it was to reduce length of stay and discharge patients early. The model was run interactively with the Steering Group and the results used to inform them at a Stakeholder Day, at which focus groups developed a Local Services framework for emergency care. A Treatment Center was subsequently built in the grounds of a larger hospital to provide community diagnostic services and to fast-track routine surgical cases.

This model was used to inform strategic decisions. The numerical results of the model were indicative of trends and were relative rather than absolute. Moreover, the qualitative insights which the model provided (and indeed the value of the mapping process as a communication tool) were found to be as useful as the numerical results. This approach is now being used by Brailsford and Lattimer in a national study of workforce change in unscheduled care, in which many of the same strategic issues arise.

3.4 Commentary on Examples

The models in this section share several features. Firstly, all models were developed in close collaboration with clinical experts. All Level 1 models used clinically recognized stages to model disease progression. Secondly, many of the models were used to address a public health policy issue, and therefore were aimed at regional or even national decision makers rather than local hospital managers. Thirdly, the models required a lot of detailed data. The data were derived from a variety of sources, including the literature, but also data collected for a previous study as well as data collected routinely in the course of providing health care. Some model parameters were unknown and therefore multiple scenarios were run exploring a range of potential outcomes, depending on the value of these parameters, basically constituting a sensitivity analysis. Several of the models required some degree of nonstandard technical innovation in order to adapt the chosen simulation approach to the specific needs of the model or allow the model to be run in a reasonable length of time. In some cases, there was an attempt to develop a generic model that could be reused to model similar systems. The last two models were developed to study emergency care, but their scope included many other systems because it was almost impossible to draw a well-defined boundary around any subset of the modeled system. Many of the other models had to place rather arbitrary limits on the scope of the system to be modeled. Finally, almost all of the models were developed by academics for research and/or student projects, rather than consultants or management engineers doing routine analysis for the institutions.

4 Opportunities

As we noted earlier, Americans spent approximately $2.3 trillion in 2007 on health care. The budget of the NHS in the UK was approximately £120 billion. With expenditures this large, even very modest improvements in cost can amount to billions of dollars in savings. In a landmark study (Institute of Medicine 2000), the Institute of Medicine estimated that between 44,000 and 98,000 deaths were due to medical errors in 1997. A 1% drop in the medical error rate can save hundreds and perhaps thousands of lives. Clearly, there is a strong incentive to find ways to improve the functioning of various parts of the health care systems. Unfortunately, even very small subsystems within the health care system are so complex and stochastic that it is difficult or impossible for anyone, even managers with many years of experience who are intimately familiar with these systems, to predict the effect of changes in the structure and processes within them. So, there is an opportunity to apply models to predict the behavior of health care systems and perhaps to optimize such systems.

While no modeling paradigm fits every system structure and analysis objective, simulation is an ideal tool for most health care systems. As the examples in Section 3 have shown, health care systems are complex and involve relationships that are difficult to describe mathematically. For example, in the ICU model in Section 3.2.2, what is the relationship between cost per day and the number of nurses permanently based in the ICU? Since the number of nurses is a discrete variable, it seems reasonable to imagine that this relationship is discontinuous and nonlinear, but it also depends upon other system characteristics and cannot easily be expressed as a mathematical function. However, using standard discrete-event modeling techniques, this relationship can be evaluated using a simulation.

The stochastic variables include such things as patient arrival times, patient diagnoses, treatment choice, duration and efficacy, workforce availability, error occurrence, cost, environmental factors and human decision making. These variables are intimately concerned with the behavior of the system and must be considered. On the other hand, their actual effect on the system is very complex and nonlinear, and usually cannot be expressed as simple mathematical or statistical models. The modeling approach must be able to capture the relationships between these variables and the system state at a much more elemental level. Simulation is just such a tool. Indeed, any system can be modeled and simulated if its components and processes can be described at a basic level.

Since health care systems have many stakeholders with differing goals, any analysis will need to evaluate multiple measures of system performance. For example, an emergency department has patients, physicians and hospital administration as stakeholders. Patients are concerned with minimizing the length of time between arrival and the start of treatment. Physicians are interested in the delays in diagnosis due to waiting for laboratory or imaging results. Administrators consider system capacity and overall cost to be important. The model should be able to predict all of these performance measures simultaneously, and evaluate their trade-offs.

The various simulation tools (DES and SD) offer a flexible modeling approach. A wide range of system models can be developed to represent almost any system structure imaginable. The modeling world-view of DES is that the system is composed of entities that have attributes and interact only at discrete points in time when an event occurs to change the system state. For a more in-depth description of this world-view, see Chapter 2 of Fishman (2001) or Schriber and Brunner (1998). This is a very general model structure, and has proven useful to represent a huge assortment of systems over the past 50 years. Although we will discuss some needed improvements in available software, the modeling approach is robust enough to represent virtually any health care system imaginable.

The use of simulation for health care systems is further promoted by the wide variety of simulation software available. For the adventurous, general purpose programming languages such as Java, C/C++ and FORTRAN can be used, supported by libraries such as Simkit (Buss 2001) or SIMLIB (Law and Kelton 2000). Numerous libraries are available for many programming languages. This approach offers the advantages that the developer can combine the simulation library with other available libraries to support such things as database access, graphics and numerical computations, and if needed, the developer can modify the basic simulation code. This can be especially useful if changes are needed to the fundamental simulation code in order to overcome technical challenges such as those mentioned in Section 3.1. At the other end of the scale, simulation model development environments are available to support graphical interactive model development. This software requires the user to be familiar with modeling concepts but does not require programming knowledge. The disadvantages of this approach are that the user must stick with the capabilities provided, and thus must often change the model to conform to what the software will represent, and that these packages are proprietary and often quite expensive, thus discouraging their use for single projects. Swain (2007) provides a survey of some, but not nearly all, available simulation software.

The goal of most modeling efforts is to search for optimal decisions among a collection of possible choices. Simulation methodology evaluates the performance of decisions involving system design, strategy or operating policy. Using simulation, the search for optimal decisions, especially optimal combinations of system parameters, often must be done manually and can be laborious due to the large number of possible decisions. To some extent, this objection has been alleviated by optimization features that are available on many commercial packages. These features allow some types of decisions to be searched during the simulation runs to find those that result in improved performance. April et al. (2003) and Olafsson and Kim (2002) provide a brief surveys of simulation optimization methods.

Another feature that favors simulation for health care problems is that the models and methodology are relatively easy to describe and explain to administrators, clinicians and other persons not familiar with the development and use of models for decision making. Many simulation packages have animation capabilities which allow the user to view the simulation as it runs. Animations give the viewer a sense of the reality of the model and develop confidence in the model. Other modeling approaches, such as mathematical optimization and probability models, that consist of manipulating functions and solving systems of equations are much more difficult to describe and explain to managers who are not comfortable with mathematics at this level.

While simulation uses very sophisticated technology for sampling random variates, representing the system and analyzing input and output data; in a sense, it is easier to apply by persons with relatively little training in engineering, mathematics, statistics and similar disciplines. To build a simulation model, the modeler must understand the system components and their interactions, and must be able to describe these within the modeling world view, but this can be done by a person who has relatively little mathematical sophistication. In contrast, linear programming models and Markov chain models are much more abstract and require more advanced knowledge of mathematics to build and interpret. Many mathematical probability models, in particular, require the modeler to modify the model and solution procedure for every change in the model assumptions. For example, a formula is available to compute the mean waiting time in the simple stationary M/M/1 queueing model (Gross and Harris 1974, Chapter 2), but if the distribution of either the interarrival time or service time is changed from the exponential distribution, a new solution process must be adopted. Indeed, if both the interarrival and service time distribution are non-Markovian, then no analytical solution exists. Using discrete-event simulation to estimate mean waiting time for a single-server, FCFS queueing system such as any of these, one merely changes the algorithm used to sample the interarrival and service time random variates. These algorithms are available from public sources. All other computations are unchanged. Other than possibly needing to make longer runs for some versions of the model, the analyst uses the exact same code and procedures to analyze all models. In this sense, simulation is much more versatile than alternative methodologies. Finally, simulation is a mature methodology that has been widely accepted in many other areas of business, industry and government. Many organizations, especially large organizations, in areas such as transportation, communications, manufacturing, services and finance, use simulation routinely as an aid in designing new products and systems.

5 Challenges

If simulation is so well-suited to solve systems analysis problems in health care, why hasn't it been used more and why isn't it a routine part of analysis prior to major decisions? We will explore these questions in this section. In particular, we will look at the difficulties that must be overcome in order to have a successful simulation modeling project in health care.

5.1 Project Management Challenges

Project management challenges involve the problems associated with initiating and managing the modeling project. These challenges relate to the demand for projects using simulation, support for the project by management, team collaboration difficulties, problems selecting appropriate measures of success, and cultural differences between health care practitioners and simulation analysts.

Management knowledge and support. Before a technology can be adopted widely, there must be a demand for it. In this case, the demand must come from executives and managers in various parts of the health care systems or from public officials. Most health care managers are unaware of the power of modeling and simulation to understand system behavior and predict the likely effect of various decisions. In the health care management literature, i.e., the publications read daily by executives and line managers in health care systems, there is almost no mention of the use of tools like simulation to analyze and evaluate decision alternatives. So, the primary challenge to the use of simulation in health care systems is to educate the potential users about the powerful capabilities of the technology.

Team management. Most large simulation projects, including all projects discussed in Section 3, are pursued by a team, usually consisting of simulation experts, domain experts and supporting members. This setup is normally used because no one person has all of the expertise in the operations of the system and the application of simulation to do the entire job herself. In addition, systems are usually too complex to allow one person to develop the model, collect and analyze data, run the simulation and do the other tasks involved. This is especially true in health care models. Very few simulation experts have detailed knowledge of health care systems, and even fewer health care managers and practitioners know how to use simulation technology. Working in teams to develop models of health care systems is especially challenging because of the different cultures involved. Most simulation practitioners are educated in engineering, computer science, statistics or mathematics. Most managers and clinicians in health care are educated in medicine as patient care providers (doctors, nurses, etc.). The fields of engineering and health care have different concepts, goals, and vocabularies. Indeed, they have very different ways of viewing the problems to be solved. Engineers think in terms of the delivery system, with an emphasis on *the system.* Health care providers usually focus on the *individual patient.* Normally, they view the delivery system—the dance of people and resources that delivers health care to the patient—as a given, something that they accept as it is. If there are problems with getting resources in a timely manner, their view is that it is somebody else's responsibility. Their world is focussed on the patients and clinical procedures. On the other hand, the engineer sees the people, resources and processes—the physical and virtual system—and takes the clinical procedures as fixed. A model of the system must accommodate both views and model the interplay between clinical activities, the resources and the people. This means that each team member must work to learn the world view and language of the others so they can integrate their knowledge.

Selecting performance measures. In health care modeling projects, it is often hard to select appropriate measures of effectiveness. We alluded to this problem earlier when we noted that health care systems have a variety of stakeholders, each with their own goals. Some have operational goals, such as to maximize patient throughput. Others have financial goals, as, for example, to minimize the present value of cost over a period of 5 years. In private hospitals and clinics, the goals often reflect an interest in making the enterprise maximally profitable. In public hospitals and clinics, the appropriate goals are not so clear. Certainly, operational efficiency is important, but it is also important to maintain financial viability. It can be a challenge to agree on the appropriate objectives, and therefore, the most appropriate data to collect from the particular health care system and the simulation. It is helpful that data collection can be changed late in the modeling and simulation project, but it is also desirable to have a firm idea of the modeling goals early in the process.

5.2 Modeling Challenges

Many of the challenges to health care simulation involve modeling directly. Health care systems are highly dynamic, involve human behavior in an intrinsic way and have control mechanisms that make adjustments to accommodate changing conditions. Models must be able to accurately capture these characteristics.

Modeling human behavior. As we discussed earlier, many of the activities in health care delivery involve human interaction, both with machinery and with other people. Patients respond to treatments and make choices for further care. Staff respond to policies for patient care or infection control. These responses can, and often are, not rational or at least, not directed to achieving the hospital's goals. For example, nurses sometimes choose to not follow a directive if they feel that the time requirement is excessive and the policy reduces the quality of care (whether it actually does or not). Patient compliance with physician orders and medication is an issue in health care, and models will usually need to include compliance, especially for the policies the model seeks to evaluate.

Providers and public officials are often more interested in how the delivery system design affects the health of patients through the quality of care. Physicians make decisions regarding further testing, diagnosis, and treatment for patients based upon the information available from current examinations and tests. These decisions are important because they influence the behavior of the system going forward by determining the additional tests and treatments to be scheduled. Instead of using surrogates such as patient delays to evaluate patient response, models should represent the relationship between system operations and patient response directly. Thus, the model should be able to provide an accurate representation of the patient's symptoms upon presentation as well as how those symptoms work through the physician's decision making process to initiate further activities.

Health care systems can respond to extraordinary conditions by altering the system's configuration and structure. For example, if an emergency department becomes overcrowded, the director might close the department, diverting patients to other EDs in the area, possibly causing those patients to suffer additional complications due to the delay in treatment. Or, the director might reconfigure the ED by creating space in a hallway for overflow patients and assigning staff from other areas to attend the patients there. Another choice could to be to speed the discharge of some patients who have minor problems. Each of these responses to the overcrowding has a significant impact on the functioning of the ED and makes a rather fundamental change to the structure of the ED. Moreover, this response is the result of a decision by a person. The model should be able to capture this type of response.

Rapid modeling tools for health care systems. Health care facilities operate on a fast schedule. Today's problem is soon forgotten because another problem will present itself tomorrow. Managers in health care are busy identifying and solving problems in succession. Problems that yield to quick solutions get solved; others are put aside. Frequently, managers have difficulty distinguishing temporary operational problems from strategic policy-oriented problems. Thus, an issue involving ICU overcrowding will be dealt with as a temporary operational problem, rather than a policy issue, even though it recurs periodically. To keep the managers' focus on the problem, they need to quickly see potential solutions. This is where a simulation tool that allows rapid modeling would be useful.

As Davies and Davies (1994) have pointed out, in some health care models, some entities such as patients often need to belong to multiple queues simultaneously and wait for multiple resources, possibly responding to the first resource that becomes available. These systems also use priorities to determine how clinicians and other workers will be preempted and reallocated when a new patient arrives or other events occur. While these features can be built into any model if an appropriate simulation library and programming language is used, the simulation software needs to make it natural and easy to implement models having these structural characteristics.

Other characteristics that are desirable in health care specific modeling software include health care domain specific processes and data. For example, easy access to ICD-9 diagnosis and procedure codes with frequency data for various locations that can be used to sample incoming diagnoses would be valuable. Access to generic data would also be useful for cases where provider-specific data is not available. Software modules to utilize this data to sample diagnoses and other purposes should be provided so they do not have to be built from scratch. Pre-built but configurable processes such as ED triage and registration should be provided so the model building process can consist mainly of assembling modules. Of course, like any other special purpose simulation tool, a health care modeling package should include facilities for model animation, graphical presentation of data and flexible reporting that can be configured to look similar to existing hospital reports.

All simulation software is designed to allow the user to evaluate alternative system designs or other decisions related to the system. It is important, therefore, that the software allow the alternative decisions to be represented fairly easily, so the simulation can implement the alternatives with minimal effort. Some alternatives, which can be implemented as parameter changes, are easy to implement simply by changing some elements in the input file. On the other hand, some alternatives that represent more fundamental changes in system structure are much more difficult to implement. For example, if alternatives involve changing system policies, such as allowing nurse practitioners to perform additional duties that would normally be done by physicians, or modify care processes, the actual logic of the model must be modified to implement the alternative policy. The simulation software should be designed so these more fundamental alternatives can be implemented by users if possible. This would encourage health care managers to re-use the model for additional decision evaluation beyond the original scope.

5.3 Challenges from the Health Care System

Lack of data. All simulations need system-specific data in order to validate the model and to have a model that is valid for the current environment. Data availability is a challenge in every health care modeling project. Health care systems keep copious amounts of data. Indeed, it has been said that health care delivery primarily involves data collection and storage, and only a minority of time is spent on actual diagnosis and treatment. However, the data collected is for clinical, administrative and legal uses, not for modeling. Data collected for clinical uses include all medically relevant information and often include process data such as times when treatments begin. For example, in a hospital, nurses record the times when medications are administered in order to monitor and assure that they are given on the prescribed schedule.

Some data is collected just for legal and risk management purposes to protect the hospital or clinic in the case of an accusation of negligence or poor quality care. Most emergency departments record each patient's arrival time and time of the start of the first examination by each physician so the patient's waiting time before being seen by a physician can be computed. In a lawsuit, this delay would be relevant.

Hospitals and certain other providers collect data to monitor and evaluate existing processes, but not to analyze existing systems. For example, emergency departments record patient arrival time and discharge time in order to compute their total time in the ED. This is a performance measure for the ED, but it is not useful in a simulation because this would be a parameter the simulation seeks to estimate. The simulation needs more basic data. Data requirements include activity times such as the durations of patient-physician encounters or the time to deliver a breathing treatment to an asthma patient. This data is seldom collected because managers do not realize the need but also because of resistance from the staff to being measured.

If data are manually collected, the cost can be excessive. Moreover, conditions within and surrounding the health care system under study can change so rapidly that the data are no longer relevant when the model is ready to use. It seems that the only truly complete solution to this problem involves creating a system of ongoing data collection. We will examine this topic in Section 6.

Lack of standardization. While most health care subsystems have much in common, virtually all of them have some unique characteristics. These unique characteristics might involve the layout of a department or the presence or absence of a particular resource such as a lab or imaging machine, or they might involve unique process structures. For example, in one ED, breathing treatments might be administered by a nurse, whereas, in another ED, these treatments might be required to be administered by a respiratory therapist. As a result of these differences, it is very hard or impossible to develop a generic model of a system that can just be configured and used routinely. Generic models would greatly reduce the cost of developing and applying the simulation, and they would make it possible for a manager, rather than a modeler, to do the analysis using simulation. Instead, every model must be modified to fit the current system, resulting in increased cost and time for the project.

Complexity. We have already mentioned how complex health care systems are. As in any modeling effort, the difficulty of developing the model increases exponentially with the complexity of the system. If health care systems are compared to, say, manufacturing systems, many components and interactions are not immediately visible to the observer. Many processes are developed ad-hoc and are either not documented or poorly documented. A great deal of time and effort can be required to describe and model these processes at an appropriate level of detail. In addition, many health care systems are in fact multiple systems functioning relatively independently. For example, in a pre-operative unit, anesthesiology, nursing and surgical staff each constitutes an independent system with their own procedures but the systems normally perform them with little coordination. This structure also greatly increases the modeling difficulties.

6 Overcoming the Challenges

Often when discussing simulation in health care, the goal is limited to solving a particular problem using this methodology. In this section, we would like to raise that goal to one of having simulation (and perhaps other analytical operations research methodologies) used routinely in health care management to improve the quality of decision making, system design, and ultimately, patient care. Most comments in this section could just as easily apply to other operations research tools. Since this paper is about simulation, we will just say "simulation" and leave it to the reader to determine if other OR tools apply.

If the effort over the past half century by thousands of professionals to develop and refine simulation methodology and its applications is to pay off maximally, simulation needs to become a standard tool in the health care decision process. Obviously, simulation is not the preferred analytical tool in every decision, but it should at least be given consideration and used when appropriate and feasible. To this end, we have a long way to go since simulation, when it is used, is used mainly in a research setting in large medical centers, and funds are almost always provided by government research grants. Indeed, this is the case with most of the examples in Section 3. This indicates that most health care managers are not aware that tools of this sort are available, or they do not see sufficient value in them to invest the money, or they cannot find a company or consultant to perform the analysis, or they cannot do the analysis because some required component, such as data, is not available. Many problems in health care are well suited for simulation, but the environment is not.

One approach to this conundrum is to think of this as a marketing problem in which the product to be marketed is problem analysis using simulation and the marketplace is health care management. In order to promote the sale and use of a product in a marketplace, certain conditions must be met. Normally, three things must be done to launch a successful product:

1. Develop the product to meet the needs of the marketplace.
2. Grow demand for the product in the marketplace.
3. Work with the customer to purchase and implement the product.

6.1 Develop the Product

Simulation is certainly a mature and successful technology for problem analysis, but it is still not ready for routine use in the health care marketplace. The current problems have been outlined in Section 5. Here, we will offer some suggestions about how these problems might be overcome.

The simulation community needs to develop software that satisfies demand in this marketplace, mainly by developing specialized tools for modeling and simulating health care systems. These tools need to be modular and standards-based so they can be used and reused by a broad group of analysts and consultants. At present, most health care simulation models are implemented using generic tools, so all health care-specific aspects of the models must be developed from basic components—entities and resources—and their health care-specific attributes and behaviors must be created for each individual model. The simulation modeler needs to believe that he or she can quickly and reliably develop the model, and the new tools should support this capability.

Many other areas, such as manufacturing and communications, have tools that were developed specifically for modeling those types of systems. Like these other tools, the need is not for a tool that allows an untrained analyst to build models, but rather one that provides support for a trained simulation modeler in building health care models quickly. Since all models will potentially require customization, the tool needs to give the modeler fine-grained control, but make definition of common entities such as patients, physicians, nurses, imaging machines, beds, etc. quick and easy. These components should be predefined and configurable. The goal of having such a tool is to enable the modeler to build a simulation prototype within one to two weeks to show the stakeholders, and not only provide a starting point for the path to a final model that can be used for actual analysis, but also to keep the managers interested in the project.

One package built for modeling health care systems is MedModel (Price and Harrell 1999). This package has been used to develop a number of DES models, but it is limited in its features because it is basically a repurposed version of ProModel, which is a modeling tool for production systems. As such, MedModel requires a physical model layout and primarily models the physical operations of health care clinics and similar systems, rather than the conceptual procedures and processes. A more appropriate tool would allow models of non-physical systems to be developed in a natural way.

Currently, most published health care modeling is done on a special project basis by researchers. We believe that most simulation modeling projects in the U.S. are done by academics. In the UK, consultants do a lot of health care modeling, but in both countries and the rest of the world, simulation is frequently overlooked when systems are being analyzed and modified. If public health agencies, hospitals and other health care systems are to use this technology routinely, a workforce must be available to deliver the service, i.e., perform the systems analysis, modeling and simulation. At present, there are not enough people trained in modeling and simulation, and particularly in modeling health care systems, to support this objective. The educational community needs to offer more courses in operations research and systems analysis graduate programs, at the master's degree level, in health care modeling and simulation. The graduates of these programs need to have the specific skills to work on system modeling problems when they graduate. As such, they should have a good working knowledge of modeling and simulation, and also a thorough understanding of the structure and workings of the health care systems in their country of interest. These programs could also be developed to train current simulation modelers in the specifics of health care systems and to train current health care managers in modeling and simulation.

The simulation community needs to shift its view from the producer's perspective to the consumer's. If simulation is to be not only accepted but demanded, the users—health care managers—must see value in the product. That is, they must believe that the product will return information for decision making that justifies the cost. The simulation community must fully understand this viewpoint in order to design simulation products for health care that will be successful in the marketplace.

6.2 Grow Demand for Simulation

Although we have not done a statistical study, it is apparent to us that there is very little awareness of simulation methods in the health care sector. We explored this topic in Section 5.1, but it is clear that the job of convincing health care managers to utilize this methodology in their projects would be greatly facilitated by their having an understanding of the existence and usefulness of simulation. This could be accomplished by developing materials on simulation that showcase its useful application in health care and provide a realistic idea of how to use this methodology in problem analysis and decision making. Some sample models that managers can download and experiment with would also be helpful. This material could then be presented, perhaps through guest lectures, to students in public health and health administration curricula, as well as other venues such as professional meetings and focussed conferences. A major research project to do exactly this is currently under way in the UK. RIGHT (Research Into Global Healthcare Tools, `http://www-edc.eng.cam.ac.uk/right/`) is a collaborative study, one of whose outputs will be a web-based method selection tool, a paper-based Workbook explaining the methods in lay terms, and a website with successful case studies.

Another way to promote the use of simulation is by publishing articles in trade publications that are read by line managers as well as top executives of health care organizations. These articles need to be problem-focussed, but also show clearly how simulation is a technology that is uniquely suited to analyze some of the most important problems these managers face. Unlike academic journals, trade publications often promote the use of specific products and solutions, so this would be an ideal place for the simulation community to promote the use of simulation. Finally, there are many websites dedicated to health care management. This is yet another opportunity to publish informative articles about simulation and health care. These articles should focus on how simulation can solve problems by providing insight into system behavior that cannot be gained from other analyses.

Consulting companies have excellent access to health care executives because hospitals and other health care providers make frequent use of consultants. Encouraging health care consultants to make more use of simulation and promote its use by their clients is another way the value and benefits of simulation can be communicated to the health care community. It is also important to provide a means for consultants and other non-academics to publish their successful simulation applications in short articles, perhaps through a column on simulation successes in health care or a blog on the topic.

6.3 Work with the Customer

Simulation projects often take months to complete (if they are ever really completed), and encounter multiple problems during their progress. This will be discouraging to managers, and will lead them to conclude that while simulation is a valuable tool, it is just too hard to use. In this section, we want to explore some ways to overcome this problem.

Virtually every simulation project in health care reports major problems finding appropriate data. Hospital managers and other health care system administrators can be encouraged to collect and store the more detailed data that is necessary in simulation modeling. This collection and storage should be automated so it does not impose a burden on the workers. As electronic medical records and online clinical ordering systems are adopted, this provides a great opportunity to collect some activity times automatically. It might be necessary to consider modifying some models to use activities whose epochs are marked by times that can be recorded automatically by these online systems. Other innovative methods for collecting data, such as using RFID or sonic technology, are available and should also be considered. Of course, it will be hard to convince health care managers to invest the money to collect this data before models are built requiring it, but hopefully as some institutions make the investment and reap the benefits, others will decide to follow. For those cases where data is not available for the specific system being modeled, it would be useful to collect and share data from other systems. For example, data for the activity times of various surgeries, categorized by procedure code, patient demographics and other attributes, could be used to estimate activity time duration distributions for surgery, similar to the work in Section 3.2.1. This type of data could be anonymized and contributed to a central repository where it would be available to anyone developing a model of surgical suite operations. Going one step further, the data could be used to fit distributions for surgery times, and the distributions could be published and used in those cases where more specific data are not available.

On the output side, it is important that the models produce results that are relevant and important to health care managers. Engineers and analysts tend to think in terms of waiting time and resource utilization. Hospital administrators and clinicians think in terms of financial impact and quality of care. Obviously, these are very closely related metrics, but a simulation output that just presents waiting times and utilization will not be taken seriously by hospital administrators. Perhaps all simulations should produce reports that show financial results such as revenue and cost, unless there is a good reason to exclude these measures.

Ideally, some models could be implemented by starting with a generic model and just configuring the model to represent the specific system under consideration. For example, a generic model of an emergency department could be configured by providing parameters such as the arrival rates, numbers of clinical personnel by job type, numbers of examination rooms, and so forth. Some work toward this goal has been done by Gunal and Pidd (2007), and appears promising. If such generic models can be developed and shown to provide useful predictions, they can greatly simplify the process of model development and use.

Similar to generic models, many health care systems use common components such as laboratories and imaging centers. Models of these subsystems can be developed as modules for reuse, simplifying the process of model development. To carry this idea further, a model building process could just consist of assembling the component modules and providing appropriate parameters. In engineering and architecture, this is the approach a draftsperson frequently takes to create complex drawings. Modules could be stored in a central repository online, making them available to any potential user. This would be difficult at first, and would require simulation professionals to establish interface standards for models, but the goal would justify the effort, in our opinion.

In recent years, much work has been done on Web-based simulation (Miller et al. 2000). The concept here is that a simulation can be distributed among various locations on the Internet, accessing data and model components from remote servers. This technology opens many attractive options for health care modelers. For example, some model components, such as a model of a laboratory or imaging department, could be developed by experts or even vendors and implemented and maintained on their server. The process of model building would then involve locating and incorporating these specific components into the larger model. This approach, known as service-oriented architecture, is becoming popular in other areas of internet technology and, if implemented, could improve the ease of developing complex simulation models of health care systems.

Generally, the simulation community and the health care management community need to reduce the difficulty, and hence cost, of developing health care simulation models. One activity that would promote this goal is to develop rapid modeling tools, especially tools with simple user interfaces that implement generic models of common systems and can be used by health care analysts and managers, rather than simulation analysts. Another effort could be aimed at promoting standards in health care processes and terminology. Standardization would encourage the development and use of generic models and reusable modules.

Finally, and perhaps most importantly, the simulation community needs to really take an interest in all aspects of health care, and not only learn the terminology and concepts of diagnosis and care delivery, but also understand the culture of the profession. Simulation modeling projects require a team approach, and it is important to master how to create and manage these teams so they are effective. If simulation experts are able to understand the attitudes and thought processes of the health care experts, smooth functioning and effective management of the team will result, and ultimately the probability of success will increase.

7 Conclusion

Much has been written in recent years about the inefficiency, cost and error rates of the current health care systems. Simulation can play an important role in improving these operations and, literally, saving lives. In a recent book, Gawande (2007) persuasively argued that, while there is a role for basic research and improved therapies in medicine, much greater benefits can be gained by simply improving the way that existing therapies are delivered to patients. Simulation is a tool that should be an important component in the effort to find these improvements.

Ayers (2007) shows how regression and randomized trials, two statistical methods that have been around for many decades and are routinely used by researchers, have found new applications in business, sports, education, health care and government to improve decision making and radically reshape many of the decision processes in these and other areas. These applications have been driven by the availability of large amounts of data stored electronically, cheap high-speed computing, and the Internet, which effectively is a distributed computing platform where not only data can be stored and retrieved, but experiments can be run and results can be quickly distributed. The statistical methodology, regression and randomized trials, is not new. But, some decision makers have found new applications for it and have reaped extraordinary results and rewards.

We see simulation as having a similar opportunity. Although useful research is continuing on simulation methodology, it is not a requirement for successful application in health care. What is needed is for decision makers to recognize that the quality of decisions can be greatly improved by utilizing simulation (as well as other OR tools). But, much work needs to be done before this can happen.

We believe that academic programs in operations research, especially those that emphasize simulation, should examine their curricula and see if programs of study can be developed that integrate simulation and health care. Simulation educators, researchers and practitioners who want to work in health care need to develop a greater focus on problem solving and work to understand the world of health care, especially the culture of the people who work in that world daily. The benefits of simulation should be promoted by talking, writing and blogging about successful simulation projects. These projects do not have to be huge, but they should clearly show the feasibility and benefits of using simulation to advance the established goals of the health care systems.

Additional work is needed to prepare for routine use of simulation in health care. New modeling tools are needed that allow useful models to be built quickly and applied effectively. These tools will shorten the development time, improve model reliability and reduce cost. Data repositories and data collection systems that gather input data for simulations need to be developed in collaboration with health care software providers.

Ayers (2007) has pointed out that the availability of high-speed computing and cheap large data stores have driven down the cost of statistical analysis to almost nothing. The simulation community needs to find other ways to drive down the cost of developing, implementing and using simulation. One way is by developing generic models for common health care systems along with software tools to manage and apply these models. If successful, this could create a new role for simulation analysts that involves configuring and applying standard generic simulations.

Simulation methodologies will certainly continue to advance in the future. So will methods to manage health care systems, but citizens will continue to demand better performance from their health care systems. It is our hope that some of the ideas in this article will contribute to improving the delivery of health care with the use of simulation.

Acknowledgments The authors wish to thank Christos Alexopoulos, David Goldsman, and James R. Wilson for their work organizing and editing this volume. Our appreciation goes to Christos for suggesting the topic of this paper. We wish to thank David Block MD, PhD for reviewing an early version of the manuscript and providing very insightful comments on the topic, especially the cultural environment of health care. Dr. Seila wishes to thank his wife, Jodie, who is a registered nurse and has worked in both hospital and physician practice settings, for giving him an intimate view of health care delivery that he could not have gained otherwise. Dr. Seila also wishes to thank George Fishman for introducing him to discrete-event simulation, directing his dissertation research and providing guidance and encouragement over the past thirty years.

References

April, J., F. Glover, J. P. Kelly, and M. Laguna. 2003. Practical introduction to simulation optimization. In *Proceedings of the 2003 Winter Simulation Conference*, pp. 71–78. Piscataway, New Jersey: Institute of Electrical and Electronics Engineers. Available online via www.informs-cs.org/wsc03papers/010.pdf [accessed October 21, 2008].

Ashby, M., M. Miller, D. Ferrin, and T. Flynn. 2007. Simulating the patient move: Transitioning to a replacement hospital. In *Proceedings of the 2007 Winter Simulation Conference*, pp. 1562–1565. Piscataway, New Jersey: Institute of Electrical and Electronics Engineers. Available online via www.informs-sim.org/wsc07papers/192.pdf [accessed October 1, 2008].

Ayers, I. 2007. *Super Crunchers*. New York: Bantam.

Brailsford, S. C., V. A. Lattimer, P. Tarnaras, and J. A. Turnbull. 2004. Emergency and on-demand health care: Modelling a large complex system. *Journal of the Operational Research Society* 55:34–42.

Buss, A. 2001. Discrete event programming with Simkit. *Simulation News Europe* 32/33:15–24. Available online at diana.nps.edu/Simkit/ [accessed October 22, 2008].

Centers for Disease Control and Prevention. 2004. Chlamydia screening among sexually active young female enrollees of health plans – United States, 1999–2001. *Morbidity and Mortality Weekly Report* 53(42): 983–985. Available online at www.cdc.gov/mmwr/ preview/ mmwrhtml/mm5342a1.htm [accessed October 22, 2008].

Dangerfield, B. C. 1999. System dynamics applications to European health care issues. *Journal of the Operational Research Society* 50:345–353.

Davies, R., and S. C. Brailsford. 2004. Screening for diabetic retinopathy. In *Handbook of OR/MS Applications in Health Care*, International Series in Operations Research and Management Science, pp. 493–518. Norwell, Massachusetts: Kluwer.

Davies, R., S. C. Brailsford, P. J. Roderick, C. R. Canning, and D. N. Crabbe. 2000. Using simulation modelling for evaluating screening services for diabetic retinopathy. *Journal of the Operational Research Society* 51:476–484.

Davies, R., and H. Davies. 1994. Modelling patient flows and resource provision in health systems. *Omega* 22:123–131.

Davies, R., R. M. O'Keefe, and H. T. O. Davies. 1993. Simplifying the modeling of multiple activities, multiple queuing and interruptions: a new low-level data structure. *ACM Transactions on Modeling and Computer Simulation* 3:332–346.

Evenden, D., P. R. Harper, S. C. Brailsford, and V. Harindra. 2005. System dynamics modelling of chlamydia infection for screening intervention planning and cost benefit estimation. *IMA Journal of Management Mathematics* 16:265–279.

Evenden, D., P. R. Harper, S. C. Brailsford, and V. Harindra. 2007. Improving the cost-effectiveness of chlamydia screening with targeted screening strategies. *Journal of the Operational Research Society* 58:1400–1412.

Fishman, G. S. 1973. *Concepts and Methods in Discrete Event Digital Simulation*. New York: John Wiley & Sons.

Fishman, G. S. 1978. *Principles of Discrete Event Simulation*. New York: John Wiley & Sons.

Fishman, G. S. 2001. *Discrete-Event Simulation: Modeling, Programming, and Analysis*. New York: Springer-Verlag.

Fone, D., S. Hollinghurst, M. Temple, A. Round, N. Lester, A. Weighman, K. Roberts, E. Coyle, G. Bevan, and S. Palmer. 2003. Systematic review of the use and value of computer simulation modelling in population health and health care delivery. *Journal of Public Health Medicine* 25:325–335.

Gawande, A. 2007. *Better*. New York: Picador.

Griffiths, J. D., N. Price-Lloyd, M. Smithies, and J. E. Williams. 2005. Modelling the requirement for supplementary nurses in an intensive care unit. *Journal of the Operational Research Society* 56:126–133.

Gross, D., and C. M. Harris. 1974. *Fundamentals of Queueing Theory*. New York: John Wiley & Sons.

Gunal, M. M., and M. Pidd. 2007. Moving from specific to generic: Generic modelling in health care. In *Proceedings of the 2007 INFORMS Simulation Society Research Workshop*. Available online at www.informs-cs.org/2007informs-csworkshop/19.pdf [accessed October 21, 2008].

Harindra, V., J. M. Tobin, and G. Underhill. 2002. Opportunistic chlamydia screening: Should positive patients be screened for co-infections? *International Journal of STDs and AIDS* 13:821–825.

Harper, P. R., and A. K. Shahani. 2002. Modelling for the planning and management of bed capacities in hospitals. *Journal of the Operational Research Society* 53:11–18.

Health Protection Agency. 2004. HIV and other sexually transmitted infections in the United Kingdom in 2003. Annual Report, November 2004.

Institute of Medicine. 2000. *To Err is Human. Building a Safer Health System*. Washington, DC: National Academy Press.

Jackson, R. R. P. 1964. Appointment systems in hospitals and general practice. *Operational Research Quarterly* 15:219–237.

Jun, J. B., S. H. Jacobson, and J. R. Swisher. 1999. Application of discrete-event simulation in health care clinics: a survey. *Journal of the Operational Research Society* 50:109–123.

Klein, R., M. D. David, S. E. Moss, B. E. Klein, and D. L. De Mets. 1985. The Wisconsin epidemiologic study of diabetic retinopathy. A comparison of retinopathy in younger and older onset diabetic patients. *Advances in Experimental Medical Biology* 189:321–335.

Lane, D. C., C. Monefeldt, and J. V. Rosenhead. 2000. Looking in the wrong place for healthcare improvements: A system dynamics study of an accident and emergency department. *Journal of the Operational Research Society* 51:518–531.

Law, A. M., and W. D. Kelton. 2000. *Simulation Modeling and Analysis*, 3rd edition. New York: McGraw-Hill.

Miller, J., A. F. Seila, and X. Xiang. 2000. The JSIM web-based simulation environment. *The International Journal of Grid Computing: Theory, Methods and Applications*, Special issue on Future Generation Computer Systems 17(2):119–133.

Nance, R. E. 1993. A history of discrete event simulation programming languages, *ACM SIGPLAN Notices* 28(3):149–175.

NHS, NSC Diabetic Retinopathy Screening. 2001. Resources for a national diabetic risk-reduction programme. Available online at www.diabetic-retinopathy.screening.nhs.uk/costings.html [accessed October 21, 2008].

OECD. 2007a. OECD health data 2007: How does the United Kingdom compare? Available online at www.oecd.org/dataoecd/54/45/39092215.pdf [accessed October 20, 2008].

OECD. 2007b. OECD health data 2007: How does the United States compare? Available online at www.oecd.org/dataoecd/54/44/39092204.pdf [accessed October 20, 2008].

Olafsson, S., and J. Kim. 2002. Simulation optimization. In *Proceedings of the 2002 Winter Simulation Conference*, pp. 79–84. Piscataway, New Jersey: Institute of Electrical and Electronics Engineers. Available online at www.informs-cs.org/wsc02papers/011.pdf [accessed October 21, 2008].

Price, R. N., and C. Harrell. 1999. Healthcare simulation modeling and optimization using MedModel. In *Proceedings of the 1999 Winter Simulation Conference*, pp. 215–219. Piscataway, New Jersey: Institute of Electrical and Electronics Engineers. Available online at www.informs-cs.org/wsc99papers/028.PDF [accessed October 21, 2008].

Pritsker, A. A. B. 1998. Organ transplantation allocation policy analysis. *ORMS Today*, August, 1998. Available online at www.lionhrtpub.com/orms/orms-8-98/ transplant.html [accessed October 22, 2008].

Pritsker, A. A. B., O. P. Daily, and K. D. Pritsker. 1996. Using simulation to craft a national organ transplantation policy. In *Proceedings of the 1996 Winter Simulation Conference*, pp. 1163–1169. Piscataway, New Jersey: Institute of Electrical and Electronics Engineers.

Pritsker, A. A. B., D. L. Martin, J. S. Reust, M. A. Wagner, J. R. Wilson, M. E. Kuhl, M. D. Allen, O. P. Daily, A. M. Harper, E. B. Edwards, L. E. Bennett, J. P. Roberts, and J. F. Burdick. 1995. Organ transplantation policy evaluation. In *Proceedings of the 1995 Winter Simulation Conference*, pp. 1314–1323. Piscataway, New Jersey: Institute of Electrical and Electronics Engineers.

Royston, G., A. Dost, J. Townshend, and H. Turner. 1999. Using system dynamics to help develop and implement policies and programmes in healthcare in England. *System Dynamics Review* 15:293–313.

Schriber, T., and D. T. Brunner. 1998. How discrete event simulation software works. In *Handbook of Simulation: Principles, Metholology, Advances, Applications, and Practice*, pp. 765–811. New York: John Wiley & Sons.

Seila, A. F., V. Ceric, and P. Tadikamalla. 2003. *Applied Simulation Modeling*. Belmont, CA: Brooks/Cole.

Sterman, J. D. 2000. *Business Dynamics*. Columbus, OH: McGraw-Hill/Irwin.

Swain, J. J. 2007. New frontiers in simulation: Biennial survey of discrete-event simulation software tools. *ORMS Today* 34(5):23–43. Available online at `www.lionhrtpub.com/orms/orms-10-07/frsurvey.html` [accessed October 22, 2008].

Townshend, J. R. P., and H. S. Turner. 2000. Analysing the effectiveness of chlamydia screening. *Journal of the Operational Research Society* 51:812–824.

UK Department of Health. 2005. National chlamydia screening programme. Available online at `www.dh.gov.uk` [accessed October 22, 2008].

Future Trends in Distributed Simulation and Distributed Virtual Environments

Steffen Straßburger, Thomas Schulze, and Richard Fujimoto

Abstract This paper reports the main results of a peer study on future trends in distributed simulation and distributed virtual environments. The peer study was based on the opinions of more than 60 experts which were collected by means of a survey and personal interviews. The survey collected opinions concerning the current state-of-the-art, relevance, and research challenges that must be addressed to advance and strengthen these technologies to a level where they are ready to be applied in day-to-day business in industry. The most important result of this study is the observation that as research areas, both distributed simulation and distributed virtual environments are attributed a high future practical relevance and a high economic potential. At the same time the study shows that the current adoption of these technologies in the industrial sector is rather low. The study analyzes reasons for this observation and identifies open research challenges.

1 Introduction

This paper reports the main results of a peer study on future trends in distributed simulation (DS) and distributed virtual environments (DVE)—two of the most-important areas in the discipline of simulation. The survey, as well as this entire study, was initiated by the Fraunhofer IFF in Magdeburg and was conducted by an independent research team under the leadership of the authors.

The survey was aligned with an attempt to establish an "Innovation and Research Center for Distributed, Interoperable Virtual Reality and Simulation" in Magdeburg, Germany, which follows a German national funding initiative.[1] Although this initiative provided the background and motivation for this study, its results are independent and are therefore expected to be of value to the entire DS/DVE community.

S. Straßburger (✉)
School of Economic Sciences, Ilmenau University of Technology, 98684 Ilmenau, Germany
e-mail: Steffen.Strassburger@tu-ilmenau.de

[1] "Centers for Innovation Competence" is a funding instrument of the German Federal Ministry of Education and Research exclusively targeted towards the former East German states. Its intention is to establish internationally recognized research centers in dedicated basic research areas.

C. Alexopoulos et al. (eds.), *Advancing the Frontiers of Simulation: A Festschrift in Honor of George Samuel Fishman*, International Series in Operations Research & Management Science 133, DOI 10.1007/b110059_11,

The intention of this survey was to assess the current status in the fields of distributed simulation and distributed virtual environments and to identify new trends and research challenges in these fields.

The motivation for this survey and study is multifaceted. On one hand, computer simulation and interactive virtual-reality-based visualizations have already established themselves as useful tools. On the other hand, there is an increasing complexity in both product development and production processes. This requires new methods for planning, evaluating, and controlling the underlying systems.

Technologies such as DS and DVE (which are already used rather frequently in the defense sector) could also be key enablers for addressing complexity issues in non-military applications. They can be the basis for simulating complex systems by integrating heterogeneous sub-components which cannot be executed as a monolithic application on one computer. They can connect all involved stakeholders even if they are located on different sites around the world.

The survey collected opinions concerning the current state-of-the-art, relevance, and the research challenges that must be addressed to advance and strengthen these technologies to a level where they are ready to be applied in day-to-day business.

Portions of this paper appeared in Straßburger et al. (2008).

2 Survey on Future Trends in Distributed Simulation and Distributed Virtual Environments

This section reports in detail about the results of the survey conducted on the topic of future trends in distributed simulation and distributed virtual environments.

2.1 Introduction

The survey was officially conducted during the period from September 15, 2007 until October 15, 2007.[2] The survey was designed to be distributed and completed in electronic form. The survey was mainly targeted towards experts in the fields of DS/DVE. Invitations to complete the survey were therefore distributed through relevant conference distribution lists like the Winter Simulation Conference (WSC), the Principles of Advanced and Distributed Simulation Conference (PADS), the IEEE International Symposium on Distributed Simulation and Real-Time Applications (DS-RT), and the Annual Conference of the German Simulation Society (ASIM). Furthermore the survey invitation was posted on the homepage of the Simulation Interoperability Standards Organization (SISO) and distributed to its members. It was also distributed to members of the SimSummit organization whose membership includes government, industry, and academia organizations concerned with Modeling and Simulation, especially for defense applications.

[2] Returned surveys were accepted until the cut-off date of November 1, 2007.

The survey questionnaire was completed by 61 individuals. The majority of the respondents (67%) classified themselves as working in research organizations,while 20% were from industry and 10% from defense (Fig. 1). As the main intention of this study was to focus on the research aspects of the DS/DVE fields, the distribution of participants agrees with our objective.

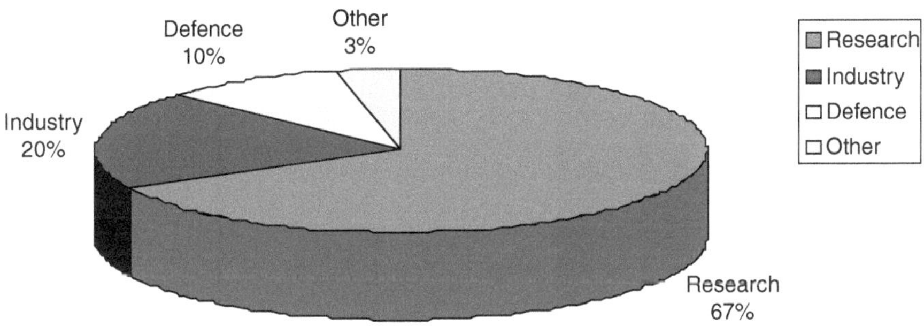

Fig. 1 Classification of participants with regard to their organization

The participants were also asked to classify themselves with regard to their relationship to distributed simulation and distributed virtual environments (Fig. 2). The responses show that 92% of the participants are directly involved with these topics, either as researcher/developer (79%) or as practitioners (13%).

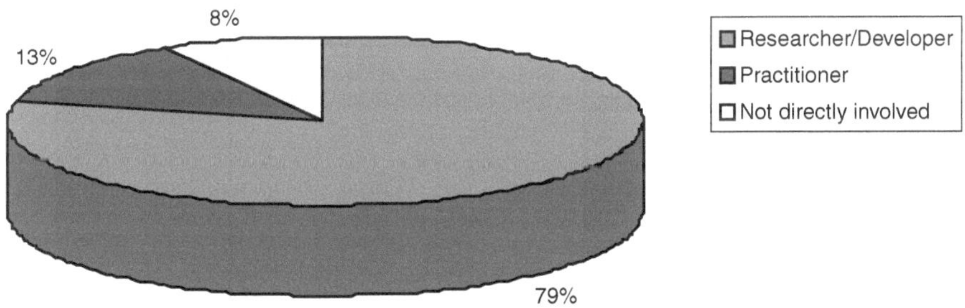

Fig. 2 Relationship of participants to DS/DVE technologies

The answers of the participants can thus be expected to give substantiated statements towards the state-of-the-art of the research in the fields as well as towards open research questions.

The actual survey consisted of two main parts, which will both be discussed in detail in the next section. While the first part addressed the relevance of DS/DVE technologies today and in the future, the second part focused on the open research challenges and the latest trends in these areas.

2.2 Survey Evaluation

This section reports and discusses the results of each question of the survey. All questions which asked for a rating or classification typically operated on a scale from 0 (none) to 5 (very high). Exceptions to this rule are pointed out in the text. As the survey design intentionally included several open questions and possibilities to comment, the answers to those questions are clustered and also reported.

2.2.1 Part 1: Evaluation of the Relevance of the Technologies (DS/DVE) Today and in the Future

Part 1 of the survey was intended to address the relevance of DS/DVE technologies today and in the future. Its intention was to assess the relevance of these technologies for practical applications. Furthermore, these questions attempt to identify those fields where the experts expect these technologies to have the highest economical potential and to identify the kind of applications that will derive the most benefit.

Question 1.1: Please rate the future relevance of the following potential applications of the DS/DVE technologies for improving internal processes within companies (including their suppliers) or other organizations. Please give grades in the range from (5) = highest relevance to (0) = no relevance!

Question 1.1 queried the future relevance of DS/DVE technologies to potential applications focusing on the *process side* of companies and other organizations. The questions suggested some examples for potential applications of DS/DVE technologies for improving processes and asked the experts to rate them. The results arc shown in Fig. 3.

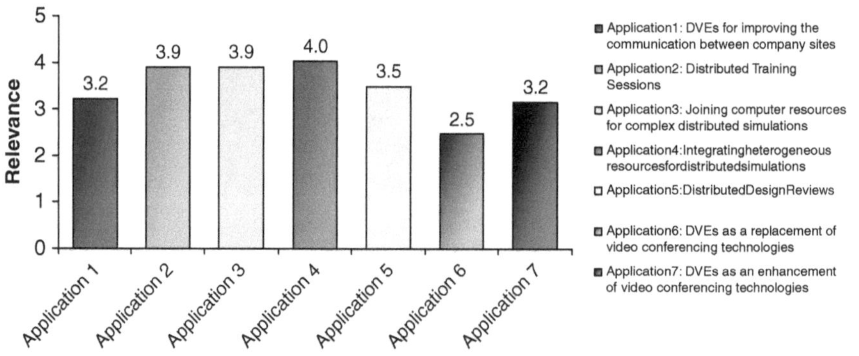

Fig. 3 Potential applications for process improvements

The diagram shows that from the suggested choices applications 2, 3, and 4 are attributed the highest relevance. The highest rated application (no. 4) suggests the application of distributed simulation to integrate heterogeneous resources and is followed by no. 3 that involves the application of distributed simulation to join computer resources for complex distributed simulations. Both types of scenarios can be anticipated to play an important role in companies that have become sufficiently complex that they require the integration of simulation applications spanning several geographically distinct locations.

Application no. 2 which involves distributed training sessions is also assigned a high future relevance. This is easily understandable, since distributed virtual training sessions are a straightforward means of joining geographically distributed experts to conduct training or to provide remote assistance.

These three applications are followed by application 5 ("Distributed Design Reviews") which receives a ranking of 3.5 indicating that many participants still consider this a relevant topic for the future.

The usage of DVEs as a replacement of video conferencing technologies is not considered as being highly relevant for the future; however, DVEs are attributed some relevance as an enhancement of video conferencing technologies. The future could therefore bring about some mixed form of DVEs and video conferencing, e.g., by combining the traditional video feed with some interactive 3D view of the items to be discussed, or by integrating the video feed into the virtual world.

Question 1.2: In which areas do you see additional relevant applications of the technologies for improving internal processes within companies (including their suppliers) or other organizations?

Question 1.2 was designed as an open-ended question and queried for any additional relevant applications for improving processes within organizations. The answers have been clustered into the four categories: integration aspects, product development, production, and miscellaneous. The following briefly summarizes the responses.

1. Integration aspects
   - Distributed simulation (DS) allows geographic separation of simulation tools (and resources) from the points where they are needed.
   - Different company locations and their suppliers can join their simulation resources and thus improve their cooperative processes (e.g., supply chains).
   - Distributed Design (not only Distributed Design Reviews) but the entire design process is an application for both DS and DVE.
   - DVEs can enable better communication between various project participants (manager, architects, designer, end-user, ...).
2. Product development
   - Remote product testing, e.g., in settings where the product simulator is at one location different from the product testers, can be an important future application of DS/DVEs. This applies for both product development as well as for product marketing.
   - Knowledge protection in multiple-component product models can be assured with distributed simulation technology. This is highly relevant for joint product

development where multiple companies develop individual components of products. Typically, companies want information concerning their component to be protected while still allowing it to be used in conducting simulations of the entire product. Distributed simulation offers the potential for companies to protect their intellectual property by providing their component behavior as a black box model that can be integrated in a distributed simulation of the overall product.
- Virtual assembly/product integration before physical prototypes exist.

3. Production
   - Distributed planning of manufacturing lines (joint planning of suppliers and OEMs).
   - Direct process control of automated production processes (simulation-based Command&Control center).
4. Miscellaneous
   - Distributed Virtual Environments as market places for subcontractors on the Internet.
   - 3D web: Current 2D web will be combined with DVE capabilities (e.g., as known from applications such as Second Life).
   - Real-time decision making using DS/DVE.
   - Analysis and feedback for developing interpersonal skills for management and teamwork.
   - Changing current business processes to achieve greater orientation around end-user needs.

Question 1.3: How do you rate the relevance of the technologies distributed simulation and distributed virtual environments for improving the life cycle of future products (e.g., for the product development, the product operation, or product maintenance)?

While the first two questions focused on the process side of organizations, question 1.3 queried for the relevance of DS/DVE technologies in improving the life cycle of future *products*. The participants were asked to rate relevance on a scale from 0 (none) to 5 (very high).

The mean value of all answers is 3.9, indicating high relevance. This question was answered by 98% of the participants. The standard deviation of all answers is rather low (0.9), indicating a high level of agreement among the participants.

The comments entered by the majority of the participants indicate that they can envision DS/DVE technologies in many areas of the life cycle of future products, including the design, testing, acquisition, training, and maintenance of products. A special interest is attributed to the product development and testing phases of the product life cycle.

Arguments given in favor of this statement include the general trend towards globalization. As products are often composed of parts developed and manufactured by multiple enterprises, integrating simulation models of such parts is a key point to produce realistic simulations of the entire product. This obviously applies especially to very complex products in high tech industries such as automotive, aircraft, and aerospace.

Several comments decidedly attribute DVE technology a high relevance for distributed design reviews in globally distributed enterprises, a finding already confirmed by question 1.1.

Several people also indicated their conviction that distributed virtual training technologies will play an important future role for the operation and maintenance life cycle phase of a product.

A few people expressed their doubts that DS/DVE technologies will see use in applications that go beyond usage in a specific product life cycle phase. This argumentation is quite reasonable when considering the current state-of-the-art in existing base technologies and IT tools. There certainly is a lack of easy-to-use and easy-to-adapt standards for such solutions. Only if such standards will be created and accepted by leading tool vendors and an accompanying methodology is established can the usage of DS/DVE technologies become commonplace during the entire life cycle of future products.

Question 1.4: How would you rate the current adoption of the technologies in industry and defense?

This question attempted to capture the *current* relevance of DS/DVE technologies. The experts were queried for their opinion concerning the current adoption (i.e., the practical application) of these technologies in industry and in defense. As a majority of the experts work in research institutions, their answers can, of course, only constitute a somewhat subjective judgment of practical adoption. Still the responses provide an impression about the ratio by which industry and defense differ in their usage of DS/DVE technologies.

Fig. 4 shows the mean values of the responses concerning DS and DVE technology adoption in each sector (industry/defense), respectively. As expected, it can be seen that the military domain already makes good use (between medium and high) of distributed simulation and distributed virtual environments. The reasons for this are quite obvious as military training and acquisition are typical applications which are expected to rely highly on DS/DVE technologies. Also, domain standards such as HLA and DIS originate in the defense community.

Even though there is already some good adoption of DS/DVE technologies in the defense community, it is interesting that the degree of adoption in military applications is not rated between "very high" and "high". This may indicate the existence of technological barriers (thus indicating more research is needed), as well as organizational issues that prevent more widespread usage. This is supported by some respondents who indicated that HLA—which should constitute the leading edge of standards in the military DS/DVE domain—is said to have suffered from some degree of fragmentation in the US Department of Defense, the original creator of HLA, resulting in the creation of additional standards (e.g., TENA). It is certainly

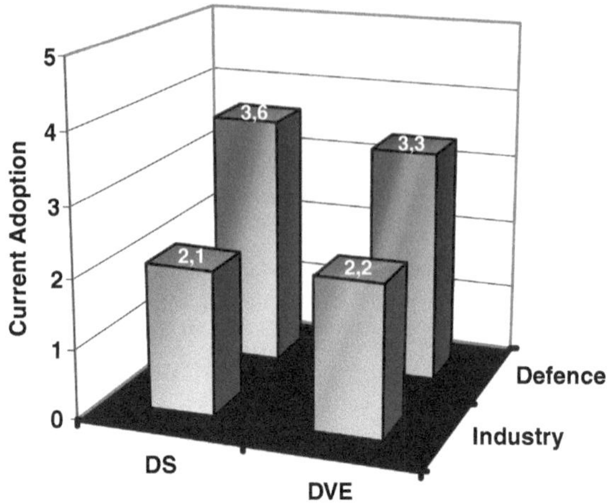

Fig. 4 Current adoption of DS/DVE technologies in industry and defense

true that even in defense today, use of DS/DVE technologies has not yet reached its fullest potential.

The low usage of DS/DVE in industry, combined with the answers from questions 1.1 through 1.3 suggest that industrial usage may be limited by a lack of good technical solutions and the need for more basic research in this area. It may, for instance, indicate that existing solutions are focused on the needs of the defense community and may not sufficiently take into account requirements in industry.

Some participants also indicated that DS/DVE technologies are missing recognition in industry. This is not only due to technical reasons which may prevent the usage, but also a problem of acceptance and appreciation. Prejudices exist in industry concerning cost as well as the invasive and disruptive character of DVEs. These problems must be overcome, along with the issue of establishing clear business cases to cover the return on investment (ROI).

This lies in line with comments on needed improvements towards technical issues such as the need for seamless integration of DS/DVE with existing industrial IT infrastructures and their applications and processes.

Question 1.5: Which economical potential do you see in the technologies? Please give us your opinion which areas might have the highest economical potential.

Question 1.5 attempts to estimate the economical potential of DS/DVE technologies. The answers cannot provide any quantitative prediction; instead, they can only give a qualitative assessment across the opinions of the respondents.

This average value for the economic potential is 3.7. Considering our scale from 0 to 5 which rates 3 as “medium” and 4 as “high”, this is a rather high rating. This confirms that our experts believe that DS/DVE technologies do have a significant economical potential. The standard deviation for this value is low (0.98) compared to that of other questions. This indicates good agreement among the participants on this issue.

The experts were also asked to identify areas where they expect the highest economic potential. Many participants indicated the defense sector as the one with the highest economic potential. Applications here include mission training and rehearsal, decision support, and technology acquisition. This rating is obvious, as the defense sector is already aligned with these technologies and the military constitutes the world's largest contractor in the simulation market.

Besides defense, many participants identified the manufacturing sector as well as product development as areas with high economic potential. Furthermore, the entire area of distance learning technologies (which includes advanced distributed training solutions) is mentioned by many participants as having high economic potential. Several participants also identified new emerging markets with an increasing economic potential. These markets include the areas of emergency and security management, homeland security, and global environmental problems. Further responses include the gaming industry and consumer applications as well as product marketing.

A few participants also argued that the economic potential is limited because of the expense related to using these technologies. This is an issue that certainly needs to be addressed, especially if the technologies are to be used by small companies that cannot afford high investments in hardware or software.

Question 1.6: Which future developments do you expect in the cooperation between OEMs and their suppliers which could make the application of technologies like distributed simulation and distributed virtual environments inevitable? Which other technologies might be required?

This question attempted to stir up thinking in non-technical directions in order to identify societal or industrial trends which could influence DS/DVE technology usage. Some participants commented on the phrasing of the question, especially the word "inevitable" which has a rather strict meaning. This phrasing was chosen on purpose to encourage thinking in controversial directions. One sample train of thought envisioned during the setup of the questionnaire was along the lines that "if the future shortage of fuel and fossil resources will lead to less travel", then "there will be an absolute necessity for more virtual cooperation in DVEs."

Not all responses in fact commented on future developments in the cooperation between OEMs and their suppliers. Those which have done so can be clustered into the categories "business environment" and "success stories."

The category "business environment" includes three main subcategories:

- Globalization: Economic incentives and increasing competition has led to the forming of industrial clusters (aka globalization). This most obviously requires increased cooperation among OEMs and builds a demand for DS/DVE technologies.

- IPR protection: With globalization and increasing cooperation among companies comes an increased need for the protection of intellectual property rights (IPR). This will increase the need for secure component-based distributed simulation in which companies can join their component models without revealing their internal knowledge and expertise.
- Try-before-buy: There will be an increased usage of the try-before-buy approach, i.e., selecting the right sub-components for integration in the final product from a given range of options by investigating the component's functionality and interplay with the final product using DS/DVE technologies.

The answers in the cluster "success stories" indicate quite consistently that such stories are needed to overcome psychological barriers. Also, convincing pilot applications are needed to demonstrate the positive effects of these technologies. It was further commented that technological advances must be triggered by a customer demand.

Most respondents did not in fact comment on the future developments in the cooperation between OEMs and suppliers, but commented on technical factors which will positively influence the adoption of DS/DVE. Again, these answers have been clustered into two categories, "ready and robust solutions" and "technological advances."

In the cluster "ready and robust solutions," several respondents commented that reliable standards are indispensible for a more widespread application of DS/DVE technologies by OEMs and their suppliers. Furthermore the issue of semantic interoperability must be addressed, at least within a given application domain. This requires standardized ontologies which provide out-of-the-box semantic interoperability, not only syntactic connectivity which can already be achieved today. Also, *trustworthy* tools which address these issues are needed.

In the cluster of technological advances, several commented on the need for secure high bandwidth and fast network/communication technologies to enable OEM-supplier cooperation based on DS/DVE technology. In addition, the adoption and the emulation of game technologies were suggested to simplify the use and to reduce the entry barriers which today prevent people from applying those technologies.

An interesting comment suggested the introduction of an interim technology between traditional 2D (paper) documents and 3D environments in order to help people to become accustomed to the new 3D technologies. Furthermore, many commented on the need to reduce the cost for equipment needed to adopt these technologies.

Question 1.7: We are interested in your opinion about the distributed virtual online community "Second Life." Q1.7.1: Will the concepts applied there get any industrial relevance? Q1.7.2: Which weaknesses do you see (technical, conceptual …)?

The virtual online community Second Life[3] has received significant media coverage in the past and can be considered a trendy end-user version of a DVE. Second Life (SL) is available free of charge to any interested end-user. The creator, Linden Lab, provides client software which is available for Windows, Linux, and Mac operating systems. Second Life's vision is to create a second reality in which real users (represented by avatars) can interact in a persistent three-dimensional world.

The number of registered SL users is in the range of 11.5 million, but it has to be noted that only a fraction of this number is regularly active in SL. Users in SL can buy "land" in the virtual world in order to create their own "properties." The creation of 3D geometries for this property can be done using primitives (cubes, spheres, cones, ...) and simple tools provided by SL. On the technical side, SL is hosted by a powerful central server pool comprising several thousands of computers. Each server performs the physics simulation (collisions, interactions,...) of a dedicated region of SL's virtual world whereas the clients simply visualize the server data. Assets (e.g., 3D objects) created by users are stored in a separate server farm, currently comprising 24 terabytes of storage capacity.

Question 1.7 of the survey tries to look behind the hype and questions the suitability of the concepts behind SL for serious industrial applications (Q1.7.1) and identifies its weaknesses (Q1.7.2).

The responses towards the industrial relevance of SL are (with very few exceptions) quite consistent. The consensus is that its current and future industrial relevance lies mainly in areas such as advertising, marketing, and entertainment. Its potential for improving communications is specifically considered relevant for the communication between companies and their customers. Only very few respondents see potential for using SL for improving communication *within* companies.

More-serious industrial applications are questioned by many respondents. This unsuitability is attributed to technical as well as to conceptual issues. On the technical side it was pointed out that fidelity and resolution in SL are not appropriate for "serious" applications. SL is considered less effective than video conferences and not effectively usable for collaborative design and development efforts.

Specific conceptual problems are seen concerning privacy issues. SL provides too little user verification and options for separating serious usage from personal activities. One participant answered exemplarily that "you cannot keep naked avatars out of your business meeting." However, it was also stated that other DVE products are emerging which may better address the needs of business users.

Question 1.7.2 elaborated on opinions towards the technical and conceptual weaknesses in SL. On the technical side, several again cited a lack of fidelity and resolution as well as missing security mechanisms. This includes missing user verification and control.

[3] Second Life is developed and distributed by the company Linden Lab. For more information please refer to http://secondlife.com/

Further, in this line the issue of scalability and bandwidth requirements and the lack of compressing technologies are considered as technical weaknesses. The scalability issue arises from the central server pool in which each server performs the physics simulation for a certain region in Second Life. This limits the number of clients which can concurrently stay in a region.

In addition to these issues the respondents mentioned the limited graphics capabilities of SL as a problem. This includes the inability to build content with standard 3D tools and the missing ability to import content created with such tools into SL.

Several people also criticized the fact that SL is not based on open standards and has closed interfaces. This limits the possibility to integrate it with other industrial applications. Also, the available protocols and scripting capabilities, e.g., for component motion, were considered insufficient by some respondents.

On the conceptual side the interactions of the kind enabled by SL are most beneficial when done in real time between participants (like a telephone call). However, synchronizing participation across global time zones makes this difficult (imagine a conference call between Japan, the USA, and Germany—who stays up late and who comes in early?). This is a problem which is valid for any DVE-based communcation.

Specific critiques on the conceptual side of SL include its unfocused character—it provides no appropriate tools for (say) engineering tasks, or any other serious industrial task.

Another noted problem addresses the relation between the virtual money in SL and real money. This relation depends on a single company which makes it a difficult base for conducting real business.

In summary, the majority of the survey respondents consider SL as a DVE for entertainment and social interaction (for which it in fact was created). Its value for industry is mostly limited to company presentations and marketing purposes. Conceptual and technical issues currently prevent more-serious industrial usage.

Question 1.8: Could you name a potential non-military "Killer Application," i.e., an application which obviously requires DS or DVE technologies to implement it, and which has a significant practical relevance?

This question attempted to identify potential "breakthrough" applications which would push the DS/DVE market forward. Answers were restricted to non-military domains, since the military domain already has some very convincing use cases for those technologies.

The answers have been clustered into the following categories:

1. Decision Support Systems for homeland security/catastrophes/crisis situations: Many answers of the participants fell into this category. The envisioned decision support systems are complex and networked IT systems which provide the operator/decision maker with simulation support for such crisis simulations. Possible application scenarios include the simulation of the effects of a crisis (e.g., on complex critical infrastructures) and potential countermeasures, e.g.,

a module for the virtual training (and possibly remote assistance) for rescue teams.

As this simulation capability is required *on demand* this creates the need for powerful simulation capabilities (possibly on massively parallel architectures) and the integration of heterogeneous components for the simulation (live data feeds, data bases, geographical information systems, and multiple command and control centers).

2. Virtual Training Applications: As virtual training applications are quite successful applications for DVEs in the military domain it is obvious to look for applications of this technology in industrial domains as well. Nominations in this category suggest virtual training in truly geographically and internationally distributed contexts (e.g., for the International Space Station, ISS) and training applications which combine one or multiple users with real and simulated equipment.
3. Space Exploration: Nominations in this category suggest the application of DS/DVE as a Command and Control tool for remote operations, especially in space exploration missions. With these technologies it becomes possible to send humans virtually where they cannot go physically.
4. Virtual Meetings: As this is one of the most obvious DVE applications several people nominated this as a killer application. Such virtual meetings are suggested for technological development and design teams, for project progress meetings, as well as for social interaction and entertainment.
5. Industrial Supply Chain Simulation: The necessity for distributed simulation applications for global supply chain simulation and optimization is motivated by the need for know-how protection. As the individual members of a supply chain typically do not want to reveal knowledge about their internal processes to the others, a traditional (monolithic) supply chain simulation model cannot be built. DS provides a means for each participant in the supply chain to submit their supply chain node model as a black box into the overall distributed supply chain simulation.

The following nominations have been suggested by very few (or even single) users, but they are included here as they suggest some interesting applications apart from the mainstream.

6. Emulation: Applications in this area require the coupling of real equipment with simulated parts of reality. The purpose here is typically to test the equipment. Complex application scenarios contain multiple real controllers which are connected to a DVE allowing testing by multiple people.
7. Virtual Travel at Street Level: This nomination is a derivation of existing DVE concepts like Second Life. Here, it is suggested to create a DVE which models the *real earth*. In the DVE one could then meet and travel in a synthetic 3D environment as realistic as the images from Google Earth or Google Street View.
8. Real Estate and Home Design: This nomination suggests the usage of DVE technology to enable customers to visualize and virtually enter future houses in the actual environment in which they will be built. It is argued that if a potential buyer of an existing property could visualize the (future) house within its actual

environment or the potential buyer of a new house could "see" it and use VR to select colors, designs, styles, and landscaping, the resulting increase in sales would be significant.

9. Cultural Education: This nomination suggests creating history and art museums as a DVE in which artifacts are displayed in the real context in which they existed. An alternative may combine traditional museums with real objects with an artificial environment generating distributed mixed reality systems.
10. Sales Activities: Virtual shopping malls in which 3D products with dynamic properties can be tested are suggested as a future sales instrument. In the case of car sales this may, for instance, enable the customer to virtually enter the car, view it from all angles, change its configuration, and even drive it on any road in the world.

2.2.2 Part 2: Research Challenges and Trends

Whereas the first part of the survey focused on the relevance of DS/DVE technologies for practical applications, the second part analyzes open research challenges in this area as well as current and possible future trends.

Question 2.1: Which research activities are you and your institution currently conducting in the field of distributed simulation and distributed virtual environments?

This question was intended to identify topics that are currently under investigation by researchers active in the DS/DVE fields. The answers can be clustered into the categories "application areas for DS/DVE technologies," "research in base technologies," and "interdisciplinary activities." Some exemplary and typical nominations are mentioned below for each of the categories.

Participants who are working in application areas for DS/DVE technologies mentioned homeland security, emergency management, manufacturing and logistics, military simulation (training, weapons), and complex technical or natural systems (particle, material, climate) as application areas in which they are actively applying DS or DVE approaches.

Those who conduct research towards developing base technologies perform research in the following exemplary areas:

- effects of wide area network latency of real-time and interactive distributed simulations
- combination of discrete-event simulations with DS and DVE
- fundamental interoperability mechanisms
- synchronization algorithms
- distributed haptic DVEs

Interdisciplinary activities which are addressed by a number of participants include the integration of game technology with advanced simulation technologies (e.g., to leverage the strengths of each), the work on simulation-based Command and Control (C2) systems, as well as agent-based approaches for decision making in DS/DVE.

Question 2.2: What would you consider as the specific strengths/unique selling points of your institution in the context of the fields under investigation?

This question has been included for reasons of identifying success factors for research in DS/DVE. As this question provided highly individualized answers, they cannot be summarized in greater detail here.

Something that became obvious is that the interdisciplinary character of the research activities is often considered a specific strength, either through interdisciplinary research groups or through close cooperation with industry.

Question 2.3: How would you rate the maturity and practical relevance of the following standards/protocols?

This question asked for a rating of the maturity and the practical relevance of selected standards and protocols which each play an important role in the DS/DVE market. Fig. 5 lists the mean values determined from the survey responses. (See the Appendix for a list of the various protocols.)

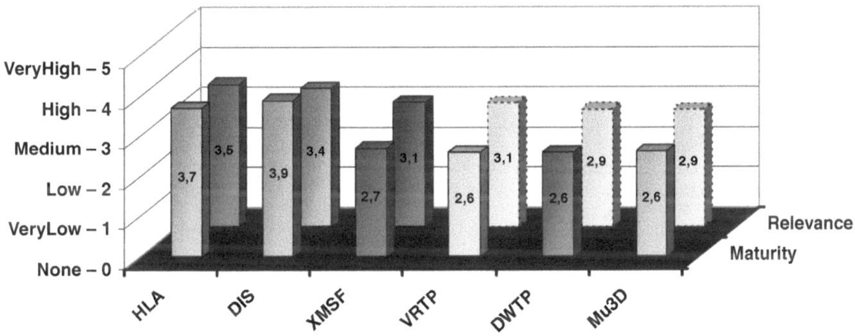

Fig. 5 Maturity and relevance of DS/DVE standards and protocols

While the answers concerning the maturity of the standard can be considered quite objective, the values for the practical relevance are only significant for HLA, DIS, and (to a certain degree) XMSF. This is due to the fact that the sample size varies quite significantly in the rating of the standards/protocols, i.e., if a standard was not known to a participant, then it did not receive a rating. HLA and DIS are known and have been rated by more than 80% of the participants. XMSF was still known by approximately 50% of the participants.

VRTP, DWTP, and Mu3D on the other hand were only known by a minority of the participants (VRTP: 34%, DWTP: 21%, Mu3D: 18%). Therefore the ratings and statements concerning the practical relevance of these protocols may be questionable.

The main conclusions which can be drawn from the results shown in Fig. 5 are that HLA and DIS are the leading standards in the DS/DVE sector and that both already have a rather high maturity. On the other hand, their practical relevance is rated between medium and high (3.5 and 3.4 respectively), a value which is relatively high, but might be expected to be even higher considering that both standards have been on the market for more than 10 years (HLA) or 15 years (DIS). The other standards, especially VRTP, DWTP, and Mu3D, are attributed a rather low maturity and in fact only have a minor practical relevance (as explained above).

For XMSF, the evaluation of the rating and the comments indicates that while the concepts of XMSF did have quite a good recognition in the community, XMSF itself, however, does not seem to be supported any longer as an ongoing activity.

The participants were also asked to comment on the specific weaknesses which they would attribute to the standards and protocols. The following lists the weaknesses attributed to HLA and DIS.

Weaknesses of HLA:

- No load-balancing as part of the standard
- Poor scalability
- Too much reliance on peer-to-peer structures; large DVEs may be better with client-server structures where multiple servers are peers of each other
- Covers only syntactic interoperability (not semantic)
- Standard is too "heavy," i.e., very complex, difficult to learn, and thus time-consuming to adopt and use

Weaknesses of DIS:

- PDU broadcast concept allows no interest management (publish/subscribe) and no load balancing
- Limited conceptual versatility (i.e., only applicable to real-time simulations)
- Restriction to a single domain (military training simulations)
- Limitations of the standard lead to proprietary modifications and custom implementations that do not allow re-use outside original application

In conclusion, the most critical issue concerning DIS is its limitation towards a certain niche of the simulation market (real-time, mostly military training applications) and its broadcast mechanism. HLA as the current state-of-the-art standard is attributed to at least the perception of limited scalability and the fact that it only addresses the syntactic, but not the semantic, interoperability issues.

Question 2.4: How would you rate the maturity of the following underlying base technologies needed to implement DS/DVE applications and their significance for advancing the fields of DS/DVE?

This question is an attempt to identify the most promising research areas concerning base technologies for DS/DVE applications. To do so the survey participants were asked to rate the current maturity of certain base technologies as well as their significance for advancing the fields of DS/DVE. The results are displayed in Fig. 6.

To interpret the diagram with the objective of identifying the interesting and promising research areas, one should look for the base technologies which currently have the lowest degree of maturity, but still have a high rating with regard

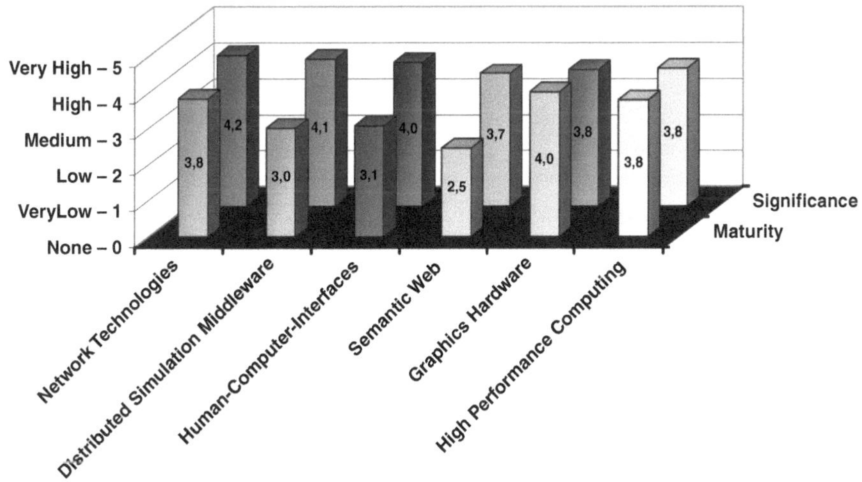

Fig. 6 Maturity and significance of base technologies

to significance for advancing the fields of DS/DVE. In this sense, this indicates that distributed simulation middleware, human-computer-interfaces, and the semantic web (as a placeholder for approaches supporting semantic interoperability) are the fields rated as the most promising areas of research.

On the other hand, as the maturity rating of graphics hardware is already high, there are no breakthroughs in this area to be expected, unless some completely new paradigm develops (e.g., "no triangles needed any more"). A similar observation applies to the area of network technologies.

Besides these statistics it is also interesting to look at the comments provided by the respondents concerning the improvements needed within each of the base technologies. Table 1 lists the most important nominations of the respondents.

Question 2.5: How would you rate the overall maturity of the technologies and solutions developed in the fields of distributed simulation and distributed virtual environments?

This question asked for the opinion of the participants about the overall maturity of the technologies and solutions in the DS/DVE areas. The possible answers included the following options:

4 – Very mature and already applied for many practical applications
3 – Mature, but not applied widely yet
2 – In the process of maturing
1 – Technologies exist, but still have significant weaknesses
0 – Academic research/prototypes

Table 1 Required improvements for each of the base technologies for DS/DVE applications

Network technologies	Distributed simulation middleware
• Lower latency and better bandwidth • Robustness and fault tolerance • Security • Quality of Service (QoS) specifications	• Plug-and-Play capabilities • Standardization (also: interoperability between different standards) • Semantic connectivity • Ubiquity (accessible anywhere with any device)
Human-computer-interfaces	**Semantic web**
• Multimodal interfaces, including haptics, without data gloves • More immersive environments • Focus change to enhanced reality instead of virtual reality required • Usability improvements • Human-centered interfaces	• More mature ways than current ontologies for defining semantics (ontologies cannot be the only answer) • Standardization of terms of reference for certain domains • Ways to transform current know-how stored in the WWW into a semantic web knowledge
Graphics hardware	**High-performance computing**
• Better physics integration • Promotion of standardization	• Involvement of all heterogeneous nodes into a universal grid • Demonstration of application potential to broader community • Tools to use high-performance computing in engineering software environments

The mean value of all the responses according to this scale was 2.1. Our interpretation for this value is that DS/DVE technologies have been around for some time; however, there are still weaknesses and technological issues which need to be resolved and more basic research is needed to bring them to a widespread and cost-efficient usage.

Some comments provided by the participants indicate that there are no standardized products and off-the-shelf solutions available upon which to build. A few have also commented that there are some proprietary application fields (especially military) within which technologies are already quite mature.

Question 2.6: Which research challenges in the research fields (DS/DVE) would you consider "Grand Challenges" which must be solved in order to advance the field significantly?

This question tries to identify the "Grand Challenges" which, if solved, will advance the fields of DS/DVE significantly. Grand Challenges are generally considered very complex problems in a certain research field, for which currently no solution exists. On the other hand, a problem for which it is known (or generally believed) that no such solution can ever exist (e.g., a problem which a mathematical proof shows to be unsolvable) cannot be considered a Grand Challenge.[4]

The answers provided by the respondents of the survey do not all qualify as Grand Challenges; however, the most-interesting and appealing problems are listed and briefly discussed in the following.

1. Solving the intrinsic conflict between the desire for high interactivity/response times and the need for maintaining consistency in DVEs

   A well-known problem of DVEs is the need to maintain consistency among all users. Consider, for instance, a situation in which a user tries to grab a virtual part in his local representation of the DVE. To make sure that he can in fact grab it and that no other user can do so at the same time, the DVE would need to implement some locking mechanism which must synchronize with all participants before the user can actually grab the part. This would, in the worst case, cause a delay of two times the travel time which a message needs on the network between this participant and the participant with the slowest network connection. This is in strong conflict with the objective of giving the user a fast response time, i.e., to be realistic there should be no noticeable delay between his intention of grabbing the part and the action taking place.

2. Easy-to-use synchronization algorithms which idealistically solve the "zero lookahead problem"

   Efficient synchronization algorithms are the key factor for any parallel or distributed simulation. The topic of synchronization algorithms has been on the research agenda for quite a number of years. Several protocols (conservative, optimistic, hybrid) have been developed in the past which all work better or worse depending on the type of application. The implementation and usage of optimistic protocols is quite complicated and cannot be done easily in any commercial off-the-shelf simulation software. Conservative protocols, on the other hand, are rather simple to use. However, their performance depends highly on how much lookahead[5] can be extracted from the participating simulations (the more the better). Often the participating simulations of the DS are so closely interconnected that a lookahead value of zero is required. In that case the conservative protocols lead to a serialization of the entire distributed simulation with the obvious severe performance implications. This is the intrinsic problem that needs to be solved in this Grand Challenge.

---

[4] This excludes futuristic visions—like the "Holodeck" known from the science fiction show "Star Trek"—from the list, as it would require quantum leaps in physics which no one truly believes to be feasible.

[5] Lookahead is a well-known term in the DS area which refers to the amount of time which a simulation can look into its future. It is a guarantee of how far ahead of time the simulation will generate any messages/events for other simulations.

3. True plug-and-play simulation capabilities

   So far today no standard (not even HLA) has enabled simulation packages to be coupled in a true plug-and-play fashion. What is needed is a standard approach to couple the distributed models and gain acceptance for this standard by industry (simulation tool vendors as well as end-users). Only this can lay the basis for effortless integration of distributed and independently developed simulators. This also requires approaches for interoperability between multi-level resolution models, i.e., models which operate on different levels of detail.

4. Automatic or semi-automatic semantic interoperability between domains (ontologies, standard reference models, metamodels)

   This Grand Challenge could be considered a sub-challenge of the previous one. In order to achieve plug-and-play interoperability between simulations, some methodology for gaining semantic interoperability between domains (or even within a domain) must be established. So far, standards like HLA have mainly addressed the syntactic interoperability between simulation systems.

5. DVEs of the future

   Several suggestions have been made towards requirements for the DVE of the future. They include:

   - Improvement of network performance to allow realistic interaction between attendees of a DVE meeting, plus new technologies to allow the interaction with the environment realistically in terms of Computer-Human-Interaction.
   - Realistic real-time visualization with full account of underlying physics and integration of voice and sounds
   - Living Dynamic Worlds, i.e., the creation of a world that is constantly active and evolving, even if there are no human players participating
   - Use of city-sized large-scale mobile nodes in reasonable speed
   - Platform and information/data handling technologies to support multi-user, multi-role, multi-viewpoint simulations
   - Unification of discrete and continuous simulation theory and practice

   While on the first view this suggestion of a Grand Challenge does not seem to be directly related to the DS/DVE fields, a solution to this issue would, of course, also have significant impact on the way distributed simulations could be treated.

Question 2.7: Which findings and results would you expect from an external research group in order to advance the field significantly?

This question was naturally (and rightfully so) motivated by the attempt of the Fraunhofer IFF as the initiator of this study to determine the expectations of the research center which it is about to establish. The answers to this question apply for any research group that wants to make a “difference,” i.e., to advance the field significantly.

First, the community obviously expects solutions to the Grand Challenges introduced in the previous section. Second, the community expects the *unification* of research, development, utilization, and education in DS/DVE.

The setting of a research and development agenda of research for short and long-term goals in these areas is expected. This includes the identification of important trends, the creation of a forum for prime players to interact and collaborate, the definition of reference models, and the definition of standard approaches.

Finally, the research group is also expected to expand the predominant application fields of DS/DVE to industry, design, manufacturing, and the consumer sector by demonstrating how research translates into real use and by integrating its findings with a variety of needs in industry, manufacturing, health care, security, environment, and education.

Question 2.8: Would you consider collaboration in distributed virtual environments a viable topic for the future or will personal meetings with physical attendance of all participants always be preferable?

This question was intended to validate whether or not the participants considered DVE-based collaboration viable at all. The question was answered by more than 90% of the participants. Among those who answered, 95% considered collaboration in DVEs as a viable topic for the future. This confirms again that research in this area continues to have a high relevance in the future.

With regard to the second part of this question, statistics were collected concerning whether meetings with physical attendance of the participants would always be preferable. Seventyfour percent answered that at least *sometimes* physical meetings are indeed preferable. Only 23% insisted that personal meetings are *always* preferable.

Question 2.9: Can you name interesting trends, solutions, and actors in the areas of DS/DVE which you would consider drivers in these fields?

This last main question of the survey gave the participants the opportunity to name anything which they considered drivers influencing the future of DS/DVE. The answers were structured into trends, solutions, and actors and are reported accordingly.

- Trends
  - Increasing popularity of personal computing devices
  - Service-oriented architectures
  - Ambient networks
  - Open source solutions
  - Rising importance of homeland security and critical infrastructure protection
  - Ubiquity of visual media
  - Expectation of instant easy communication (cell phone, I-pod, email, ...)
  - Augmented reality systems
  - Introduction of haptic and other multimodal interfaces

- Solutions
  - SISO draft standard for simulation package interoperability
  - German Armed Forces' Simulation & Test Environment (SuT Bw)
  - VBS2.com (Virtual Battlespace 2)
  - HLA (High Level Architecture for Modeling and Simulation)
  - MDA (Model Driven Architecture)
  - DEVS (Discrete Event System Specification)
- Actors/Participants
  - Gaming industry
  - Defense agencies (US DOD, German Armed Forces' IT-Agency)
  - SISO (Simulation Interoperability Standards Organization)
  - Marketing decision makers

3 Summary and Conclusions

This study has investigated future trends in the fields of distributed simulation and distributed virtual environments. It was designed as a peer study taking into account the opinions of more than 60 leading experts in the fields under investigation. This has been accomplished in the forms of executing a survey, conducting interviews with selected experts, and evaluation of related publications.

The survey results show that both DS and DVE are characterized as having a high practical relevance for improving both the processes within organizations and the overall product life cycle of future products. The greatest practical relevance is considered to be in the areas of joining and integrating (possibly heterogeneous) computer resources for conducting complex distributed simulations as well as in the execution of distributed training sessions.

Important applications are also considered to be in the areas of production planning and control, product development, and the general integration of geographically distributed computing resources for stakeholders. DS/DVE technologies are also attributed to having a considerable economic potential.

The survey indicates that the current adoption of DS/DVE technologies in industry today is limited. While the defense sector already makes better usage of those technologies, a lower industrial usage may be attributed to the need to articulate a clear business case for the adoption of the technologies. Although there is a high economic potential and a high practical relevance of certain applications, there are a limited number of success stories and publications articulating the return on investment in using these technologies.

On the other hand, technological immaturities in these technologies exist and have been reported in this study, preventing widespread usage of both technologies. These immaturities help explain the different levels of usage of DS/DVE in industry vs. defense: Because the existing solutions and standards are focused on the needs of the defense community, they may not take into account commercial requirements to a sufficient degree. This is in large part confirmed by the study, e.g., by revealing that there are no plug-and-play capable standards for industrial usage of DS/DVE or that there is no automatic interoperability between domains because of the lack of semantic interoperability.

As this is the case, any industrial usage must overcome the need to perform a costly integration of the needed tools, a fact that prevents a more widespread usage even if the practical application (e.g., a distributed design review) exists. That there still are significant technical and conceptual weaknesses is also confirmed by the assessment of the study that DS/DVE technologies are generally considered to still be "in the process of maturing."

Further, certain Grand Challenges have been identified that will, if solved, significantly leverage and stimulate the usage of DS/DVE. These include

- solutions to the intrinsic conflict in DVEs between the desire for high interactivity and the need to maintain consistency
- easy-to-use synchronization algorithms for DS which solve the "zero lookahead" problem
- true plug-and-play simulation capabilities
- (semi-) automatic semantic interoperability between domains

This study has also revealed that there are, in fact, already some instances of DVEs which are quite successful outside the defense sector. One example of such a DVE is Second Life, used for social interaction and entertainment. Their industrial usage, however, is typically limited to marketing purposes for larger companies. More serious industrial usage is prevented by conceptual and technical weaknesses.

New applications areas that could drive the DS/DVE market forward include areas such as decision support systems for homeland security and crisis management, virtual training applications, space exploration, and virtual meetings.

Some trends which can influence the development of the DS/DVE fields are the increasing popularity of personal computing devices, the existence of ambient networks, as well as the expectation of instant and easy communications. While this may lead to new forms of accessing and using these technologies, this also imposes new research requirements, as solutions and algorithms for the special requirements of this form of usage need to be developed. As an example, algorithms for participating in DVEs under the special requirements and conditions of a mobile phone (low power consumption profile, unreliable communication, limited display size, easy interaction mechanisms) would be required.

As for the industries and participants which will drive future innovation in these fields, it is certain that the defense and gaming industries will have leading roles. The gaming industry in general already has some very good proprietary solutions for implementing DVEs. However, their decision makers have little interest in revealing their solutions or in standardization efforts. On the contrary, the quality of their solutions contributes largely to the success of their products and is thus almost always considered as proprietary intellectual property.

On the other hand, this study also shows a growing interest and need for these technologies in other industries. Especially any high-tech industry (e.g., the automotive and aeronautics industries, as well as manufacturing) will see an increasing demand for their application. This increasing demand is directly derived from the trend toward globalization. As products are often composed of parts developed and manufactured by multiple enterprises, DS and DVE technologies will make significant contributions in product development as well as in production.

Based on this growing interest, standardization bodies such as the Simulation Interoperability Standards Organization through its product development groups (PDGs) have begun to develop standardized solutions for enabling better DS/DVE interoperability in selected non-military applications. Examples include the Commercial Off-the-Shelf Simulation Package Interoperability (CSPI) and the Core Manufacturing Simulation Data PDGs.

References and Selected Bibliography

Broll, W., and D. Schick. DWTP—A basis for networked VR on the Internet. VRML 98: Third Symposium on the Virtual Reality Modeling Language.

Brutzman, D. nd. Virtual Reality Transfer Protocol (VRTP). Available online via `http://web.nps.navy.mil/~brutzman/vrtp` [accessed December 14, 2007].

Fujimoto, R. 2000. *Parallel and Distributed Simulation Systems*. Wiley Interscience.

Galli, R., and Y. Luo. 2000. Mu3D: a causal consistency protocol for a collaborative VRML editor. In *Proceedings of the 5th Symposium on Virtual Reality Modeling Language (Web3D-VRML)*. Monterey, CA, USA.

IEEE 1516-2000: IEEE Standard for Modeling and Simulation (M&S) High Level Architecture (HLA)—Framework and Rules.

IEEE 1516.1-2000: IEEE Standard for Modeling and Simulation (M&S) High Level Architecture (HLA)—Federate Interface Specification.

IEEE 1516.2-2000: IEEE Standard for Modeling and Simulation (M&S) High Level Architecture (HLA)—Object Model Template (OMT) Specification.

Moves Institute. nd. The Extensible Modeling and Simulation Framework (XMSF). Monterey, CA, USA. Available online via `https://www.movesinstitute.org/xmsf/xmsf.html` [accessed December 14, 2007].

SimSummit Survey on US DoD M&S Management/Leadership August 1, 2007. Available online via `http://www.sim-summit.org/Final1Aug2007.pdf` [accessed December 14, 2007].

Straßburger, S. 2001. *Distributed Simulation Based on the High Level Architecture in Civilian Application Domains*. SCS-Europe BVBA.

Straßburger, S., T. Schulze, and R. Fujimoto. 2008. Future trends in distributed simulation and distributed virtual environments: Results of a peer study. In *Proceedings of the 2008 Winter Simulation Conference*, eds. S. J. Mason, R. R. Hill, L. Mönch, O. Rose, T. Jefferson, and J. W. Fowler, pp. 777–785. Piscataway, New Jersey: Institute of Electrical and Electronics Engineers.

Appendix

Some Terms of Reference—This section defines some general terms as well as standards and protocols used frequently within this study.

DIS: Distributed Interactive Simulation is an IEEE standard for conducting real-time platform-level wargaming simulations across multiple host computers and is used worldwide especially by military organizations.

DS: Distributed Simulation.

DVE: Distributed Virtual Environment.

DWTP: The Distributed Worlds Transfer and Communication Protocol is a protocol for shared virtual environments on the internet (Broll and Schick).

HLA: The High Level Architecture is a general-purpose architecture for distributed computer simulation systems. The creation of HLA was initiated by the U.S. Defense Modeling and Simulation Office with the main objective of supporting interoperability and reuse of simulations. Unlike DIS, HLA provides support for a wide range of simulation applications. Communication between HLA simulations is managed by a runtime infrastructure (RTI). The three key components of HLA (framework and rules (IEEE 1516-2000), federate interface specification (IEEE 1516.1-2000), object model template (IEEE 1516.2-2000)) are standardized in the IEEE 1516 standard series.

IEEE: Institute of Electrical and Electronics Engineers.

Mu3D: The Multi-User 3D protocol is an XML-based protocol for exchanging interaction data in distributed 3D applications. Its core functionality provides a causal consistency protocol for collaborative VRML editors (Galli and Luo 2000).

PADS: Parallel and Distributed Simulation.

PDU: Protocol Data Unit.

RTI: Runtime Infrastructure.

SISO: Simulation Interoperability Standards Organization.

SIW: Simulation Interoperability Workshop.

UDP: User Datagram Protocol.

VRTP: The Virtual Reality Transfer Protocol was an attempt to provide client, server, multicast streaming, and network-monitoring capabilities in support of networked 3D graphics and large-scale virtual environments (Brutzman).

XMSF: The Extensible Modeling and Simulation Framework is a set of Web-based technologies, applied within an extensible framework that enables a new generation of modeling and simulation applications to emerge, develop, and interoperate (The Moves Institute).

Survey on Future Trends in Distributed Simulation and Distributed Virtual Environments

Survey Background

The intention of this survey is to assess the current status in the fields of distributed simulation (DS) and distributed virtual environments (DVE) and to identify new trends and research challenges in these fields. The results of the survey will be summarized and published in a Peer Study.

Participants of the survey will receive a **free copy** of the final version of this study.

The motivation for this survey and study is multifaceted. On one side computer simulation and interactive virtual reality based visualizations have already established themselves as adequate tools in the past. On the other side there is an increasing complexity of both product development and production processes. This requires new methods for planning, evaluating, and controlling the underlying systems.

Technologies like distributed simulation and distributed virtual environments (which are already used frequently in the defence sector) could also be key enablers for addressing the complexity in non-military applications. They can be the basis for simulating complex systems by integrating heterogeneous sub-components which cannot be executed as a monolithic application on one computer. They can connect all involved stakeholders even if they are located on different sites around the world.

The following questions try to collect opinions on the state-of the art, the relevance, and research challenges which must be addressed to advance and strengthen these technologies to a level where they are ready to be applied in day-to-day business.

About the initiators

The study has been initiated by the Fraunhofer IFF in Magdeburg and is conducted by an independent team of researchers, including Prof. Richard Fujimoto (Georgia Tech, Atlanta), Prof. Steffen Strassburger (Technical University of Ilmenau, Germany) and Prof. Thomas Schulze (University of Magdeburg, Germany).

The study is aligned with an attempt to establish an "Innovation and Research Center for Distributed, Interoperable Virtual Reality and Simulation" in Magdeburg, Germany which follows a national German funding scheme. Once established, this center seeks international cooperation with leading scientists in the field.

Personal Data

Please provide the following personal information.

<table>
<tr><td>Name:</td><td rowspan="4">Address:</td></tr>
<tr><td>Position:</td></tr>
<tr><td>Organization:</td></tr>
<tr><td>Type of Organization: - Please Select -</td></tr>
<tr><td colspan="2">Relationship to DS/DVE technologies: - Please Select -</td></tr>
</table>

Please return the completed survey in electronic form to survey@tu-ilmenau.de no later than October 15, 2007.

Part 1: Evaluation of the Relevance of the Technologies (DS/DVE) Today and in Future

Q1.1: Please rate the future relevance of the following potential applications of the DS/DVE technologies for improving internal *processes* within companies (including their suppliers) or other organizations.

Please give grades in the range from (5) = highest relevance
to (0) = no relevance!

Grade	Applications
	Distributed interactive virtual environments which are used for improving the communication between different sites of a company
	Distributed training sessions (e.g., to join geographically distributed trainees and trainers)
	Joining computer resources to conduct complex distributed simulations
	Integrating heterogeneous resources to perform distributed simulations
	Distributed Design Reviews
	Distributed virtual environment as a replacement of video conferencing technologies
	Distributed virtual environment as an *enhancement* of video conferencing technologies

Q1.2: In which areas do you see additional relevant applications of the technologies for improving internal *processes* within companies (including their suppliers) or other organizations?

Q1.3: How do you rate the relevance of the technologies distributed simulation and distributed virtual environments for improving the life cycle of future *products* (e.g. for the product development, the product operation, or product maintenance)?

Very High ☐ High ☐ Medium ☐ Low ☐ Very Low ☐ None ☐

Please explain your selection briefly.

Q1.4: How would you rate the current adoption of the technologies in industry and defence?

Adoption of	In Industry	In Defence
Distributed Simulation	Please Select	Please Select
Distributed Virtual Environments	Please Select	Please Select

Comments (if any).

Q1.5: Which economical potential do you see in the technologies? Please give us your opinion which areas might have the highest economical potential.

Economic Potential: Please Select

Areas:

Q1.6 Which future developments do you expect in the cooperation between OEMs and their suppliers which could make the application of technologies like distributed simulation and distributed virtual environments inevitable? Which other technologies might be required?

Q1.7 We are interested in your opinion about the distributed virtual online community "Second Life".

Q1.7.1 Will the concepts applied there get any industrial relevance?

Q1.7.2 Which weaknesses do you see (technical, conceptual ...)?

Q1.8 Could you name a potential non-military "Killer Application", i.e. an application which obviously requires DS or DVE technologies to implement it, and which has a significant practical relevance?

Part 2: Research Challenges and Trends

Q2.1 Which research activities are you and your institution currently conducting in the field of distributed simulation and distributed virtual environments?

Q2.2 What would you consider as the specific strengths / unique selling points of your institution in the context of the fields under investigation?

Q2.3 How would you rate the maturity and practical relevance of the following standards/protocols?

HLA (High Level Architecture for Modeling and Simulation)
Maturity: - Please Select -
Practical Relevance: - Please Select -
Weaknesses (please describe):

DIS (Distributed Interactive Simulation)
Maturity: - Please Select -
Relevance: - Please Select -
Weaknesses (please describe):

XMSF (Extendible Modeling and Simulation Framework)
Maturity: - Please Select -
Relevance: - Please Select -
Weaknesses (please describe):

VRTP (Virtual Reality Transfer Protocol)
Maturity: - Please Select -
Relevance: - Please Select -
Weaknesses (please describe):

DWTP (Distributed Worlds Transfer and Communication Protocol)
Maturity: - Please Select -
Relevance: - Please Select -
Weaknesses (please describe):

Mu3D (Multi-User 3D Protocol)
Maturity: - Please Select -
Relevance: - Please Select -
Weaknesses (please describe):

Q2.4 How would you rate the maturity of the following underlying base technologies needed to implement DS/DVE applications and their significance for advancing the fields of DS/DVE?

Network technologies (including hardware and protocols)
Maturity: - Please Select -
Significance for advancing the fields: - Please Select -
Required improvements:

Distributed Simulation Middleware:
Maturity: - Please Select -
Significance for advancing the fields: - Please Select -
Required improvements:

Human-Computer-Interfaces
Maturity: - Please Select -
Significance for advancing the fields: - Please Select -
Required improvements:

Semantic Web
Maturity: - Please Select -
Significance for advancing the fields: - Please Select -
Required improvements:

Graphics (hardware and algorithms)
Maturity: - Please Select -
Significance for advancing the fields: - Please Select -
Required improvements:

High Performance Computing
Maturity: - Please Select -
Significance for advancing the fields: - Please Select -
Required improvements:

Q2.5 How would you rate the overall maturity of the technologies and solutions developed in the fields of distributed simulation and distributed virtual environments?

- ☐ Very mature and already applied for many practical applications
- ☐ Mature, but not applied widely yet
- ☐ In the process of maturing
- ☐ Technologies exist, but still have significant weaknesses
- ☐ Academic research/prototypes

Comments (if any):

Q2.6 Which research challenges in the research fields (DS/DVE) would you consider "Grand Challenges" which must be solved in order to advance the field significantly?

Q2.7 Which findings and results would you expect from an external research group in order to advance the field significantly?

Q2.8 Would you consider collaboration in distributed virtual environments a viable topic for the future or will personal meetings with physical attendance of all participants always be preferable?

Viable Topic	- Please Select -
Personal meetings are preferable	- Please Select -

Q2.9 Can you name interesting trends, solutions, and actors in the areas of DS/DVE which you would consider drivers in these fields?

Trends:

Solutions:

Actors:

Q2.10 Are you aware of any other reliable data collections / studies / publications relevant for this survey?

Combined Screening and Selection of the Best with Control Variates

Shing Chih Tsai, Barry L. Nelson, and Jeremy Staum

Abstract Nelson and Staum derived ranking-and-selection (R&S) procedures that employ control-variate (CV) estimators instead of sample means to obtain greater statistical efficiency. However, control-variate estimators require more computational effort than sample means, and effective controls must be identified. In this paper, we present a new CV screening procedure to avoid much of the computation cost along with a better paired CV model than that of Nelson and Staum. We also present a two-stage CV combined procedure that captures the ability to eliminate inferior systems in the first stage and the statistical efficiency of control variates for selection in the second stage. Some guidelines about control-variate selection and an empirical evaluation are provided.

1 Introduction

In simulation research and applications, ranking-and-selection procedures (R&S; see for instance Bechhofer et al. 1995) have proven to be quite useful for finding the system design that is the best, or near the best, where the "best" system is the one with the largest or smallest expected performance measure. However, R&S procedures are only recommended when the number of alternative designs is relatively small and the designs are not functionally related. For instance, the typical indifference-zone (IZ) selection procedure will require large numbers of observations to deliver the desired correct-selection guarantee when output variances or the number of systems are large. To solve this problem, Nelson et al. (2001) proposed a combined procedure that uses the subset selection approach to eliminate some noncompetitive systems in the first stage; it then applies a standard IZ selection procedure in the second stage. In this way, sampling cost can be saved while still maintaining the ease of implementation and statistical efficiency.

S.C. Tsai (✉)
Department of Industrial and Information Management, National Cheng Kung University, Tainan, Taiwan
e-mail: str6969@mail.ncku.edu.tw

C. Alexopoulos et al. (eds.), *Advancing the Frontiers of Simulation: A Festschrift in Honor of George Samuel Fishman*, International Series in Operations Research & Management Science 133, DOI 10.1007/b110059_12, © Springer Science+Business Media, LLC 2009

In almost all R&S procedures, sample means of the outputs are used as the estimators of the expected performance. Nelson and Staum (2006) derived R&S procedures that employ control-variate estimators instead of sample means. Controls are random variables in the simulation that are correlated with the output of interest, but whose expected values are known (Lavenberg and Welch 1981). These control-variate procedures can be more statistically efficient than the sample-means-based procedures. However, control-variate estimators require more computational effort than sample means, and effective controls must be identified.

One of our goals is to propose a new control-variate (CV) screening procedure to decrease the computation cost and still obtain the statistical efficiency. A superior paired CV model is provided and compared to the paired model in Nelson and Staum (2006). We also propose a two-stage procedure that captures the ability to screen out inferior systems and the statistical efficiency of CVs for selection: We use a screening procedure with CVs to eliminate obviously noncompetitive systems in the first stage and then apply a selection-of-the-best-with-control-variates procedure to the surviving subset of systems in the second stage. Nelson and Staum (2006) showed that the screening threshold with CVs is expected to be tighter than with sample means when the correlation between the output and control is not too small. Therefore, the expected subset size is correspondingly smaller. For the selection-of-the-best-with-control-variates procedure, Nelson and Staum (2006) also showed that we can expect a smaller sample size than Rinott's (1978) procedure even when the correlation between the output and control is modest. Thus the sample size of the CV selection procedure is typically smaller than that of Rinott's (1978) procedure, which is based on sample means. Since the CV screening procedure is better than the standard screening procedure based on sample means, and the CV selection procedure is better than the selection procedure based on sample means, we can expect that a combined CV procedure is better than a combined procedure based on sample means. In this paper we develop the theory and procedures to support such a combined approach.

The remainder of this paper is organized as follows: In Section 2, we outline the generic combined procedure. Sections 3–5 review CV estimators and several CV R&S procedures. We present the improved paired CV model and a new CV screening procedure in Section 4. Section 6 contains some guidance for selecting control variates in this context. In Section 7, we present the CV combined procedure in detail. The paper ends with an empirical evaluation, including a queueing example, performed to compare the two combined procedures (Sections 8 and 9), and conclusions in Section 10. All proofs are relegated to the Appendix.

2 Generic Combined Procedure

In the CV combined procedure, we apply the CV selection-of-the-best procedure to the subset of systems chosen by the CV screening procedure to acquire both statistical and computational efficiency. The generic combined procedure is as follows. In the remainder of the paper we fill in specific pieces of this procedure.

1. For each system, obtain a small number of observations of the system performance measure and the controls. Then form CV estimators of each system's mean and calculate an estimator of the variance of each CV estimator.
2. Apply a CV screening procedure to eliminate inferior systems based on the information acquired in the first step.
3. If only one system survives, then stop and return that one as the best system. Otherwise, calculate the total number of observations needed for each system to detect a specified practically significant difference in performance with the desired confidence level.
4. Take additional observations from each surviving system and form CV estimators. Then select the system with the best CV estimator.

3 Screening Procedure with Individual Control Variates

In this section we briefly provide the definitions and notation that will be used throughout the paper and review the screening procedure with individual control variates in Nelson and Staum (2006). The following description is based on Nelson and Staum (2006).

3.1 *Individual Control-Variate Estimators*

Let X_{ij} be the jth simulation observation from system i, for $i = 1, 2, \ldots, k$. We assume it can be represented as

$$X_{ij} = \mu_i + (\boldsymbol{C}_{ij} - \boldsymbol{\xi}_i)'\boldsymbol{\beta}_i + \eta_{ij}, \tag{1}$$

where the $q_i \times 1$ vector $\boldsymbol{C}_{ij}$ is called the *control* and is assumed to be multivariate normal, while $\{\eta_{ij},\ i = 1, 2, \ldots, k,\ j = 1, 2, \ldots, n\}$ are mutually independent and $\{\eta_{ij},\ j = 1, 2, \ldots, n\}$ is a set of independent and identically distributed (i.i.d.) $\mathrm{N}(0, \tau_i^2)$ random variables. For each system $i = 1, 2, \ldots, k$, the controls $\{\boldsymbol{C}_{ij},\ j = 1, 2, \ldots, n\}$ are also i.i.d., are independent of $\{\eta_{ij},\ j = 1, 2, \ldots, n\}$, and have known expected value $\boldsymbol{\xi}_i$. The X_{ij} are therefore i.i.d. $\mathrm{N}(\mu_i, \sigma_i^2)$ random variables, with both μ_i and σ_i^2 unknown and (perhaps) unequal. The multiplier $\boldsymbol{\beta}_i$ is a $q_i \times 1$ vector of unknown constants that captures the relationship between the output X_{ij} and the control $\boldsymbol{C}_{ij}$, while η_{ij} represents that part of the variability in X_{ij} that is not explained by the controls.

A control-variate estimator of μ_i can be much more statistically efficient than the sample mean of the X_{ij}. We review some basic properties of the CV estimator under Model (1) below. The development is based on Nelson (1990), Nelson and Hsu (1993), and Nelson and Staum (2006).

Let

$$\boldsymbol{X}_i(n) = \begin{pmatrix} X_{i1} \\ X_{i2} \\ \vdots \\ X_{in} \end{pmatrix} \quad \text{and} \quad \boldsymbol{C}_i(n) = \begin{pmatrix} \boldsymbol{C}'_{i1} \\ \boldsymbol{C}'_{i2} \\ \vdots \\ \boldsymbol{C}'_{in} \end{pmatrix}$$

be vectors of the outputs and controls across all n observations from system i. Define the sample means of the outputs and controls as

$$\bar{X}_i(n) = \frac{1}{n}\sum_{j=1}^{n} X_{ij} \quad \text{and} \quad \bar{\boldsymbol{C}}_i(n) = \frac{1}{n}\sum_{j=1}^{n} \boldsymbol{C}_{ij}.$$

We append "(n)" to quantities defined across n observations.

To define the CV point estimator, let

$$\boldsymbol{L}'_i(n) = \left[(\boldsymbol{C}_{i1} - \bar{\boldsymbol{C}}_i(n)), (\boldsymbol{C}_{i2} - \bar{\boldsymbol{C}}_i(n)), \ldots, (\boldsymbol{C}_{in} - \bar{\boldsymbol{C}}_i(n))\right].$$

If $\boldsymbol{1}_{n\times 1}$ is a column vector whose n elements all equal one, then the CV point estimator of μ_i is

$$\begin{aligned}\widehat{\mu}_i(n) &= \left[\frac{1}{n}\boldsymbol{1}'_{n\times 1} - \left(\bar{\boldsymbol{C}}_i(n) - \boldsymbol{\xi}_i\right)'\left(\boldsymbol{L}'_i(n)\boldsymbol{L}_i(n)\right)^{-1}\boldsymbol{L}'_i(n)\right]\boldsymbol{X}_i(n) \\ &= \bar{X}_i(n) - \left(\bar{\boldsymbol{C}}_i(n) - \boldsymbol{\xi}_i\right)'\widehat{\boldsymbol{\beta}}_i\end{aligned}$$

(Nelson 1990). It is known that under Model (1)

$$\mathrm{E}[\widehat{\mu}_i(n)] = \mu_i \quad \text{and} \quad \mathrm{Var}[\widehat{\mu}_i(n)] = \left(\frac{n-2}{n-q_i-2}\right)\frac{\tau_i^2}{n}$$

where $\tau_i^2 = (1 - R_i^2)\sigma_i^2$ and R_i^2 is the square of the multiple correlation coefficient between X_{ij} and $\boldsymbol{C}_{ij}$ (Lavenberg and Welch 1981).

We need to know the distribution of $\widehat{\mu}_i(n)$ and an estimator of its variance to derive R&S procedures. For each system $i = 1, 2, \ldots, k$, let

$$\boldsymbol{A}_i(n) = \begin{pmatrix} 1 & (\boldsymbol{C}_{i1} - \boldsymbol{\xi}_i)' \\ 1 & (\boldsymbol{C}_{i2} - \boldsymbol{\xi}_i)' \\ \vdots & \vdots \\ 1 & (\boldsymbol{C}_{in} - \boldsymbol{\xi}_i)' \end{pmatrix}.$$

If $\boldsymbol{I}$ is the identity matrix of rank q_i, then define

$$\widehat{\tau}_i^2(n) = \frac{1}{n-q_i-1} \boldsymbol{X}_i(n)' \left[\boldsymbol{I} - \boldsymbol{A}_i(n) \left(\boldsymbol{A}_i'(n) \boldsymbol{A}_i(n) \right)^{-1} \boldsymbol{A}_i'(n) \right] \boldsymbol{X}_i(n)$$

$$= \frac{1}{n-q_i-1} \sum_{j=1}^{n} \left[X_{ij} - \widehat{\mu}_i(n) - (\boldsymbol{C}_{ij} - \boldsymbol{\xi}_i)' \widehat{\boldsymbol{\beta}}_i(n) \right]^2 \tag{2}$$

as the residual variance estimator. Further, let

$$\widehat{\Delta}_i^2(n) = \frac{1}{n} + \frac{1}{n-1} \left(\bar{\boldsymbol{C}}_i(n) - \boldsymbol{\xi}_i \right)' \boldsymbol{S}_{\boldsymbol{C}_i}^{-1}(n) \left(\bar{\boldsymbol{C}}_i(n) - \boldsymbol{\xi}_i \right) \tag{3}$$

where $\boldsymbol{S}_{\boldsymbol{C}_i}(n)$ is the sample variance-covariance matrix of $\boldsymbol{C}_{ij}$. Then we have the following key result:

Lemma 1 (Nelson and Hsu 1993, Theorem 4.1) *If Model (1) pertains, then conditional on $\boldsymbol{C}_1(n), \boldsymbol{C}_2(n), \ldots, \boldsymbol{C}_k(n)$, the following properties hold:*

P1: $\widehat{\mu}_i(n) \sim \mathrm{N}(\mu_i, \widehat{\Delta}_i^2(n)\tau_i^2)$, $i = 1, 2, \ldots, k$.

P2: $\widehat{\tau}_i^2(n) \sim \dfrac{\tau_i^2 \chi_{n-q_i-1}^2}{n-q_i-1}$ *and is independent of* $\widehat{\mu}_i(n)$, *for* $i = 1, 2, \ldots, k$, *where* $\chi_{n-q_i-1}^2$ *is a chi-squared random variable with* $n - q_i - 1$ *degrees of freedom.*

P3: *If* $\{\eta_{ij}, i = 1, 2, \ldots, k, j = 1, 2, \ldots, n\}$ *are mutually independent, then* $\{\widehat{\mu}_i(n), \widehat{\tau}_i^2(n), i = 1, 2, \ldots, k\}$ *are mutually independent.*

Property P3 requires that the η_{ij} are independent for all systems i as well as for all observations j. In practice P3 will hold either if all systems are simulated independently, or if common random numbers (CRN) are used but the dependence due to CRN is entirely explained by the controls. CRN is a technique that tries to induce a positive correlation between the outputs of different systems by using the same pseudorandom numbers to simulate each alternative system and therefore reduce the variance of the difference between them.

3.2 Screening with Individual Control Variates

We will assume that unknown to us $\mu_k \geq \mu_{k-1} \geq \cdots \geq \mu_1$ and that bigger is better. The goal of the procedure is to find a subset I that contains system k with prespecified confidence. We also assume that Model (1) holds but relax the assumption that $\boldsymbol{C}_{ij}$ has to be multivariate normal. Let $t_{p,\nu}$ represent the p quantile of the t distribution with ν degrees of freedom.

Procedure 1 (Individual CV Screening Procedure)

1. *Choose the confidence level* $1 - \alpha > 1/k$.

2. *Obtain* $n_i > q_i + 2$ *observations from system* $i = 1, 2, \ldots, k$ *and form CV estimators* $\widehat{\mu}_i(n_i),\ i = 1, 2, \ldots, k$.
3. *Let* $t_i = t_{(1-\alpha)^{1/(k-1)}, n_i - q_i - 1}$ *and create the subset*

$$I_{\text{Indiv}} = \{i : \widehat{\mu}_i(n_i) - \widehat{\mu}_\ell(n_\ell) \geq -W_{i\ell}, \forall \ell \neq i\}, \tag{4}$$

where

$$W_{i\ell} = \sqrt{t_i^2\, \widehat{\Delta}_i^2(n_i)\widehat{\tau}_i^2(n_i) + t_\ell^2\, \widehat{\Delta}_\ell^2(n_\ell)\widehat{\tau}_\ell^2(n_\ell)}.$$

Nelson and Staum (2006) proved that $\Pr\{k \in I_{\text{Indiv}}\} \geq 1 - \alpha$ when Model (1) holds even if we relax the assumption that $\boldsymbol{C}_{ij}$ is multivariate normal. Nelson and Staum (2006) also showed that very little correlation between the output and control is required for the subset size with CVs to be smaller than that with sample means.

The advantage of this procedure is that we just need to compute k CVs. Its disadvantage is that the assumption in Model (1) that there is no dependence between residuals across systems induced by CRN will not hold in practice. Therefore Nelson and Staum (2006) proposed a screening procedure with paired control variates, which we improve upon in the next section.

4 Screening Procedures with Paired Control Variates

In this section we briefly review the paired control variate model of Nelson and Staum (2006) and propose a more-general model on which a new procedure is based. An adjustment is also provided to reduce the computation cost and retain the benefit of paired CV estimators.

Nelson and Staum (2006) use a paired CV model to avoid the assumption that the controls entirely explain the dependence induced by CRN. To do this, they form pairwise differences across systems: $X_j(i,\ell) = X_{ij} - X_{\ell j}, \boldsymbol{C}_j(i,\ell) = \boldsymbol{C}_{ij} - \boldsymbol{C}_{\ell j}, \mu_{i\ell} = \mu_i - \mu_\ell$, and $\boldsymbol{\xi}_{i\ell} = \boldsymbol{\xi}_i - \boldsymbol{\xi}_\ell$, for $i \neq \ell$. Since they need the outputs and the controls to be paired across systems, the number of observations must be equal for each system in the same pair, and the number of controls for each system in the same pair should also be equal. For convenience we let n be the common number of replications and q be the common number of controls for each system. Then we assume that a model like Model (1) holds:

$$X_j(i,\ell) = \mu_{i\ell} + (\boldsymbol{C}_j(i,\ell) - \boldsymbol{\xi}_{i\ell})'\boldsymbol{B}(i,\ell) + \varepsilon_j(i,\ell), \tag{5}$$

where $\{\varepsilon_j(i,\ell),\ j = 1, 2, \ldots, n\}$ is a set of i.i.d. $\mathrm{N}(0, \tau_{i\ell}^2)$ random variables. The $q \times 1$ vector $\boldsymbol{C}_j(i,\ell)$ is assumed multivariate normal. For each pair of systems $i, \ell = 1, 2, \ldots, k, i \neq \ell$, the controls $\{\boldsymbol{C}_j(i,\ell), j = 1, 2, \ldots, n\}$ are also i.i.d., are independent of $\{\varepsilon_j(i,\ell), j = 1, 2, \ldots, n\}$, and have known expected value $\boldsymbol{\xi}_{i\ell}$.

Unlike Model (1), Model (5) can hold even when η_{ij} and $\eta_{\ell j}$ are dependent. However, this model may break down when CRN causes $\boldsymbol{C}_{ij} = \boldsymbol{C}_{\ell j}$ for all j, which cancels the controls. To avoid this, we present a different model to explain the relationship between the controls and the outputs. We assume the following new model holds:

$$X_j(i,\ell) = \mu_{i\ell} + (\boldsymbol{C}_{ij} - \boldsymbol{\xi}_i)'\boldsymbol{\beta}_i - (\boldsymbol{C}_{\ell j} - \boldsymbol{\xi}_\ell)'\boldsymbol{\beta}_\ell + \eta_j(i,\ell), \tag{6}$$

where $\eta_j(i,\ell) = \eta_{ij} - \eta_{\ell j}$ and $\{\eta_j(i,\ell),\ j = 1,2,\ldots,n\}$ is a set of i.i.d. $\mathrm{N}(0,\sigma^2_{i\ell})$ random variables. The $(q_i + q_\ell) \times 1$ vector $(\boldsymbol{C}'_{ij}, \boldsymbol{C}'_{\ell j})'$ is assumed multivariate normal. For each pair of systems $i, \ell = 1,2,\ldots,k$, $i \neq \ell$, the controls $\{\boldsymbol{C}_{ij},\ j = 1,2,\ldots,n\}$ and $\{\boldsymbol{C}_{\ell j},\ j = 1,2,\ldots,n\}$ are also i.i.d., are independent of $\{\eta_j(i,\ell),\ j = 1,2,\ldots,n\}$, and have known expected values $\boldsymbol{\xi}_i$ and $\boldsymbol{\xi}_\ell$. Like Model (5), Model (6) can also hold even when η_{ij} and $\eta_{\ell j}$ are dependent. For all $i \neq \ell$, we let $\widehat{\mu}_{i\ell}(n)$ be the corresponding CV estimator of $\mu_{i\ell}$ under Model (6), and define $\widehat{\tau}^2_{i\ell}(n)$ and $\widehat{\Delta}^2_{i\ell}(n)$ in analogy to Equations (2) and (3).

We now assume that Model (6) holds in order to execute the All Pairs screening procedure. Before describing the procedure, we present an argument for the superiority of Model (6) over Model (5):

- Model (6) is a more-general model that is equivalent to Model (5) when $\boldsymbol{\beta}_i = \boldsymbol{\beta}_\ell$.
- Suppose Model (5) holds, but we compute $\widehat{\mu}_{i\ell}(n)$ assuming Model (6) holds. Then $\widehat{\mu}_{i\ell}(n)$ is still unbiased. However, $\mathrm{Var}[\widehat{\mu}_{i\ell}(n)]$ will be inflated because of the loss of degrees of freedom (from $n - q - 1$ to $n - 2q - 1$). The resulting inflation of variance will not be substantial when n is not too small.
- Suppose Model (6) holds with $\boldsymbol{\beta}_i \neq \boldsymbol{\beta}_\ell$, but we compute $\widehat{\mu}_{i\ell}(n)$ assuming Model (5) holds. Then $\widehat{\mu}_{i\ell}(n)$ will be biased and $\mathrm{Var}[\widehat{\mu}_{i\ell}(n)]$ will be increased, especially when $\boldsymbol{\beta}_i$ is very different from $\boldsymbol{\beta}_\ell$ (see the Appendix).
- Under Model (6) the number of controls for each system in the same pair is not required to be equal. Therefore, we gain potential benefits in terms of CV selection (notice that the degrees of freedom is $n - q_i - q_\ell - 1$ in general).
- Model (6) makes the All Pairs screening procedure below more compatible with the CV selection procedure (see Section 5) that relies on Model (1). In fact Model (1) implies Model (6). Therefore we do not have to be concerned about any incongruity in the CV combined procedure.

We form the following All Pairs screening procedure based on Model (6).

Procedure 2 (All Pairs Screening Procedure)

1. *Choose the confidence level* $1 - \alpha > 1/k$.
2. *Obtain* $n > \max_{i\neq\ell}(q_i + q_\ell + 2)$ *observations from each system and form the* $k(k-1)/2$ *CV estimators* $\widehat{\mu}_{i\ell}(n)$ *for all* $i \neq \ell$.
3. *Let* $t_{i\ell} = t_{1-\alpha/(k-1),n-q_i-q_\ell-1}$ *and create the subset*

$$I_{\text{All Pairs}} = \left\{ i : \widehat{\mu}_{i\ell}(n) \geq -t_{i\ell}\,\widehat{\Delta}_{i\ell}(n)\widehat{\tau}_{i\ell}(n),\ \forall \ell \neq i \right\}.$$

Nelson and Staum (2006) proved that $\Pr\{k \in I_{\text{All Pairs}}\} \geq 1 - \alpha$ when Model (5) holds. This procedure is also valid when Model (6) pertains and $(\boldsymbol{C}'_{ij}, \boldsymbol{C}'_{\ell j})'$ is not required to be multivariate normal. The advantage of this procedure is that we do not have to be concerned about the dependence remaining in the residuals due to CRN. Its disadvantages are that we have to compute $k(k-1)/2$ CV estimators and that the procedure uses the conservative Bonferroni inequality. Therefore, we propose a new procedure that requires less computation and creates a subset $I \supseteq I_{\text{All Pairs}}$, and therefore we can guarantee that $\Pr\{k \in I\} \geq 1-\alpha$. To accomplish this we choose some system K^* that is very likely to be the best system, and then perform screening with paired CVs just against K^*.

In the following "Best Bet" screening procedure, we use Model (6) and denote the system with the largest $\widehat{\mu}_i(n)$ as K^*.

Procedure 3 (Best Bet Screening Procedure)

1. *Choose the confidence level* $1-\alpha > 1/k$.
2. *Obtain* $n > \max_{i \neq \ell}(q_i + q_\ell + 2)$ *observations from each system and form the k CV estimators* $\widehat{\mu}_i(n)$, $i = 1, 2, \ldots, k$.
3. *Let* K^* *be the index of the system with the largest* $\widehat{\mu}_i(n)$, *that is,* $K^* = \operatorname{argmax}_i \widehat{\mu}_i(n)$, *and then form the* $k-1$ *paired CV estimators* $\widehat{\mu}_{iK^*}(n)$ *for all* $i \neq K^*$.
4. *Create the subset*

$$I_{\text{Best Bet}} = \left\{ i : \widehat{\mu}_{iK^*}(n) \geq -t_{iK^*} \widehat{\Delta}_{iK^*}(n) \widehat{\tau}_{iK^*}(n) \right\} \cup \left\{ K^* \right\}.$$

The advantage of this procedure is that it can decrease the computation cost and achieve the desired statistical efficiency as well. The subset size will be close to that of the All Pairs screening procedure, because there is a large correlation between $\widehat{\mu}_{i\ell}(n)$ and $\widehat{\mu}_i(n) - \widehat{\mu}_\ell(n)$. The disadvantage is that it needs to compute $2k-1$ CV estimators, which is more than the individual CV screening procedure (k). However, it still saves computation cost compared with the All Pairs screening procedure ($k(k-1)/2$), when the number of alternatives is large, and it avoids the assumption that CVs explain all the dependence induced by CRN.

Remark 1 The system with the largest sample mean is also a potential best system, so we could let $K^* = \operatorname{argmax}_i \bar{X}_i(n)$, and then do screening with paired CVs just against K^*. This procedure can save a great deal of computation cost because we only need to compute $k-1$ paired CV estimators. Unfortunately, the subset formed by this procedure may be much larger than that formed by the All Pairs screening procedure, because there is not necessarily much correlation between $\widehat{\mu}_{i\ell}(n)$ and $\bar{X}_i(n) - \bar{X}_\ell(n)$.

5 Selecting the Best with Control Variates

In this section we briefly review the selection-of-the-best-with-control-variates procedure in Nelson and Staum (2006). The following description is based on Nelson and Staum (2006). Under Model (1), we adopt the indifference-zone (IZ) formulation in which we require a guaranteed probability of selecting system k whenever the difference $\mu_k - \mu_{k-1} \geq \delta$, where the indifference-zone parameter $\delta > 0$ is set to the smallest difference the analyst feels is worth detecting. We also assume that all systems have the same number of controls q. The procedure is as follows:

Procedure 4 (Selecting the Best with Controls)

1. *Choose the indifference-zone parameter* $\delta > 0$, *confidence level* $1 - \alpha > 1/k$ *and choose* $\alpha_0, \alpha_1 > 0$ *such that* $\alpha = \alpha_0 + \alpha_1$.
2. *For each system* $i = 1, 2, \ldots, k$, *obtain* $n_0 > q + 2$ *observations and calculate* $\widehat{\tau}_i^2(n_0)$.
3. *For each system* $i = 1, 2, \ldots, k$, *set the total sample size*

$$N_i = \min_{n \geq n_0} \left\{ n : \left(\frac{n-q}{q} \right) \left(\frac{n\delta^2}{h^2 \widehat{\tau}_i^2(n_0)} - 1 \right) \geq \mathcal{F}_{q,n-q}^{(\gamma)} \right\}$$

where $h = h_{k,1-\alpha_1,n_0-q-1}$ *is Rinott's (1978) constant,* $\mathcal{F}_{q,n-q}^{(\gamma)}$ *is the* γ *quantile of the F distribution with* $(q, n-q)$ *degrees of freedom, and*

$$\gamma = \begin{cases} (1-\alpha_0)^{\frac{1}{k}}, & \text{if the systems are simulated independently} \\ 1 - \alpha_0/k, & \text{otherwise.} \end{cases}$$

4. *Collect* $N_i - n_0$ *observations from system* i *and form the CV estimators* $\widehat{\mu}_i(N_i)$ *for* $i = 1, 2, \ldots, k$.
5. *Select system* $B = \operatorname{argmax}_i \widehat{\mu}_i(N_i)$, *and form the* $(1-\alpha)100\%$ *simultaneous confidence intervals*

$$\mu_i - \max_{\ell \neq i} \mu_\ell \in \left[-\left(\widehat{\mu}_i(N_i) - \max_{\ell \neq i} \widehat{\mu}_\ell(N_\ell) - \delta \right)^-, \left(\widehat{\mu}_i(N_i) - \max_{\ell \neq i} \widehat{\mu}_\ell(N_\ell) + \delta \right)^+ \right] \quad (7)$$

for $i = 1, 2, \ldots, k$, *where* $-y^- = \min\{0, y\}$ *and* $y^+ = \max\{0, y\}$. *Furthermore,*

$$\Pr\left\{ \mu_B - \max_{\ell=1,\ldots,k} \mu_\ell \geq -\delta \right\} \geq 1 - \alpha, \quad (8)$$

that is, with high confidence, the mean of the selected system is within δ *of the mean of the truly best system.*

Nelson and Staum (2006) proved that $\Pr\{B = k\} \geq 1 - \alpha$ whenever Model (1) holds and $\mu_k - \mu_{k-1} \geq \delta$. Regardless of the configuration of the true means, the confidence intervals (7) have coverage probability at least $1 - \alpha$ by Theorem 1 of Nelson and Matejcik (1995), while Inequality (8) follows from Corollary 1 of Nelson and Goldsman (2001).

6 Control-Variate Selection for Screening

In this section we provide some guidance for selecting control variates. When CRN is involved in the screening procedure, can we take advantage of CRN when choosing control variates, or should we just select favorable control variates to minimize the variance of each CV point estimator individually? For the screening procedure with individual control variates, we use the screening threshold $W_{i\ell}$ in Equation (4) as the measure of chosen subset size. The smaller $W_{i\ell}$ is, the more difficult it is for system i to survive in the subset. To make the expected subset size as small as possible, we select favorable control variates to minimize $\mathrm{E}[W_{i\ell}^2]$ because $W_{i\ell}$ is nonnegative. To simplify the analysis, suppose we choose first-stage sample sizes and controls for each system such that $n_i = n$ and $q_i = q$, for all i. Consequently, for all $i = 1, 2, \ldots, k$, $t_i = t$, and we know that

$$\begin{aligned}\mathrm{E}[W_{i\ell}^2] &= \mathrm{E}\left[t_i^2\, \widehat{\Delta}_i^2(n_i)\widehat{\tau}_i^2(n_i) + t_\ell^2\, \widehat{\Delta}_\ell^2(n_\ell)\widehat{\tau}_\ell^2(n_\ell)\right] \\ &= \mathrm{E}\left[t^2\, \widehat{\Delta}_i^2(n)\widehat{\tau}_i^2(n)\right] + \mathrm{E}\left[t^2\, \widehat{\Delta}_\ell^2(n)\widehat{\tau}_\ell^2(n)\right] \\ &= t^2\, \left(\mathrm{Var}[\widehat{\mu}_i(n)] + \mathrm{Var}[\widehat{\mu}_\ell(n)]\right) \\ &= t^2 \left(\frac{n-2}{n-q-2}\right)\left(\frac{\mathrm{Var}[\eta_{ij}] + \mathrm{Var}[\eta_{\ell j}]}{n}\right).\end{aligned}$$

Clearly we would like to minimize the variance of each CV point estimator. In other words, we should choose control variates for each system to obtain the greatest variance reduction individually. Notice that $\mathrm{E}[W_{i\ell}^2]$ is unaffected by CRN; therefore, CRN is irrelevant with respect to the individual CV screening procedure. Añonuevo and Nelson (1988), Bauer and Wilson (1992), and Nelson (1989) give algorithms for selecting good control variates individually. However, CRN can affect $\mathrm{Cov}[\eta_{ij}, \eta_{\ell j}]$, which represents the benefits of CRN that we cannot capture in the individual CV screening procedure. This is the disadvantage inherent in the screening procedure with individual control variates.

For the screening procedure with paired control variates under Model (6), the expectation of the square of the screening threshold from Step 3 of Procedure 2 is

$$\begin{aligned}\mathrm{E}\left[t_{i\ell}^2\, \widehat{\Delta}_{i\ell}^2(n)\widehat{\tau}_{i\ell}^2(n)\right] &= t_{i\ell}^2\, \mathrm{Var}[\widehat{\mu}_{i\ell}(n)] \\ &= t_{i\ell}^2 \left(\frac{n-2}{n-2q-2}\right)\frac{\mathrm{Var}[\eta_j(i,\ell)]}{n} \\ &= t_{i\ell}^2 \left(\frac{n-2}{n-2q-2}\right)\frac{\mathrm{Var}[\eta_{ij} - \eta_{\ell j}]}{n} \\ &= t_{i\ell}^2 \left(\frac{n-2}{n-2q-2}\right)\left(\frac{\mathrm{Var}[\eta_{ij}] + \mathrm{Var}[\eta_{\ell j}] - 2\mathrm{Cov}[\eta_{ij}, \eta_{\ell j}]}{n}\right)\end{aligned}$$

which directly incorporates the reduced variance by applying CVs to the paired observations. The more positive correlation that remains in the residuals across systems induced by CRN, the larger $\text{Cov}[\eta_{ij}, \eta_{\ell j}]$ will be. The paired CV procedure exploits this dependence so that it can perform better than the individual CV screening procedure, especially when n is much larger than q. Therefore, we should choose control variates that can be paired across systems i and ℓ to minimize the variance of $\widehat{\mu}_{i\ell}(n)$, and these may be different than what we would choose to minimize the variance of each individual CV estimator. As a result it is possible that the CVs chosen for system i could be different when system i is paired with each system $\ell = 1, 2, \ldots, k, \ell \neq i$ (in which case Model (6) would be modified to allow the selected controls to depend on the pair (i, ℓ)).

7 Combined Procedure

In the combined procedure, we apply a screening procedure with control variates to eliminate noncompetitive systems in the first stage. Then in the second stage the CV selection-of-the-best procedure is applied to the surviving systems to pick the best system, while still gaining the desired overall confidence level. Here are some key observations:

- We spend α_0 of the overall allowable error α for incorrect selection on the first screening stage, and $\alpha_1 + \alpha_2$ on the second selection-of-the-best stage.
- If we use the individual CV screening procedure in the first stage, then a multiplicative approach is applied, i.e., $1 - \alpha = (1 - \alpha_0)(1 - \alpha_1 - \alpha_2)$.
- If we use the paired CV screening procedure in the first stage, then an additive approach is applied, i.e., $1 - \alpha = 1 - \alpha_0 - \alpha_1 - \alpha_2$.
- We set the appropriate critical constant t_i of each system $i = 1, 2, \ldots, k$ in the CV screening procedure for k systems, n_i first-stage samples, q_i control variates, and confidence level $1 - \alpha_0$.
- We set the appropriate critical constant h of each system $i = 1, 2, \ldots, k$ in the CV selection-of-the-best procedure for k systems, n_i first-stage samples, q_i control variates, and confidence level $1 - \alpha_1$.
- We set the appropriate critical constant γ in the CV selection-of-the-best procedure for k systems, confidence level $1 - \alpha_2$, and depending on whether or not the systems are simulated independently or with CRN.

In the procedure below we assume that $n_i - q_i$ is the same for each system $i = 1, 2, \ldots, k$ and mention the necessary adjustment for unequal $n_i - q_i$ in Remark 2. The following is a procedure that combines the individual CV screening procedure with the CV selection-of-the-best procedure.

Procedure 5 (Individual CV Combined Procedure)

1. *Select overall confidence level* $1 - \alpha > 1/k$, *indifference-zone parameter* $\delta > 0$, *number of systems* k, *and first-stage sample size* $n_i > q_i + 2$ *from system* $i = 1, 2, \ldots, k$. *Set* $t_i = t_{(1-\alpha_0)^{1/(k-1)}, n_i - q_i - 1}$ *and* $h = h_{k, 1-\alpha_1, n_i - q_i - 1}$, *which is Rinott's constant (see Wilcox 1984 or Bechhofer et al. 1995 for tables).*

2. *Obtain n_i observations from each system and calculate $\widehat{\mu}_i(n_i)$, $\widehat{\Delta}_i^2(n_i)$ and $\widehat{\tau}_i^2(n_i)$, $i = 1, 2, \ldots, k$. We also create the subset*

$$I = \{i : \widehat{\mu}_i(n_i) - \widehat{\mu}_\ell(n_\ell) \geq -W_{i\ell}, \forall \ell \neq i\},$$

where

$$W_{i\ell} = \sqrt{t_i^2\,\widehat{\Delta}_i^2(n_i)\widehat{\tau}_i^2(n_i) + t_\ell^2\,\widehat{\Delta}_\ell^2(n_\ell)\widehat{\tau}_\ell^2(n_\ell)}.$$

3. *If I contains a single index, then stop and return that system as the best. Otherwise, for all $i \in I$, compute the second-stage sample size*

$$N_i = \min_{n \geq n_i}\left\{n : \left(\frac{n - q_i}{q_i}\right)\left(\frac{n\delta^2}{h^2\widehat{\tau}_i^2(n_i)} - 1\right) \geq \mathcal{F}^{(\gamma)}_{q_i, n-q_i}\right\}$$

where

$$\gamma = \begin{cases} (1-\alpha_2)^{\frac{1}{k}}, & \textit{if the systems are simulated independently} \\ 1 - \alpha_2/k, & \textit{otherwise.} \end{cases}$$

Notice: $1 - \alpha = (1 - \alpha_0)(1 - \alpha_1 - \alpha_2)$ (multiplicative approach).

4. *Take $N_i - n_i$ additional observations from all systems $i \in I$ and form the CV estimators $\widehat{\mu}_i(N_i)$ for $i \in I$.*
5. *Select the system $B = \text{argmax}_I \widehat{\mu}_i(N_i)$ as best from all systems $i \in I$.*

Theorem 1 *If Model (1) holds, then the individual CV combined procedure selects a system B such that $\Pr\{B = k\} \geq 1 - \alpha$ whenever $\mu_k - \mu_{k-1} \geq \delta$. For any configuration of the means, the following hold with probability greater than or equal to $1 - \alpha$:*

- *For all $i \in I$,*

$$\mu_i - \max_{\ell \in I, \ell \neq i} \mu_\ell \in \left[-\left(\widehat{\mu}_i(N_i) - \max_{\ell \in I, \ell \neq i} \widehat{\mu}_\ell(N_\ell) - \delta\right)^-, \left(\widehat{\mu}_i(N_i) - \max_{\ell \in I, \ell \neq i} \widehat{\mu}_\ell(N_\ell) + \delta\right)^+\right]. \quad (9)$$

- *The mean of the system we select will be within δ of the mean of the truly best system in I with probability $\geq 1 - \alpha$, that is,*

$$\Pr\left\{\mu_B - \max_{\ell \in I, \ell \neq B} \mu_\ell \geq -\delta\right\} \geq 1 - \alpha. \quad (10)$$

Remark 2 Suppose that $n_i - q_i$ is different across systems. This causes the first-stage residual-variance estimators $\widehat{\tau}_1^2(n_1), \widehat{\tau}_2^2(n_2), \ldots, \widehat{\tau}_k^2(n_k)$ to have different degrees of freedom. One approach is to use the adjusted constant

$$h = h_{2,(1-\alpha_1)^{1/(k-1)},\min_i\{n_i-q_i-1\}}$$

which is valid when the degrees of freedom are unequal (Boesel, Nelson, and Kim 2003).

Remark 3 We can combine the paired CV screening procedure with the CV selection-of-the-best procedure. When we use the Best Bet screening procedure, we need to change Step 2 to the following:

2. *Obtain* $n > \max_{i\neq\ell}(q_i + q_\ell + 2)$ *observations from each system and form the k CV estimators* $\widehat{\mu}_i(n)$, $i = 1, 2, \ldots, k$. *Let* K^* *be the index of the system with the largest* $\widehat{\mu}_i(n)$, *that is,* $K^* = \text{argmax}_i \widehat{\mu}_i(n)$, *and then form the* $k-1$ *paired CV estimators* $\widehat{\mu}_{iK^*}(n)$ *for all* $i \neq K^*$. *Then we let* $t_{iK^*} = t_{1-\alpha_0/(k-1), n-q_i-q_{K^*}-1}$ *and create the subset*

$$I_{\text{Best Bet}} = \left\{ i : \widehat{\mu}_{iK^*}(n) \geq -t_{iK^*}\,\widehat{\Delta}_{iK^*}(n)\widehat{\tau}_{iK^*}(n) \right\} \cup \left\{ K^* \right\}.$$

An additive approach is applied ($1-\alpha = 1-\alpha_0 - \alpha_1 - \alpha_2$).

Theorem 2 *If Model (1) holds, then the paired CV combined procedure selects a system B such that* $\Pr\{B = k\} \geq 1 - \alpha$ *whenever* $\mu_k - \mu_{k-1} \geq \delta$, *and statistical inferences (9) and (10) still hold regardless of the configuration of the true means.*

We prove that $\Pr\{B = k\} \geq 1 - \alpha$ with independence among $\{\eta_{ij},\ i = 1, 2, \ldots, k,\ j = 1, 2, \ldots, n\}$ in the Appendix. However, experiments showed that this paired CV combined procedure can perform very well even when $\{\eta_{ij},\ i = 1, 2, \ldots, k\}$ are positively dependent.

8 Empirical Results

In this section we summarize the results of an empirical evaluation performed to compare the following procedures:

1. The combined sample-means-based procedure (NSGS) due to Nelson et al. (2001) that uses a screening procedure with sample means to eliminate non-competitive systems after the first stage of sampling, and then applies Rinott's IZ selection procedure in the second stage. This procedure allows for unknown and unequal variances across systems, but CRN is not exploited.
2. The individual CV combined procedure which we call TNS-I, and the paired CV combined procedure which we call TNS-P. These procedures allow for unknown and unequal variances across systems and the use of CRN, although TNS-I does not exploit CRN.

The systems are represented by various configurations of k normal distributions; in all cases, system k was the best (had the largest true mean). Let X_i be a simulation observation from system i, for $i = 1, 2, \ldots, k$. For simplicity, we assume that there is $q = 1$ control variate. Then we assume the output can be represented as

$$X_i = \mu_i + (C_i - \xi_i)\beta_i + \eta_i,$$

where $\{\eta_i,\ i = 1, 2, \ldots, k\}$ are $\mathrm{N}(0, \sigma_\eta^2)$ random variables. The $\{C_i,\ i = 1, 2, \ldots, k\}$ are assumed to be $\mathrm{N}(\xi_i, \sigma_c^2)$ random variables and independent of $\{\eta_i,\ i = 1, 2, \ldots, k\}$. The correlation between controls C_i and C_ℓ for $i \neq \ell$ is ρ_c. The correlation between residuals η_i and η_ℓ for $i \neq \ell$ is ρ_η. The squared correlation coefficient between X_i and C_i is $\rho^2_{(x,c)}$.

We evaluated each procedure on different configurations of the systems, examining factors including the number of systems k, the practically significant difference δ, the initial sample size n_0, the variance of the controls σ_c^2, the variance of the residuals σ_η^2, the correlation of the controls ρ_c, and the correlation of the residuals ρ_η. The larger σ_c^2 is compared with σ_η^2, the more of the variability in outputs can be explained by the controls. When $\rho_\eta \neq 0$, then Model (6) holds but Model (1) does not hold. A larger ρ_η means more dependence due to CRN is accounted for by the residuals. The configurations, the experiment design, and the results are described below.

8.1 Configurations and Experiment Design

We used the slippage configuration (SC) of the true means of the systems, in which μ_k was set to δ, while $\mu_1 = \mu_2 = \cdots = \mu_{k-1} = 0$. This is a difficult scenario for screening procedures because all the inferior systems are close to the best system. These experiments with the slippage configuration illustrated that CVs can make the screening procedure more efficient even under the most difficult situation.

We chose the initial sample size to be $n_0 = 10$, for $i = 1, 2, \ldots, k$. The mean of the controls, ξ_i, is set to be 0, for $i = 1, 2, \ldots, k$. We also set β_i to be 1, for $i = 1, 2, \ldots, k$. The number of systems in each experiment varied over $k = 2, 5, 10, 25, 100$. The indifference-zone parameter, δ, was set to $\delta = \sqrt{(\sigma_c^2 + \sigma_\eta^2)/n_0}$. For each configuration, 500 macroreplications (complete repetitions) of the entire combined procedure were performed. In all experiments, the nominal probability of correct selection was set at $1 - \alpha = 0.95$. We took $\alpha_0 = \alpha_1 = \alpha_2 = \alpha/3$ in paired CV screening cases and took $\alpha_0 = \alpha/3$, $\alpha_1 = \alpha_2 = \alpha/(3 - \alpha)$ in individual CV screening cases. For NSGS, we set $\alpha_0 = \alpha_1 = \alpha/2$. To compare the performance of the procedures we recorded the estimated probability of correct selection (PCS), the average number of samples per system (ANS), and the percentage of systems that received second-stage sampling (PSS).

8.2 Summary of Results

The PCS of the CV combined procedure was over 0.95 in all configurations. The overall experiments showed that the CV combined procedure was superior to the combined sample-means-based procedure under any configuration we examined.

We do not try to present comprehensive results from such a large simulation study. Instead, we present selected results that highlight the key conclusions. Notice that we apply Model (6) and the Best Bet screening procedure in TNS-P.

8.2.1 Effect of Number of Systems

See Table 1 for an illustration. Systems are simulated independently since NSGS and TNS-I do not exploit CRN. The goal is to compare NSGS with TNS-I when we have different numbers of systems. As k increases, the average number of samples per system increases greatly in NSGS compared to TNS-I. The percentage of systems that received second-stage sampling is smaller in TNS-I than in NSGS.

Table 1 Effect of number of systems for NSGS and TNS-I when $\sigma_c = 4$, $\sigma_\eta = 1$, $\rho_c = \rho_\eta = 0$

Number of systems	Procedure	PCS	ANS	PSS
$k = 2$	NSGS	0.98	98	0.86
	TNS-I	1	12	0.41
$k = 5$	NSGS	0.98	186	0.96
	TNS-I	1	19	0.76
$k = 10$	NSGS	0.98	234	0.97
	TNS-I	1	27	0.86
$k = 25$	NSGS	0.98	306	0.99
	TNS-I	1	34	0.92
$k = 100$	NSGS	0.99	430	0.99
	TNS-I	1	49	0.98

8.2.2 Effect of Control Variates

See Table 2 for an illustration. We know that $\rho^2_{(x,c)} = \sigma^2_c/\sigma^2_x = \sigma^2_c/(\sigma^2_c + \sigma^2_\eta)$, which represents how good this CV is. In our experiments, we fix σ^2_x to be 16. For example, $\rho^2_{(x,c)} = 0.2$ means $\sigma^2_x = 16$ and $\sigma^2_c = 3.2$. We find that the performance of the individual CV combined procedure is almost the same as NSGS when $\rho^2_{(x,c)}$ is 0.2. When $\rho^2_{(x,c)}$ is larger than 0.2, the CV combined procedure can outperform NSGS. Thus, very small $\rho^2_{(x,c)}$ is required for the CV combined procedure to be out performed by NSGS. Larger $\rho^2_{(x,c)}$ means the CVs can explain more variability of the outputs, and thereby makes the CV combined procedure more efficient.

Table 2 Effect of control variates for TNS-I in comparison with NSGS when $\rho_c = 0$, $\rho_\eta = 0$, and $k = 10$

	Procedure	PCS	ANS	PSS
$\sigma_x^2 = 16$	NSGS	1	235	0.97
$\rho^2_{(x,c)} = 0.2$	Individual CV	0.97	241	0.98
$\rho^2_{(x,c)} = 0.4$	Individual CV	1	181	0.99
$\rho^2_{(x,c)} = 0.6$	Individual CV	1	129	0.97
$\rho^2_{(x,c)} = 0.8$	Individual CV	1	68	0.99

8.2.3 Effect of Correlation

See Table 3 for an illustration. Here we compare TNS-I and TNS-P under different ρ_η. When the correlation between residuals is larger, TNS-P performs better and outperforms TNS-I easily. In Table 3, we see that the PSS of TNS-P is as low as 0.10, which shows the high efficiency of TNS-P when ρ_η is large. Notice that CRN does not affect the screening threshold for TNS-I, but it does affect the point estimator, which is why the performance of TNS-I in Table 3 varies when we have different ρ_η.

Table 3 Effect of correlation for TNS-I and TNS-P when $\sigma_c = 4$, $\sigma_\eta = 1$, and $k = 10$

Correlation	Procedure	PCS	ANS	PSS
$\rho_c = 0$	Individual CV	1	34	0.80
$\rho_\eta = 0.2$	Paired CV	1	30	0.74
$\rho_c = 0$	Individual CV	1	34	0.90
$\rho_\eta = 0.5$	Paired CV	1	26	0.55
$\rho_c = 0$	Individual CV	1	35	0.90
$\rho_\eta = 0.8$	Paired CV	1	13	0.10

9 Illustration

In this section we use a queueing example to illustrate the application of TNS-I, TNS-P, and NSGS. We use the $M/M/s/c$ model which represents a queueing system with Poisson arrivals, exponentially distributed service times, s servers, a capacity of c customers, and a first-come, first-served queueing discipline. The customers arrive with arrival rate λ. The service rate for an individual server is μ. We perform each procedure on five different configurations of the systems in which $\lambda/(s\mu) = 4/5$ where the performance measure is the steady-state mean of the waiting time in system. The capacity c is set as 15. The five configurations are shown in Table 4 along with their true expected waiting times. System 1 is obviously the best system. Note that smaller is better here.

To mitigate the initial transient bias, we initialize the simulation in steady state. That is, we calculate the steady-state distribution of the number of customers in the system, then sample the initial conditions for each replication in accordance with that steady-state distribution. An average waiting time for thirty customers is used

Table 4 The five queueing systems and their expected waiting times in steady state

System i	λ	s	μ	$\mathrm{E}[W]$
1	4	1	5	0.88
2	4	2	5/2	0.98
3	4	3	5/3	1.10
4	4	4	5/4	1.24
5	4	5	1	1.38

as the output on each replication. For TNS-I and TNS-P, we use the average service time as the control on replication j, which means

$$X_{ij} = \frac{\sum_{m=1}^{30} W_{ijm}}{30} \text{ and } C_{ij} = \frac{\sum_{m=1}^{30} S_{ijm}}{30}$$

where W_{ijm} is the waiting time in system for customer m of replication j from system i and S_{ijm} is the service time for customer m of replication j from system i. The initial sample size n_0 is set as 10 for each system. We choose the indifference zone δ to be 0.1 and CRN is applied.

Table 5 shows the results of the TNS-I, TNS-P, and NSGS procedures with 100 macroreplications and confidence level $1-\alpha = 0.95$. We also provide the estimated standard error of ANS to show that there is a significant difference.

Table 5 Results for NSGS, TNS-I, and TNS-P in 100 trials with $\delta = 0.1$, $n_0 = 10$, and $1-\alpha = 0.95$

Procedure	PCS	ANS	$\widehat{\mathrm{se}}$(ANS)	PSS
NSGS	0.97	462	13.4	1
TNS-I	1	301	10.8	0.81
TNS-P with CRN	1	207	10.4	0.68

These three procedures all exceed the desired probability of correct selection. NSGS is unable to screen out inferior systems in the first stage; therefore, its ANS is much larger than that of the other procedures. We can eliminate more systems in TNS-P than in TNS-I to further reduce the average number of samples needed by using CVs with CRN.

10 Conclusions

In this paper we presented a CV combined procedure that exploits the ability to screen out inferior systems and the statistical efficiency of control variates. We also proposed a more-general paired CV model and a new paired CV screening procedure to reduce the algorithm overhead and retain the benefits of paired CVs as well. As we showed in Sections 8 and 9, TNS-I is superior to NSGS for all the scenarios we examined. NSGS is based on the assumption that all systems are simulated independently, and TNS-I assumes that the dependence induced by CRN is entirely explained by the controls. On the other hand, TNS-P is significantly more efficient than TNS-I when the CVs do not explain all dependence due to CRN because a great deal of sampling cost can be saved. However, computational experiments show that the advantage of TNS-P over TNS-I diminishes with larger numbers of systems, and TNS-P incurs more algorithm overhead than TNS-I. As a rough rule of thumb, we use TNS-P when CRN is involved, but use TNS-I when all the systems are simulated independently.

Appendix

For the individual and paired CV combined procedures, the proofs in this appendix assume that Model (1) holds. We also assume $\mu_k \geq \mu_{k-1} \geq \cdots \geq \mu_1$ and $\mu_k - \mu_{k-1} \geq \delta$.

Model (5) and Model (6)

Here we compare Model (5), the Nelson and Staum paired CV model, and Model (6), our new paired CV model, in terms of $\mathrm{E}[\widehat{\mu}_{i\ell}(n)]$ and $\mathrm{Var}[\widehat{\mu}_{i\ell}(n)]$. We know that Model (6) tends not to lose much when the true underlying model is Model (5) because $\widehat{\mu}_{i\ell}(n)$ is still unbiased and the inflation of $\mathrm{Var}[\widehat{\mu}_{i\ell}(n)]$, due to the loss of degrees of freedom from $n-q-1$ to $n-2q-1$, will not be substantial when n is not too small. Therefore, we focus on the consequences of assuming that Model (5) holds when in fact the true model is Model (6).

We first compute $\widehat{\mu}_{i\ell}(n)$ under the assumption that Model (5) holds while Model (6) is actually the true model with $\boldsymbol{\beta}_i \neq \boldsymbol{\beta}_\ell$. The case $q_i = 1$, for all i, is sufficient to illustrate the point. For convenience, let

$$\boldsymbol{C} = \{C_{ij}, j = 1, 2, \ldots, n, i = 1, 2, \ldots, k\}$$

be the collection of all observed controls. Define

$$\widehat{B}(i,\ell) = \frac{\sum_{j=1}^{n}(X_j(i,\ell) - \bar{X}(i,\ell))(C_j(i,\ell) - \bar{C}(i,\ell))}{\sum_{j=1}^{n}(C_j(i,\ell) - \bar{C}(i,\ell))^2}.$$

We know that $\widehat{\mu}_{i\ell}(n) = \bar{X}(i,\ell) - (\bar{C}(i,\ell) - \xi_{i\ell})\widehat{B}(i,\ell)$, and $\mathrm{E}[\widehat{\mu}_{i\ell}(n)|\boldsymbol{C}] = \mathrm{E}[\bar{X}(i,\ell)|\boldsymbol{C}] - (\bar{C}(i,\ell) - \xi_{i\ell})\mathrm{E}[\widehat{B}(i,\ell)|\boldsymbol{C}]$. Therefore,

$$\mathrm{E}[\widehat{\mu}_{i\ell}(n)] = \mathrm{E}\left[\mathrm{E}[\widehat{\mu}_{i\ell}(n)|\boldsymbol{C}]\right] = \mu_{i\ell} - \mathrm{E}\left[(\bar{C}(i,\ell) - \xi_{i\ell})\mathrm{E}[\widehat{B}(i,\ell)|\boldsymbol{C}]\right]$$

and

$$\mathrm{E}[\widehat{B}(i,\ell)|\boldsymbol{C}] = \frac{\sum_{j=1}^{n} \mathrm{E}[X_j(i,\ell) - \bar{X}(i,\ell)|\boldsymbol{C}](C_j(i,\ell) - \bar{C}(i,\ell))}{\sum_{j=1}^{n}(C_j(i,\ell) - \bar{C}(i,\ell))^2}.$$

Because Model (6) holds, we know that

$$\begin{aligned}
&\mathrm{E}[X_j(i,\ell) - \bar{X}(i,\ell)|\boldsymbol{C}] \\
&\quad = (C_{ij} - \xi_i)\beta_i - (C_{\ell j} - \xi_\ell)\beta_\ell - (\bar{C}_i - \xi_i)\beta_i + (\bar{C}_\ell - \xi_\ell)\beta_\ell \\
&\quad = (C_{ij} - \bar{C}_i)\beta_i - (C_{\ell j} - \bar{C}_\ell)\beta_\ell.
\end{aligned}$$

Thus, we obtain

$$\begin{aligned}
\mathrm{E}[\widehat{B}(i,\ell)|\boldsymbol{C}] &= \frac{\sum_{j=1}^{n} \left((C_{ij} - \bar{C}_i)\beta_i - (C_{\ell j} - \bar{C}_\ell)\beta_\ell\right)(C_j(i,\ell) - \bar{C}(i,\ell))}{\sum_{j=1}^{n}(C_j(i,\ell) - \bar{C}(i,\ell))^2} \\
&= \beta_i + (\beta_i - \beta_\ell)\frac{\sum_{j=1}^{n} \left(C_{\ell j} - C_\ell\right)(C_j(i,\ell) - \bar{C}(i,\ell))}{\sum_{j=1}^{n}(C_j(i,\ell) - \bar{C}(i,\ell))^2}. \qquad (11)
\end{aligned}$$

Therefore, the bias is $\mathrm{E}[\widehat{\mu}_{i\ell}(n)] - \mu_{i\ell}$, which is equal to

$$\begin{aligned}
&-\mathrm{E}\left[(\bar{C}(i,\ell) - \xi_{i\ell})\mathrm{E}[\widehat{B}(i,\ell)|\boldsymbol{C}]\right] \\
&= -\mathrm{E}\left[\left((\beta_i - \beta_\ell)\frac{\sum_{j=1}^{n}(C_{\ell j} - \bar{C}_\ell)(C_j(i,\ell) - \bar{C}(i,\ell))}{\sum_{j=1}^{n}(C_j(i,\ell) - \bar{C}(i,\ell))^2}\right)\left(\bar{C}(i,\ell) - \xi_{i\ell}\right)\right], \qquad (12)
\end{aligned}$$

is not equal to 0 in general if $\beta_i \neq \beta_\ell$.

We now examine the impact on variance. Notice that

$$\begin{aligned}
\mathrm{Var}[\widehat{\mu}_{i\ell}(n)] &= \mathrm{Var}\left[\mathrm{E}\left[\widehat{\mu}_{i\ell}(n)|\boldsymbol{C}\right]\right] + \mathrm{E}\left[\mathrm{Var}[\widehat{\mu}_{i\ell}(n)|\boldsymbol{C}]\right] \\
&= \mathrm{Var}\left[(\bar{C}_i - \xi_i)\beta_i - (\bar{C}_\ell - \xi_\ell)\beta_\ell - (\bar{C}(i,\ell) - \xi_{i\ell})\mathrm{E}[\widehat{B}(i,\ell)|\boldsymbol{C}]\right] \\
&\quad + \mathrm{E}\left[\mathrm{Var}[\widehat{\mu}_{i\ell}(n)|\boldsymbol{C}]\right].
\end{aligned}$$

Since

$$\widehat{B}(i,\ell) = \frac{\sum_{j=1}^{n} X_j(i,\ell)(C_j(i,\ell) - \bar{C}(i,\ell))}{\sum_{j=1}^{n}(C_j(i,\ell) - \bar{C}(i,\ell))^2},$$

we have

$$\begin{aligned}
\mathrm{Var}[\widehat{B}(i,\ell)|\boldsymbol{C}] &= \mathrm{Var}[X_j(i,\ell)|\boldsymbol{C}]\frac{\sum_{j=1}^{n}(C_j(i,\ell) - \bar{C}(i,\ell))^2}{\left(\sum_{j=1}^{n}(C_j(i,\ell) - \bar{C}(i,\ell))^2\right)^2} \\
&= \frac{\mathrm{Var}[X_j(i,\ell)|\boldsymbol{C}]}{\sum_{j=1}^{n}(C_j(i,\ell) - \bar{C}(i,\ell))^2}.
\end{aligned}$$

Now

$$\begin{aligned}\text{Var}[\widehat{\mu}_{i\ell}(n)|\boldsymbol{C}] &= \text{Var}[\bar{X}(i,\ell) - (\bar{C}(i,\ell) - \xi_{i\ell})\widehat{B}(i,\ell)|\boldsymbol{C}] \\ &= \text{Var}[\bar{X}(i,\ell)|\boldsymbol{C}] + (\bar{C}(i,\ell) - \xi_{i\ell})^2\text{Var}[\widehat{B}(i,\ell)|\boldsymbol{C}] \\ &\quad -2(\bar{C}(i,\ell) - \xi_{i\ell})\text{Cov}[\bar{X}(i,\ell), \widehat{B}(i,\ell)|\boldsymbol{C}] \\ &= \text{Var}[X_j(i,\ell)|\boldsymbol{C}]\left(\frac{1}{n} + \frac{(\bar{C}(i,\ell) - \xi_{i\ell})^2}{\sum_{j=1}^n (C_j(i,\ell) - \bar{C}(i,\ell))^2}\right) \\ &\quad -2(\bar{C}(i,\ell) - \xi_{i\ell})\text{Cov}[\bar{X}(i,\ell), \widehat{B}(i,\ell)|\boldsymbol{C}].\end{aligned}$$

Further,

$$\begin{aligned}&\text{Cov}[\bar{X}(i,\ell), \widehat{B}(i,\ell)|\boldsymbol{C}] \\ &\quad = \text{Cov}[\mu_i - \mu_\ell + (\bar{C}_i - \xi_i)\beta_i - (\bar{C}_\ell - \xi_\ell)\beta_\ell + \bar{\eta}(i,\ell), \widehat{B}(i,\ell)|\boldsymbol{C}] \\ &\quad = \text{Cov}[\bar{\eta}(i,\ell), \widehat{B}(i,\ell)|\boldsymbol{C}].\end{aligned}$$

Since

$$\begin{aligned}&\widehat{B}(i,\ell) \\ &= \frac{\sum_{j=1}^n \big((C_{ij} - \bar{C}_i)\beta_i - (C_{\ell j} - \bar{C}_\ell)\beta_\ell + \eta_j(i,\ell) - \bar{\eta}(i,\ell)\big)(C_j(i,\ell) - \bar{C}(i,\ell))}{\sum_{j=1}^n (C_j(i,\ell) - \bar{C}(i,\ell))^2} \\ &= \beta_i + (\beta_i - \beta_\ell)\frac{\sum_{j=1}^n (C_{\ell j} - \bar{C}_\ell)(C_j(i,\ell) - \bar{C}(i,\ell))}{\sum_{j=1}^n (C_j(i,\ell) - \bar{C}(i,\ell))^2} \\ &\quad + \frac{\sum_{j=1}^n (\eta_j(i,\ell) - \bar{\eta}(i,\ell))(C_j(i,\ell) - \bar{C}(i,\ell))}{\sum_{j=1}^n (C_j(i,\ell) - \bar{C}(i,\ell))^2},\end{aligned}$$

we have

$$\begin{aligned}&\text{Cov}[\bar{\eta}(i,\ell), \widehat{B}(i,\ell)|\boldsymbol{C}] \\ &\quad = \text{Cov}\left[\bar{\eta}(i,\ell), \frac{\sum_{j=1}^n (\eta_j(i,\ell) - \bar{\eta}(i,\ell))(C_j(i,\ell) - \bar{C}(i,\ell))}{\sum_{j=1}^n (C_j(i,\ell) - \bar{C}(i,\ell))^2}\middle|\boldsymbol{C}\right] \\ &\quad = \frac{1}{\sum_{j=1}^n (C_j(i,\ell) - \bar{C}(i,\ell))^2}\text{Cov}\left[\bar{\eta}(i,\ell), \sum_{j=1}^n (\eta_j(i,\ell) - \bar{\eta}(i,\ell))C_j(i,\ell)\middle|\boldsymbol{C}\right] \\ &\quad = \frac{1}{\sum_{j=1}^n (C_j(i,\ell) - \bar{C}(i,\ell))^2}\left(\text{Cov}\left[\bar{\eta}(i,\ell), \sum_{j=1}^n \eta_j(i,\ell)C_j(i,\ell)\middle|\boldsymbol{C}\right]\right. \\ &\qquad\qquad \left. - \text{Cov}\left[\bar{\eta}(i,\ell), \bar{\eta}(i,\ell)\sum_{j=1}^n C_j(i,\ell)\middle|\boldsymbol{C}\right]\right)\end{aligned}$$

$$= \frac{1}{\sum_{j=1}^{n}(C_j(i,\ell) - \bar{C}(i,\ell))^2}\left(\sum_{j=1}^{n} C_j(i,\ell)\left(\frac{\mathrm{Var}\left[\eta_j(i,\ell)\right]}{n} - \mathrm{Var}\left[\bar{\eta}(i,\ell)\right]\right)\right)$$
$$= 0 \quad (\text{since } \mathrm{Var}\left[\eta_j(i,\ell)\right] = \sigma_{i\ell}^2, \forall j).$$

Then we take the expectation of $\mathrm{Var}[\widehat{\mu}_{i\ell}(n)|\boldsymbol{C}]$ to yield

$$\mathrm{E}\left[\mathrm{Var}[\widehat{\mu}_{i\ell}(n)|\boldsymbol{C}]\right] = \mathrm{Var}[\eta_j(i,\ell)]\left(\frac{n-2}{n(n-3)}\right)$$

(Lavenberg and Welch 1981). Further, from Equation (11),

$$\begin{aligned}
&\mathrm{Var}\left[(\bar{C}_i - \xi_i)\beta_i - (\bar{C}_\ell - \xi_\ell)\beta_\ell - (\bar{C}(i,\ell) - \xi_{i\ell})\mathrm{E}[\widehat{B}(i,\ell)|\boldsymbol{C}]\right] \\
&\quad = \mathrm{Var}\Bigg[(\bar{C}_i - \xi_i)\beta_i - (\bar{C}_\ell - \xi_\ell)\beta_\ell - \left(\bar{C}(i,\ell) - \xi_{i\ell}\right)\beta_i \\
&\qquad -(\beta_i - \beta_\ell)\left(\bar{C}(i,\ell) - \xi_{i\ell}\right)\frac{\sum_{j=1}^{n}\left(C_{\ell j} - \bar{C}_\ell\right)\left(C_j(i,\ell) - \bar{C}(i,\ell)\right)}{\sum_{j=1}^{n}(C_j(i,\ell) - \bar{C}(i,\ell))^2}\Bigg].
\end{aligned}$$

To simplify this expression, let

$$\Lambda(\boldsymbol{C}) = \left(\bar{C}(i,\ell) - \xi_{i\ell}\right)\frac{\sum_{j=1}^{n}\left(C_{\ell j} - \bar{C}_\ell\right)\left(C_j(i,\ell) - \bar{C}(i,\ell)\right)}{\sum_{j=1}^{n}(C_j(i,\ell) - \bar{C}(i,\ell))^2},$$

which is a function of $\boldsymbol{C}$. Then,

$$\begin{aligned}
&\mathrm{Var}\left[\mathrm{E}\left[\widehat{\mu}_{i\ell}(n)|\boldsymbol{C}\right]\right] \\
&\quad = \mathrm{Var}\left[(\bar{C}_i - \xi_i)\beta_i - (\bar{C}_\ell - \xi_\ell)\beta_\ell - (\bar{C}(i,\ell) - \xi_{i\ell})\mathrm{E}[\widehat{B}(i,\ell)|\boldsymbol{C}]\right] \\
&\quad = \mathrm{Var}\left[(\bar{C}_i - \xi_i)\beta_i - (\bar{C}_\ell - \xi_\ell)\beta_\ell - \left(\bar{C}(i,\ell) - \xi_{i\ell}\right)\beta_i - (\beta_i - \beta_\ell)\Lambda(\boldsymbol{C})\right] \\
&\quad = \mathrm{Var}\Big[(\bar{C}_i - \xi_i)\beta_i - (\bar{C}_\ell - \xi_\ell)\beta_i + (\bar{C}_\ell - \xi_\ell)\beta_i - (\bar{C}_\ell - \xi_\ell)\beta_\ell \\
&\qquad - \left(\bar{C}(i,\ell) - \xi_{i\ell}\right)\beta_i - (\beta_i - \beta_\ell)\Lambda(\boldsymbol{C})\Big] \\
&\quad = \mathrm{Var}\left[(\bar{C}_\ell - \xi_\ell)\beta_i - (\bar{C}_\ell - \xi_\ell)\beta_\ell - (\beta_i - \beta_\ell)\Lambda(\boldsymbol{C})\right] \\
&\quad = (\beta_i - \beta_\ell)^2\mathrm{Var}\left[\bar{C}_\ell - \Lambda(\boldsymbol{C})\right].
\end{aligned}$$

So, when we assume that Model (5) holds but the true model is Model (6), the variance of the CV estimator is

$$\mathrm{Var}[\widehat{\mu}_{i\ell}(n)] = (\beta_i - \beta_\ell)^2\mathrm{Var}\left[\bar{C}_\ell - \Lambda(\boldsymbol{C})\right] + \mathrm{Var}[\eta_j(i,\ell)]\left(\frac{n-2}{n(n-3)}\right). \quad (13)$$

On the other hand, when we assume Model (6) holds and Model (6) is indeed the true model,

$$\mathrm{Var}[\widehat{\mu}_{i\ell}(n)] = \mathrm{Var}[\eta_j(i,\ell)]\left(\frac{n-2}{n(n-4)}\right). \tag{14}$$

In summary, if we compute $\widehat{\mu}_{i\ell}(n)$ assuming Model (5) holds while Model (6) is the true model with $\boldsymbol{\beta}_i \neq \boldsymbol{\beta}_\ell$, then Equation (12) shows us that $\widehat{\mu}_{i\ell}(n)$ will be biased. Further, from Equations (13) and (14) we see that $\mathrm{Var}[\widehat{\mu}_{i\ell}(n)]$ will be increased, especially when $\boldsymbol{\beta}_i$ is very different from $\boldsymbol{\beta}_\ell$. This illustrates the inferiority of Model (5) relative to Model (6).

CV Combined Procedure with Individual Screening

We prove the multiplicative approach for Theorem 1, with $1-\alpha = (1-\alpha_0)(1-\alpha_1-\alpha_2)$. In the multiplicative approach, we assume that Model (1) holds. For convenience, let $\boldsymbol{C} = \{\boldsymbol{C}_{ij}, j = 1,2,\ldots,n_i, i = 1,2,\ldots,k\}$ be the collection of all observed controls, and let

$$\widehat{\tau}^2 = \{\widehat{\tau}_1^2(n_1), \widehat{\tau}_2^2(n_2), \ldots, \widehat{\tau}_k^2(n_k)\}$$

be the collection of all observed residual-variance estimators. Define

$$Z_i = \frac{\widehat{\mu}_k(n_k) - \widehat{\mu}_i(n_i) - (\mu_k - \mu_i)}{\sqrt{\widehat{\Delta}_k^2(n_k)\tau_k^2 + \widehat{\Delta}_i^2(n_i)\tau_i^2}},\quad V_i = \frac{\widehat{\mu}_k(N_k) - \widehat{\mu}_i(N_i) - (\mu_k - \mu_i)}{\sqrt{\widehat{\Delta}_k^2(N_k)\tau_k^2 + \widehat{\Delta}_i^2(N_i)\tau_i^2}},$$

$$A_i = \widehat{\Delta}_k^2(n_k)\tau_k^2 + \widehat{\Delta}_i^2(n_i)\tau_i^2,\ D_i = \widehat{\Delta}_k^2(N_k)\tau_k^2 + \widehat{\Delta}_i^2(N_i)\tau_i^2.$$

The probability of correct selection is

$$\begin{aligned}
&\Pr\left\{\widehat{\mu}_k(n_k) - \widehat{\mu}_i(n_i) \geq -W_{ki}, \forall i \neq k;\ \widehat{\mu}_k(N_k) > \widehat{\mu}_i(N_i), \forall i \in I\right\}\\
&\quad = \Pr\left\{\frac{\widehat{\mu}_k(n_k) - \widehat{\mu}_i(n_i) - (\mu_k - \mu_i)}{\sqrt{\widehat{\Delta}_k^2(n_k)\tau_k^2 + \widehat{\Delta}_i^2(n_i)\tau_i^2}} \geq \frac{-W_{ki} - (\mu_k - \mu_i)}{\sqrt{\widehat{\Delta}_k^2(n_k)\tau_k^2 + \widehat{\Delta}_i^2(n_i)\tau_i^2}}, \forall i \neq k;\right.\\
&\qquad \left.\frac{\widehat{\mu}_k(N_k) - \widehat{\mu}_i(N_i) - (\mu_k - \mu_i)}{\sqrt{\widehat{\Delta}_k^2(N_k)\tau_k^2 + \widehat{\Delta}_i^2(N_i)\tau_i^2}} > \frac{-(\mu_k - \mu_i)}{\sqrt{\widehat{\Delta}_k^2(N_k)\tau_k^2 + \widehat{\Delta}_i^2(N_i)\tau_i^2}}, \forall i \in I\right\}\\
&\geq \Pr\left\{Z_i \leq \frac{W_{ki}}{\sqrt{A_i}},\ V_i < \frac{\delta}{\sqrt{D_i}}, \forall i \neq k\right\} &(15)\\
&= \mathrm{E}\left[\Pr\left\{Z_i \leq \frac{W_{ki}}{\sqrt{A_i}},\ V_i < \frac{\delta}{\sqrt{D_i}}, \forall i \neq k \,\middle|\, \boldsymbol{C}, \widehat{\tau}^2\right\}\right]\\
&\geq \mathrm{E}\left[\Pr\left\{Z_i \leq \frac{W_{ki}}{\sqrt{A_i}}, \forall i \neq k \,\middle|\, \boldsymbol{C}, \widehat{\tau}^2\right\}\Pr\left\{V_i < \frac{\delta}{\sqrt{D_i}}, \forall i \neq k \,\middle|\, \boldsymbol{C}, \widehat{\tau}^2\right\}\right] &(16)
\end{aligned}$$

$$= \mathrm{E}\left[\mathrm{E}\left[\Pr\left\{Z_i \le \frac{W_{ki}}{\sqrt{A_i}}, \forall i \neq k \,\middle|\, \boldsymbol{C}, \hat{\tau}^2\right\}\right.\right.$$

$$\left.\left.\times \Pr\left\{V_i < \frac{\delta}{\sqrt{D_i}}, \forall i \neq k \,\middle|\, \boldsymbol{C}, \hat{\tau}^2\right\} \,\middle|\, \boldsymbol{C}\right]\right] \tag{17}$$

$$\ge \mathrm{E}\left[\mathrm{E}\left[\Pr\left\{Z_i \le \frac{W_{ki}}{\sqrt{A_i}}, \forall i \neq k \,\middle|\, \boldsymbol{C}, \hat{\tau}^2\right\} \,\middle|\, \boldsymbol{C}\right]\right.$$

$$\left.\times \mathrm{E}\left[\Pr\left\{V_i < \frac{\delta}{\sqrt{D_i}}, \forall i \neq k \,\middle|\, \boldsymbol{C}, \hat{\tau}^2\right\} \,\middle|\, \boldsymbol{C}\right]\right]. \tag{18}$$

Inequality (15) holds because $\mu_k - \mu_i \ge \delta$, $\{\forall i \in I\}$ is a smaller set than $\{\forall i \neq k\}$ which makes the condition more restrictive, and because of the symmetry of the normal distribution. Inequality (16) is an application of Slepian's inequality (cf. Tong 1980). Then since the first term in (17) is a nonnegative, real-valued function and increasing in each of $\{\hat{\tau}_1^2(n_1), \hat{\tau}_2^2(n_2), \ldots, \hat{\tau}_k^2(n_k)\}$, and the second term in (17) is nondecreasing in each of $\{\hat{\tau}_1^2(n_1), \hat{\tau}_2^2(n_2), \ldots, \hat{\tau}_k^2(n_k)\}$, and by Lemma 1 $\{\hat{\tau}_1^2(n_1), \hat{\tau}_2^2(n_2), \ldots, \hat{\tau}_k^2(n_k)\}$ are conditionally independent given $\boldsymbol{C}$, we can apply Lemma 2.4 in Tamhane (1977) to get Inequality (18). From Nelson and Staum (2006), we know that

$$\mathrm{E}\left[\Pr\left\{Z_i \le \frac{W_{ki}}{\sqrt{A_i}}, \forall i \neq k \,\middle|\, \boldsymbol{C}, \hat{\tau}^2\right\} \,\middle|\, \boldsymbol{C}\right] = \Pr\left\{Z_i \le \frac{W_{ki}}{\sqrt{A_i}}, \forall i \neq k \,\middle|\, \boldsymbol{C}\right\}$$
$$\ge 1 - \alpha_0$$

and

$$\mathrm{E}\left[\mathrm{E}\left[\Pr\left\{V_i < \frac{\delta}{\sqrt{D_i}}, \forall i \neq k \,\middle|\, \boldsymbol{C}, \hat{\tau}^2\right\} \,\middle|\, \boldsymbol{C}\right]\right] = \Pr\left\{V_i < \frac{\delta}{\sqrt{D_i}}, \forall i \neq k\right\}$$
$$\ge 1 - \alpha_1 - \alpha_2.$$

So we can conclude from Inequality (18) that the probability of correct selection is

$$\Pr\left\{\hat{\mu}_k(n_k) - \hat{\mu}_i(n_i) \ge -W_{ki}, \forall i \neq k;\ \hat{\mu}_k(N_k) > \hat{\mu}_i(N_i), \forall i \in I\right\}$$
$$\ge (1 - \alpha_0)(1 - \alpha_1 - \alpha_2) = 1 - \alpha.$$

We need to verify that Slepian's inequality can be applied for Inequality (16). It is easy to show that the $\mathrm{Cov}[Z_i, Z_j|\boldsymbol{C}, \hat{\tau}^2]$ and $\mathrm{Cov}[V_i, V_j|\boldsymbol{C}, \hat{\tau}^2]$ are nonnegative for any system $i \neq j$ (Nelson and Staum 2006). So here we only need to examine $\mathrm{Cov}[Z_i, V_j|\boldsymbol{C}, \hat{\tau}^2]$.

When $i \neq j$,

$$\begin{aligned}
&\mathrm{Cov}[Z_i, V_j|\boldsymbol{C}, \widehat{\tau}^2]\\
&= \mathrm{Cov}\left[\left.\frac{\widehat{\mu}_k(n_k)-\widehat{\mu}_i(n_i)-(\mu_k-\mu_i)}{\sqrt{\widehat{\Delta}_k^2(n_k)\tau_k^2+\widehat{\Delta}_i^2(n_i)\tau_i^2}}, \frac{\widehat{\mu}_k(N_k)-\widehat{\mu}_j(N_j)-(\mu_k-\mu_j)}{\sqrt{\widehat{\Delta}_k^2(N_k)\tau_k^2+\widehat{\Delta}_j^2(N_j)\tau_j^2}}\right| \boldsymbol{C}, \widehat{\tau}^2\right]\\
&= \frac{1}{a_i d_j}\mathrm{Cov}\left[\widehat{\mu}_k(n_k), \widehat{\mu}_k(N_k)|\boldsymbol{C}, \widehat{\tau}^2\right],
\end{aligned}$$

where $a_i = \sqrt{A_i}$, $d_j = \sqrt{D_j}$. We can factor out a_i and d_j since they are both constants when we condition on $\boldsymbol{C}$. We also know that

$$\begin{aligned}
\widehat{\mu}_k(n_k) &= \left[\frac{1}{n_k}\boldsymbol{1}'_{n_k\times 1} - \left(\bar{\boldsymbol{C}}_k(n_k)-\boldsymbol{\xi}_k\right)'\left(\boldsymbol{L}'_k(n_k)\boldsymbol{L}_k(n_k)\right)^{-1}\boldsymbol{L}'_k(n_k)\right]\boldsymbol{X}_k(n_k)\\
&= \boldsymbol{a}'\boldsymbol{X}_k(n_k)
\end{aligned}$$

and

$$\begin{aligned}
\widehat{\mu}_k(N_k) &= \left[\frac{1}{N_k}\boldsymbol{1}'_{N_k\times 1} - \left(\bar{\boldsymbol{C}}_k(N_k)-\boldsymbol{\xi}_k\right)'\left(\boldsymbol{L}'_k(N_k)\boldsymbol{L}_k(N_k)\right)^{-1}\boldsymbol{L}'_k(N_k)\right]\boldsymbol{X}_k(N_k)\\
&= \boldsymbol{b}'\boldsymbol{X}_k(N_k).
\end{aligned}$$

It follows that

$$\mathrm{Cov}[Z_i, V_j|\boldsymbol{C}, \widehat{\tau}^2] = \frac{1}{a_i d_j}\mathrm{Cov}\left[\widehat{\mu}_k(n_k), \widehat{\mu}_k(N_k)|\boldsymbol{C}, \widehat{\tau}^2\right] = \frac{1}{a_i d_j}\boldsymbol{a}'\Omega\boldsymbol{b},$$

where

$$\Omega = (\boldsymbol{E}, \boldsymbol{F}),\ \boldsymbol{E} = \mathrm{Var}[\boldsymbol{X}_k(n_k)],\ \text{and } \boldsymbol{F} = [\boldsymbol{0}]_{n_k\times(N_k-n_k)}.$$

Since $\boldsymbol{E}$ is a diagonal matrix with positive elements and $a_i d_j$ is positive, we can conclude that $\mathrm{Cov}[Z_i, V_j|\boldsymbol{C}, \widehat{\tau}^2]$ is nonnegative if $\boldsymbol{a}'\boldsymbol{B}$ is nonnegative, where the vector $\boldsymbol{B}$ is composed of the first n_k elements of the vector $\boldsymbol{b}$. We have

$$\boldsymbol{B}' = \left[\frac{1}{N_k}\boldsymbol{1}'_{n_k\times 1} - \left(\bar{\boldsymbol{C}}_k(N_k)-\boldsymbol{\xi}_k\right)'\left(\boldsymbol{L}'_k(N_k)\boldsymbol{L}_k(N_k)\right)^{-1}\left(\boldsymbol{L}'_k(n_k)+\boldsymbol{m}_k(n_k)\right)\right],$$

where

$$\boldsymbol{m}_k(n_k) = \{\bar{\boldsymbol{C}}_k(n_k)-\bar{\boldsymbol{C}}_k(N_k), \bar{\boldsymbol{C}}_k(n_k)-\bar{\boldsymbol{C}}_k(N_k), \ldots, \bar{\boldsymbol{C}}_k(n_k)-\bar{\boldsymbol{C}}_k(N_k)\}_{1\times n_k}.$$

Then

$$
\begin{aligned}
a'B &= \frac{1}{N_k} - \frac{1}{n_k}\mathbf{1}'_{n_k\times 1}\left[L_k(n_k)\left(L'_k(N_k)L_k(N_k)\right)^{-1}\left(\bar{C}_k(N_k)-\xi_k\right)\right] \\
&\quad - \frac{1}{n_k}\mathbf{1}'_{n_k\times 1}\left[m'_k(n_k)\left(L'_k(N_k)L_k(N_k)\right)^{-1}\left(\bar{C}_k(N_k)-\xi_k\right)\right] \\
&\quad - \left(\bar{C}_k(n_k)-\xi_k\right)'\left(L'_k(n_k)L_k(n_k)\right)^{-1}L'_k(n_k)\frac{1}{N_k}\mathbf{1}_{n_k\times 1} \\
&\quad + \left(\bar{C}_k(n_k)-\xi_k\right)'\left(L'_k(n_k)L_k(n_k)\right)^{-1}L'_k(n_k) \\
&\quad \times\left[L_k(n_k)\left(L'_k(N_k)L_k(N_k)\right)^{-1}\left(\bar{C}_k(N_k)-\xi_k\right)\right] \\
&\quad + \left(\bar{C}_k(n_k)-\xi_k\right)'\left(L'_k(n_k)L_k(n_k)\right)^{-1}L'_k(n_k) \\
&\quad \times\left[m'_k(n_k)\left(L'_k(N_k)L_k(N_k)\right)^{-1}\left(\bar{C}_k(N_k)-\xi_k\right)\right] \\
&= \frac{1}{N_k} - \frac{1}{n_k}\mathbf{1}'_{n_k\times 1}\left[m'_k(n_k)\left(L'_k(N_k)L_k(N_k)\right)^{-1}\left(\bar{C}_k(N_k)-\xi_k\right)\right] \\
&\quad + \left(\bar{C}_k(n_k)-\xi_k\right)'\left[\left(L'_k(N_k)L_k(N_k)\right)^{-1}\left(\bar{C}_k(N_k)-\xi_k\right)\right] \qquad (19) \\
&= \frac{1}{N_k} + \left(\bar{C}_k(N_k)-\xi_k\right)'\left[\left(L'_k(N_k)L_k(N_k)\right)^{-1}\left(\bar{C}_k(N_k)-\xi_k\right)\right] \qquad (20) \\
&= \frac{1}{N_k} + \frac{\left(\bar{C}_k(N_k)-\xi_k\right)' S^{-1}\left(\bar{C}_k(N_k)-\xi_k\right)}{N_k-1} \qquad (21) \\
&= \frac{1}{N_k} + \frac{T^2}{N_k(N_k-1)} > 0, \qquad (22)
\end{aligned}
$$

where S is the sample covariance matrix of controls from system k. That is, $L'_k(N_k)L_k(N_k) = (N_k-1)S$. Also, T^2 is the generalized T^2-statistic of controls from system k, so $T^2 = N_k(\bar{C}_k(N_k)-\xi_k)'S^{-1}(\bar{C}_k(N_k)-\xi_k)$ (Anderson 1984). Equality (19) holds since $\mathbf{1}'_{n_k\times 1}L_k(n_k) = 0$, $(L'_k(n_k)L_k(n_k))^{-1}L'_k(n_k)L_k(n_k) = \mathbf{I}$, and $L'_k(n_k)m'_k(n_k) = 0$. Equality (20) holds because $(1/n_k)\mathbf{1}'_{n_k\times 1}m'_k(n_k) = (\bar{C}_k(n_k) - \bar{C}_k(N_k))'$. Equality (21) and Equality (22) hold because of the definitions of S^{-1} and T^2. Therefore, $\mathrm{Cov}[Z_i, V_j|C, \widehat{\tau}^2]$ is positive when $i \neq j$.

When $i = j$,

$$
\begin{aligned}
&\mathrm{Cov}[Z_i, V_j|C, \widehat{\tau}^2] \\
&\quad = \frac{1}{a_i d_j}\left[\mathrm{Cov}\left[\widehat{\mu}_k(n_k), \widehat{\mu}_k(N_k)|C, \widehat{\tau}^2\right] + \mathrm{Cov}\left[\widehat{\mu}_j(n_j), \widehat{\mu}_j(N_j)|C, \widehat{\tau}^2\right]\right].
\end{aligned}
$$

We can also obtain $\mathrm{Cov}\left[\widehat{\mu}_j(n_j), \widehat{\mu}_j(N_j)|C, \widehat{\tau}^2\right] > 0$. Therefore, $\mathrm{Cov}[Z_i, V_j|C, \widehat{\tau}^2]$ is positive when $i = j$.

CV Combined Procedure with Paired Screening

We turn to the proof of Theorem 2. We can apply the additive approach to the CV combined procedure with paired screening. We assume that Model (1) holds. The probability of correct selection is

$$\begin{aligned}
&\Pr\left\{\widehat{\mu}_{ki}(n) \geq -t_{ki}\,\widehat{\Delta}_{ki}(n)\widehat{\tau}_{ki}(n), \forall i \neq k;\ \widehat{\mu}_k(N_k) > \widehat{\mu}_i(N_i), \forall i \in I\right\} \\
&\quad \geq \Pr\left\{\widehat{\mu}_{ki}(n) \geq -t_{ki}\,\widehat{\Delta}_{ki}(n)\widehat{\tau}_{ki}(n), \forall i \neq k;\right. \\
&\qquad \left. \frac{\widehat{\mu}_k(N_k) - \widehat{\mu}_i(N_i) - (\mu_k - \mu_i)}{\sqrt{\widehat{\Delta}_k^2(N_k)\tau_k^2 + \widehat{\Delta}_i^2(N_i)\tau_i^2}} > \frac{-(\mu_k - \mu_i)}{\sqrt{\widehat{\Delta}_k^2(N_k)\tau_k^2 + \widehat{\Delta}_i^2(N_i)\tau_i^2}}, \forall i \in I\right\} \\
&\quad \geq \Pr\left\{\widehat{\mu}_{ki}(n) \geq -t_{ki}\,\widehat{\Delta}_{ki}(n)\widehat{\tau}_{ki}(n), \forall i \neq k;\right. \\
&\qquad \left. V_i > \frac{-\delta}{\sqrt{\widehat{\Delta}_k^2(N_k)\tau_k^2 + \widehat{\Delta}_i^2(N_i)\tau_i^2}}, \forall i \in I\right\} \qquad (23) \\
&\quad \geq \Pr\left\{\widehat{\mu}_{ki}(n) \geq -t_{ki}\,\widehat{\Delta}_{ki}(n)\widehat{\tau}_{ki}(n), \forall i \neq k;\right. \\
&\qquad \left. V_i < \frac{\delta}{\sqrt{\widehat{\Delta}_k^2(N_k)\tau_k^2 + \widehat{\Delta}_i^2(N_i)\tau_i^2}}, \forall i \neq k;\ \widehat{\Delta}_i^2(N_i) \leq \frac{\delta^2}{h^2\,\widehat{\tau}_i^2(n_i)}, \forall i\right\} \qquad (24) \\
&\quad \geq \Pr\left\{\widehat{\mu}_{ki}(n) \geq -t_{ki}\,\widehat{\Delta}_{ki}(n)\widehat{\tau}_{ki}(n), \forall i \neq k;\ V_i < \frac{h}{\sqrt{\frac{\tau_k^2}{\widehat{\tau}_k^2(n_k)} + \frac{\tau_i^2}{\widehat{\tau}_i^2(n_i)}}}, \forall i \neq k;\right. \\
&\qquad \left. \widehat{\Delta}_i^2(N_i) \leq \frac{\delta^2}{h^2\,\widehat{\tau}_i^2(n_i)}, \forall i\right\} \\
&\quad \geq 1 - p_0 - p_1 - p_2, \qquad (25)
\end{aligned}$$

where

$$p_0 = 1 - \Pr\left\{\widehat{\mu}_{ki}(n) \geq -t_{ki}\,\widehat{\Delta}_{ki}(n)\widehat{\tau}_{ki}(n), \forall i \neq k\right\},$$

$$p_1 = 1 - \Pr\left\{V_i < \frac{h}{\sqrt{\frac{\tau_k^2}{\widehat{\tau}_k^2(n_k)} + \frac{\tau_i^2}{\widehat{\tau}_i^2(n_i)}}}, \forall i \neq k\right\}, \quad \text{and}$$

$$p_2 = 1 - \Pr\left\{\widehat{\Delta}_i^2(N_i) \leq \frac{\delta^2}{h^2\,\widehat{\tau}_i^2(n_i)}, \forall i\right\}.$$

By Lemma 1, the conditional distribution of V_i given $\{\widehat{\tau}^2, \boldsymbol{C}\}$ is standard normal. Inequality (23) holds because $\mu_k - \mu_i \geq \delta$. Inequality (24) holds because of the symmetry of the normal distribution and the probability is smaller if we require a bound on the value of $\widehat{\Delta}_i^2(N_i)$, while Inequality (25) is an application of the Bonferroni inequality.

We know that $p_0 \leq \alpha_0$, $p_1 \leq \alpha_1$, $p_2 \leq \alpha_2$ (Nelson and Staum 2006). So we can conclude that the probability of correct selection is

$$\Pr\left\{\widehat{\mu}_{ki}(n) \geq -t_{ki}\, \widehat{\Delta}_{ki}(n)\widehat{\tau}_{ki}(n), \forall i \neq k;\ \widehat{\mu}_k(N_k) > \widehat{\mu}_i(N_i), \forall i \in I\right\}$$
$$\geq 1 - \alpha_0 - \alpha_1 - \alpha_2 = 1 - \alpha.$$

References and Selected Bibliography

Anderson, T. W. 1984. *An Introduction to Multivariate Statistical Analysis*, 2nd edition. New York: John Wiley & Sons.

Añonuevo, R., and B. L. Nelson. 1988. Automated estimation and variance reduction via control variates for infinite-horizon simulations. *Computers & Operations Research* 15:447–456.

Bauer, K. W., and J. R. Wilson. 1992. Control-variate selection criteria. *Naval Research Logistics* 39:307–321.

Bechhofer, R. E., T. J. Santner, and D. Goldsman. 1995. *Design and Analysis of Experiments for Statistical Selection, Screening and Multiple Comparisons*. New York: John Wiley & Sons.

Boesel, J., B. L. Nelson, and S. Kim. 2003. Using ranking and selection to 'clean up' after simulation optimization. *Operations Research* 51:814–825.

Lavenberg, S. S., and P. D. Welch. 1981. A perspective on the use of control variables to increase the efficiency of Monte Carlo simulations. *Management Science* 27:322–335.

Nelson, B. L. 1989. Batch size effects on the efficiency of control variates in simulation. *European Journal of Operational Research* 43:184–196.

Nelson, B. L. 1990. Control-variate remedies. *Operations Research* 38:974–992.

Nelson, B. L., and D. Goldsman. 2001. Comparisons with a standard in simulation experiments. *Management Science* 47:449–463.

Nelson, B. L., and J. C. Hsu. 1993. Control-variate models of common random numbers for multiple comparisons with the best. *Management Science* 39:989–1001.

Nelson, B. L., and F. J. Matejcik. 1995. Using common random numbers for indifference-zone selection and multiple comparisons in simulation. *Management Science* 41:1935–1945.

Nelson, B. L., and J. Staum. 2006. Control variates for screening, selection, and estimation of the best. *ACM Transactions on Modeling and Computer Simulation* 16:52–75.

Nelson, B. L., J. Swann, D. Goldsman, and W. Song. 2001. Simple procedures for selecting the best simulated system when the number of alternatives is large. *Operations Research* 49:950–963.

Rinott, Y. 1978. On two-stage selection procedures and related probability-inequalities. *Communications in Statistics – Theory and Methods* A7:799–811.

Tamhane, A. C. 1977. Multiple comparisons in model I one-way ANOVA with unequal variances. *Communications in Statistics – Theory and Methods* A6:15–32.

Tong, Y. L. 1980. *Probability Inequalities in Multivariate Distributions*. New York: Academic Press.

Wilcox, R. R. 1984. A table for Rinott's selection procedure. *Journal of Quality Technology* 16:97–100.

Optimal Linear Combinations of Overlapping Variance Estimators for Steady-State Simulation

Tûba Aktaran-Kalaycı, Christos Alexopoulos, David Goldsman, and James R. Wilson

Abstract To estimate the variance parameter (i.e., the sum of covariances at all lags) of a steady-state simulation output process, we formulate an optimal linear combination of overlapping variance estimators (OLCOVE). Each variance estimator is computed from the same data set using one of the following methods: (i) overlapping batch means (OBM); or (ii) standardized time series (STS) applied to overlapping batches separately and then averaged over all such batches. Each estimator's batch size is a fixed real multiple (at least unity) of a base batch size, appropriately rounded. The overall sample size is a fixed integral multiple of the base batch size. Exploiting the control-variates method, we assign OLCOVE coefficients so as to yield a minimum-variance estimator. We establish asymptotic properties of the bias and variance of OLCOVEs computed from OBM or STS variance estimators as the base batch size increases. Finally, we use OLCOVEs to construct confidence intervals for both the mean and the variance parameter of the target process. An experimental performance evaluation revealed the potential benefits of using OLCOVEs for steady-state simulation analysis.

1 Introduction

In summarizing their recommendations on how to handle the problem of steady-state simulation analysis, Kelton et al. (2007, p. 320) make the following extraordinary statement.

> Try to get out of doing a steady-state simulation altogether by convincing yourself (or your patron) that the appropriate modeling assumptions really entail specific starting and stopping conditions.... (and don't come back here).

Although we do not disagree with this statement in its entirety, our experience has led us to a different point of view. We have found that steady-state simulation analysis is often required in the following situations: (i) the appropriate planning

---

T. Aktaran-Kalaycı (✉)
H. Milton Stewart School of Industrial and
Systems Engineering, Atlanta, GA 30332, USA
e-mail: taktaran@gmail.com

C. Alexopoulos et al. (eds.), *Advancing the Frontiers of Simulation: A Festschrift in Honor of George Samuel Fishman*, International Series in Operations Research & Management Science 133, DOI 10.1007/b110059_13, © Springer Science+Business Media, LLC 2009

horizon is long (or unknown); (ii) the initial conditions of the system are subject to uncertainty; (iii) the effects of initial conditions must be eliminated from consideration; or (iv) the system needs to tolerate sustained exposure to worst-case conditions, and we must estimate the long-run average system performance under the latter conditions. Even the most casual inspection of recent volumes of the *Proceedings of the Winter Simulation Conference*,

`www.informs-sim.org/wscpapers.html` ,

reveals that steady-state analysis is required in a variety of applications of large-scale simulation to problems in finance, health care, logistics, manufacturing, telecommunications, and transportation.

In this article we consider a univariate simulation-generated output process $\{Y_i : i = 1, 2, \ldots\}$ such that when the associated simulation model is in steady-state operation, the target process is stationary with marginal cumulative distribution function (c.d.f.) $F_Y(y) = \Pr\{Y_i \le y\}$ for all real y and for $i = 1, 2, \ldots$. Our work is motivated by the familiar problem of computing accurate point and confidence-interval estimators of the steady-state mean,

$$\mu \equiv \mathrm{E}[Y_i] = \int_{-\infty}^{\infty} y \, \mathrm{d}F_Y(y) \, .$$

If a single run of the simulation yields a time series $\{Y_i : i = 1, 2, \ldots, n\}$ of length n (also called the sample size), then the sample mean $\bar{Y}_n \equiv n^{-1} \sum_{i=1}^{n} Y_i$ is a natural point estimator for μ.

To formulate a valid confidence-interval estimator for μ, we must ultimately estimate the *variance parameter*,

$$\sigma^2 \equiv \lim_{n \to \infty} n \mathrm{Var}\big(\bar{Y}_n\big) = \sum_{\ell=-\infty}^{\infty} \mathrm{Cov}(Y_i, Y_{i+\ell}) \, . \tag{1}$$

We assume the target process $\{Y_i\}$ has a nontrivial dependency structure exhibiting short-range dependence so that $0 < \sigma^2 < \infty$, and thus μ and σ^2 are well defined. In most practical applications, the main obstacle to accurate estimation of the variance parameter is that the simulation-generated observations $\{Y_i\}$ are contaminated by initialization bias and are correlated—in short, they fail to be independent and identically distributed (i.i.d.) and thus do not constitute a random sample from the corresponding c.d.f. $F_Y(\cdot)$.

Without the property that the $\{Y_i\}$ are i.i.d., the sample variance, $S_Y^2(n) \equiv (n-1)^{-1} \sum_{i=1}^{n} (Y_i - \bar{Y}_n)^2$, can be a significantly biased estimator of σ^2, just as the sample mean $\bar{Y}_n$ can be a significantly biased estimator of μ. Unfortunately whereas deleting the observations in a suitably chosen warm-up period can effectively eliminate the bias of the sample mean $\bar{Y}_n$ as an estimator of μ, the sample variance $S_Y^2(n)$ can still be a severely biased estimator of σ^2 even when the observations

$\{Y_i : i = 1, 2, \ldots, n\}$ are taken after the simulation is warmed up. The nub of the problem is that the sample variance $S_Y^2(n)$ is only designed to estimate the steady-state marginal variance

$$\mathrm{Var}[Y_i] \equiv \mathrm{E}\big[(Y_i - \mu)^2\big] = \mathrm{Cov}(Y_i, Y_i)\ .$$

If the warmed-up (stationary) process $\{Y_i\}$ exhibits pronounced autocorrelations so that $\mathrm{Cov}(Y_i, Y_{i+\ell})$ is nonzero when the lag ℓ is nonzero, then the sample variance fails to account properly for all but one of the terms in the series representation for σ^2 appearing on the far right-hand side of Equation (1). Moreover if the covariances at nonzero lags in Equation (1) are all positive and decline slowly as $|\ell| \to \infty$ (a common occurrence for many responses of interest in heavily congested queueing simulations), then we often have $\mathrm{E}\big[S_Y^2(n)\big] \ll \sigma^2$, so that the sample variance grossly underestimates the variance parameter.

The preceding discussion can be summarized as follows: in general, using the sample variance $S_Y^2(n)$ as an estimator of the variance parameter σ^2 can give a substantially misleading representation of the precision of simulation-based point and confidence-interval estimators of both μ and σ^2. The cost of this misleading representation might range from monetary losses (for example, an invalid confidence-interval estimate of the long-run average cycle time for a product may result in unfulfilled customer orders for that product) to loss of human life (for example, an incorrect estimate of the variability of average ambulance-response times may result in increased deaths of patients needing transport to a hospital), depending on the nature of the simulation study.

The objective of our research is to develop new estimators for σ^2 with better performance (that is, lower variance, lower bias, and lower mean squared error (MSE)) than certain existing estimators in the literature. It is generally difficult to determine a warm-up period (statistics clearing time, truncation point) beyond which the simulation is (at least approximately) in steady-state operation so that the observed time series $\{Y_i : i = 1, 2, \ldots, n\}$ is (at least approximately) a realization of a stationary process (Fishman 2001). Therefore once the simulation has been warmed up, we are strongly motivated to perform a single run whose length n beyond the truncation point is sufficiently large to ensure acceptable precision in the estimation of both μ and σ^2. In particular, a single long run is generally preferable to multiple independent replications with a proportionately smaller common run length because of the difficulty and loss of sample information that is entailed in deleting observations from a warm-up period within each replication (Alexopoulos and Goldsman 2004). Given the output $\{Y_i : i = 1, 2, \ldots, n\}$ of a prolonged run of a simulation in steady-state operation, we seek to reuse that data set effectively so as to obtain different estimates of σ^2 and finally to combine those estimators to obtain an improved estimator of σ^2.

Suppose that the simulation-generated time series $\{Y_i : i = 1, 2, \ldots, n\}$ is initially organized into b adjacent nonoverlapping batches (subseries, subruns) of a base batch size m, where b is a fixed integer with $b > 1$ so that the overall sample size is $n = bm$. Given a prespecified type of variance estimator derived from the method of standardized time series (STS) such as the area estimator or the

Cramér–von Mises (CvM) estimator (Alexopoulos et al. 2007a), we can form a new estimator of σ^2 as follows:

- For a selected batch-size multiplier r (where $r \geq 1$), compute the associated batch size $M(r) = \lfloor rm \rfloor$, the greatest integer not exceeding rm.
- For $i = 1, 2, \ldots, n - M(r) + 1$, form the ith overlapping batch of size $M(r)$ consisting of the observations $\{Y_{i+j} : j = 0, 1, \ldots, M(r) - 1\}$; and compute the given variance estimator from this batch.
- Average the resulting variance estimators computed from each of the $n - M(r) + 1$ overlapping batches.

In addition to these STS-type overlapping variance estimators, we can also compute the classical overlapping batch means (OBM) variance estimator using the batch size $M(r)$ (Meketon and Schmeiser 1984).

We assume that we have selected a number of different overlapping estimators of the variance parameter, all of a single type (OBM, area, or CvM) but with different batch sizes; and we have calculated all these estimators of σ^2 from the same simulation output $\{Y_i : i = 1, 2, \ldots, n\}$. We seek an optimal linear combination of overlapping variance estimators (OLCOVE) as our final point estimator of σ^2. Standard regression techniques underlying the method of control variates (see Lavenberg and Welch 1981) are used to determine the coefficients of the variance estimators composing an OLCOVE so as to preserve low bias and minimize the variance of the final estimator of σ^2.

Because of the bias structure of the individual variance estimators composing an OLCOVE, we expect the bias of the OLCOVE to be smaller in magnitude than that of each constituent estimator using the base batch size. We consider the asymptotic behavior of an OLCOVE as the base batch size $m \to \infty$ with b fixed so that the overall sample size $n = bm \to \infty$. If the constituent estimators used to construct an OLCOVE are all first-order unbiased for σ^2 (that is, each constituent variance estimator has a bias that tends to zero faster than $1/m$ as $m \to \infty$), then that OLCOVE is also first-order unbiased. In any case, the OLCOVE will likely have lower variance than its constituent estimators, even when some of the latter estimators are correlated.

To complete our analysis of OLCOVEs, we proceed as follows:

(i) For each OBM or STS variance estimator of a given type and for a prespecified vector of batch sizes expressed as rounded multiples of the base batch size m, we establish the asymptotic joint distribution of the corresponding vector of variance estimators of that type as $m \to \infty$ with b fixed.
(ii) For each vector of variance estimators constructed as in (i), we compute the associated OLCOVE and derive its asymptotic bias and variance as $m \to \infty$; and we augment the theoretical large-sample properties of that OLCOVE with empirical small-sample results obtained by applying it to some interesting test problems.
(iii) For each OLCOVE computed as in (ii), we formulate a scaled chi-squared approximation to its distribution and exploit this distribution to construct

approximate confidence intervals for both μ and σ^2; moreover, we evaluate the performance of these confidence intervals in each of the test problems used in (ii).

The rest of this article is organized as follows. In Section 2, we summarize the relevant background on overlapping variance estimators—specifically we discuss the OBM, STS area, and STS CvM estimators. We establish the (asymptotic) theoretical properties of the proposed OLCOVEs in Section 3, where we also show how to implement an OLCOVE to minimize its variance in a general setting. In Section 3, we also provide Monte Carlo results illustrating the performance of OLCOVEs in the following test problems:

- A first-order autoregressive (AR(1)) process with lag-one correlation of 0.9; and
- The queue waiting times in an $M/M/1$ queueing system with traffic intensity of 0.8.

(To avoid the problem of initialization bias, we sample the initial condition for each of these test processes from the corresponding steady-state distribution; thus each test process is stationary. Of course this is *not* typical of practical applications.)

Section 4 addresses the costs (that is, computational requirements) associated with using OLCOVEs. In Section 5, we present a scaled chi-squared approximation to the probability density function (p.d.f.) of an OLCOVE based on a technique that is due to Satterthwaite (1941); and we use this approximation to construct approximate confidence intervals for μ and σ^2. A Monte Carlo study involving the two previously mentioned test problems is also presented in Section 5 to illustrate the adequacy of our approximation to the p.d.f. of an OLCOVE as well as the empirical coverage probabilities of our approximate confidence intervals for μ and σ^2. We discuss the shortcomings and limitations of the OLCOVE technique in Section 6, and finally in Section 7 we summarize the main findings of this work and provide recommendations for future research.

Some preliminary Monte Carlo results on linear combinations of overlapping area variance estimators are summarized in Aktaran-Kalaycı and Goldsman (2005) and in Aktaran-Kalaycı et al. (2007c). Whereas Aktaran-Kalaycı et al. (2007a) and Aktaran-Kalaycı et al. (2007b) present key theoretical properties of OLCOVEs that are based exclusively on overlapping area estimators, in this article we extend the development of OLCOVEs to encompass overlapping CvM estimators as well as the classical OBM estimators.

2 Basic Concepts and Assumptions for Steady-State Simulation Output Analysis

In this section, we introduce the notation used throughout this article; and we summarize the main results on batch-means and STS variance estimators that form the basis for our development of OLCOVEs.

2.1 Nonoverlapping Batch Means (NBM)

When the data set $\{Y_i : i = 1, 2, \ldots, n\}$ is organized into b nonoverlapping batches of size m (so that $n = mb$), the ith nonoverlapping batch consists of the observations

$$\{Y_{(i-1)m+1}, Y_{(i-1)m+2}, \ldots, Y_{im}\} \text{ for } i = 1, 2, \ldots, b\,;$$

and the corresponding batch mean is

$$\bar{Y}_{i,m} \equiv \frac{1}{m}\sum_{j=1}^{m} Y_{(i-1)m+j} \text{ for } i = 1, 2, \ldots, b\,.$$

The NBM estimator for σ^2 is given by

$$\mathcal{N}(b, m) \equiv \frac{m}{b-1}\sum_{i=1}^{b}\left(\bar{Y}_{i,m} - \bar{Y}_n\right)^2 = \frac{m}{b-1}\left(\sum_{i=1}^{b}\bar{Y}_{i,m}^2 - b\bar{Y}_n^2\right)$$

(Fishman and Yarberry 1997).

2.2 Overlapping Batch Means (OBM)

In the data set $\{Y_i : i = 1, 2, \ldots, n\}$, the ith overlapping batch of size m consists of the observations

$$\{Y_i, Y_{i+1}, \ldots, Y_{i+m-1}\} \text{ for } i = 1, 2, \ldots, n - m + 1\,;$$

and the corresponding batch mean is

$$\bar{Y}_{i,m}^{\mathrm{O}} \equiv \frac{1}{m}\sum_{j=0}^{m-1} Y_{i+j} \text{ for } i = 1, 2, \ldots, n - m + 1\,. \tag{2}$$

The OBM estimator for σ^2 is defined as

$$\mathcal{O}(b, m) \equiv \frac{nm}{(n - m + 1)(n - m)}\sum_{i=1}^{n-m+1}\left(\bar{Y}_{i,m}^{\mathrm{O}} - \bar{Y}_n\right)^2 \tag{3}$$

(Meketon and Schmeiser 1984). As elaborated in subsequent sections of this article, the OBM estimator has approximately the same bias but lower variance compared with the variance of the NBM estimator.

2.3 Standardized Time Series (STS)

The STS methodology (Schruben1983) has given rise to a large class of estimators for σ^2. In this article, we limit the discussion to area and CvM estimators that use batching and can be based on nonoverlapping or overlapping batches. The expected value of a nonoverlapping STS estimator is the same as that of its overlapping counterpart; but the variance of the latter is smaller than that of the nonoverlapped version (Alexopoulos et al. 2007a, b).

We assume that the following Functional Central Limit Theorem (FCLT) holds for the process $\{Y_i : i = 1, 2, \ldots\}$.

Assumption FCLT The sequence of random functions

$$X_n(t) \equiv \frac{\lfloor nt \rfloor (\bar{Y}_{\lfloor nt \rfloor} - \mu)}{\sigma \sqrt{n}} \quad \text{for } t \in [0, 1] \text{ and } n = 1, 2, \ldots \tag{4}$$

satisfies

$$X_n(\cdot) \underset{n \to \infty}{\Longrightarrow} \mathcal{W}(\cdot) ,$$

where: $\mathcal{W}(\cdot)$ is a standard Brownian motion process on [0, 1]; and $\underset{n \to \infty}{\Longrightarrow}$ denotes weak convergence in $D[0, 1]$, the space of functions on [0, 1] that are right-continuous with left-hand limits, as $n \to \infty$. See also Billingsley (1968) and Glynn and Iglehart (1990).

Similar to the original definition in Schruben (1983), the standardized time series from overlapping batch i is the random function

$$T^{\mathrm{O}}_{i,m}(t) \equiv \frac{\lfloor mt \rfloor \left(\bar{Y}^{\mathrm{O}}_{i,m} - \bar{Y}^{\mathrm{O}}_{i,\lfloor mt \rfloor} \right)}{\sigma \sqrt{m}} \quad \text{for } 0 \le t \le 1 \text{ and } i = 1, 2, \ldots, n - m + 1 ,$$

where for consistency with Equation (2), we define the following intermediate batch means computed from overlapping batches of size m:

$$\bar{Y}^{\mathrm{O}}_{i,j} \equiv \frac{1}{j} \sum_{\ell=0}^{j-1} Y_{i+\ell} \quad \text{for } i = 1, 2, \ldots, n - m + 1 \text{ and } j = 1, 2, \ldots, m .$$

Alexopoulos et al. (2007a) show that, under Assumption FCLT,

$$\sigma T^{\mathrm{O}}_{\lfloor sm \rfloor, m}(\cdot) \underset{m \to \infty}{\Longrightarrow} \sigma \mathcal{B}_{s,1}(\cdot) \text{ for fixed } s \in [0, b - 1] , \tag{5}$$

where: $b \equiv n/m > 1$ is a fixed ratio; and for $r \in [1, b)$ and $s \in [0, b - r]$, we let $\mathcal{B}_{s,r}(\cdot)$ denote a Brownian bridge process on the unit interval,

$$\mathcal{B}_{s,r}(t) \equiv \frac{t[\mathcal{W}(s + r) - \mathcal{W}(s)] - [\mathcal{W}(s + tr) - \mathcal{W}(s)]}{\sqrt{r}} \quad \text{for } t \in [0, 1] . \tag{6}$$

Moreover, two useful properties follow from Equation (5):

$$\sqrt{n}\left(\bar{Y}_n - \mu\right) \text{ is asymptotically } \sigma\,\text{Nor}(0,1) \text{ as } n \to \infty\,, \tag{7}$$

and

$$\sqrt{n}\left(\bar{Y}_n - \mu\right) \text{ and } \sigma T^{\mathrm{O}}_{1,n}(\cdot) \text{ are asymptotically independent as } n \to \infty\,. \tag{8}$$

2.3.1 Area Estimator

We define the area estimator computed from overlapping batch i by

$$A^{\mathrm{O}}_i(f;m) \equiv \left[\frac{1}{m}\sum_{k=1}^{m} f\left(\tfrac{k}{m}\right)\sigma T^{\mathrm{O}}_{i,m}\left(\tfrac{k}{m}\right)\right]^2 \quad \text{for } i = 1, 2, \ldots, n-m+1\,,$$

where $f(t)$ is a user-specified weight function satisfying the condition

$$\left.\begin{array}{c} \dfrac{\mathrm{d}^2}{\mathrm{d}t^2} f(t) \text{ is continuous at every } t \in [0,1]\,, \\ \mathrm{Var}\left[\int_0^1 f(t)\mathcal{B}_{0,1}(t)\,\mathrm{d}t\right] = 1\,. \end{array}\right\}$$

In this article, we consider the following weight functions: $f_0(t) \equiv \sqrt{12}$, $f_2(t) \equiv \sqrt{840}(3t^2 - 3t + 1/2)$, and $f_{\cos,j}(t) = \sqrt{8}\pi j\cos(2\pi jt)$, $j = 1, 2, \ldots$, for $t \in [0,1]$.

The *nonoverlapping (batched) area* estimator for σ^2 is

$$\mathcal{A}(f;b,m) \equiv \frac{1}{b}\sum_{i=1}^{b} A^{\mathrm{O}}_{(i-1)m+1}(f;m)$$

(Goldsman et al. 1990, Schruben 1983), which is formed from the b adjacent, nonoverlapping batches of size m.

The *overlapping (batched) area* estimator for σ^2, formed from all $n-m+1$ overlapping batches of size m, is

$$\mathcal{A}^{\mathrm{O}}(f;b,m) \equiv \frac{1}{n-m+1}\sum_{i=1}^{n-m+1} A^{\mathrm{O}}_i(f;m)$$

(Alexopoulos et al.2007a).

2.3.2 Cramér–von Mises (CvM) Estimator

The CvM estimator computed from overlapping batch i is

$$C_i^{\mathrm{O}}(g;m) \equiv \frac{1}{m}\sum_{k=1}^{m} g\left(\tfrac{k}{m}\right)\left[\sigma T_{i,m}^{\mathrm{O}}\left(\tfrac{k}{m}\right)\right]^2 \quad \text{for } i = 1, 2, \ldots, n-m+1 \,,$$

where $g(t)$ is a user-specified weight function satisfying the condition

$$\frac{d^2}{dt^2} g(t) \text{ is continuous at every } t \in [0, 1] \text{ and } \int_0^1 g(t)t(1-t)\,\mathrm{d}t = 1 \,. \tag{9}$$

In this paper, we consider the following weight functions: $g_0(t) \equiv 6$ and $g_2^{\star}(t) \equiv -24 + 150t - 150t^2$ for $t \in [0, 1]$.

The *nonoverlapping (batched) CvM* estimator for σ^2, formed from the b adjacent, nonoverlapping batches of size m, is

$$\mathcal{C}(g;b,m) \equiv \frac{1}{b}\sum_{i=1}^{b} C_{(i-1)m+1}^{\mathrm{O}}(g;m)$$

(Goldsman et al. 1999).

The *overlapping (batched) CvM* estimator for σ^2, formed from all $n-m+1$ overlapping batches of size m, is

$$\mathcal{C}^{\mathrm{O}}(g;b,m) \equiv \frac{1}{n-m+1}\sum_{i=1}^{n-m+1} C_i^{\mathrm{O}}(g;m) \tag{10}$$

(Alexopoulos et al.2007a).

2.4 Key Asymptotic Properties of Selected Variance Estimators

Table 1 contains a summary of the key asymptotic properties for the variance estimators considered so far; these properties are derived in the references cited for each estimator. The quantity

$$\gamma \equiv -2\sum_{k=1}^{\infty} k\mathrm{Cov}(Y_1, Y_{1+k})$$

introduced in Table 1 is a measure of the dependence structure of $\{Y_i\}$; see, for example, Song and Schmeiser (1995). We see from the table, for example, that

$$\mathrm{E}\big[\mathcal{A}^{\mathrm{O}}(f_0;b,m)\big] = \sigma^2 + \frac{3\gamma}{m} + o\Big(\frac{1}{m}\Big) \text{ and } \mathrm{Var}\big[\mathcal{A}^{\mathrm{O}}(f_0;b,m)\big] = \frac{0.686\sigma^4}{b} + o\Big(\frac{1}{b}\Big) \,.$$

Table 1 Approximate asymptotic bias and variance for batched estimators

Nonoverlapping	$\frac{m}{\gamma}$ Bias	$\frac{b}{\sigma^4}$ Var	Overlapping	$\frac{m}{\gamma}$ Bias	$\frac{b}{\sigma^4}$ Var
$\mathcal{A}(f_0; b, m)$	3	2	$\mathcal{A}^{\mathrm{O}}(f_0; b, m)$	3	0.686
$\mathcal{A}(f_2; b, m)$	$o(1)$	2	$\mathcal{A}^{\mathrm{O}}(f_2; b, m)$	$o(1)$	0.819
$\mathcal{A}(f_{\cos,j}; b, m)$	$o(1)$	2	$\mathcal{A}^{\mathrm{O}}(f_{\cos,1}; b, m)$	$o(1)$	0.793
$\mathcal{C}(g_0; b, m)$	5	0.8	$\mathcal{C}^{\mathrm{O}}(g_0; b, m)$	5	0.419
$\mathcal{C}(g_2^\star; b, m)$	$o(1)$	1.729	$\mathcal{C}^{\mathrm{O}}(g_2^\star; b, m)$	$o(1)$	0.777
$\mathcal{N}(b, m)$	1	2	$\mathcal{O}(b, m)$	1	1.333

3 Optimal Linear Combination of Overlapping Variance Estimators (OLCOVE)

In this section, we derive the asymptotic covariance between pairs of overlapping variance estimators of the same type that use different batch sizes—namely, pairs of overlapping CvM estimators or pairs of OBM estimators. Comparable results for pairs of overlapping area estimators are presented in Aktaran-Kalaycı et al. (2007a). These covariances are essential for the construction of OLCOVEs.

3.1 Motivation

We introduce the idea underlying an OLCOVE using a generic overlapping estimator for σ^2, denoted by $\mathcal{V}^{\mathrm{O}}(B, M)$, where: (i) the batch size M for the overlapping batches used to compute $\mathcal{V}^{\mathrm{O}}(B, M)$ is a rounded multiple of the base batch size m; and (ii) we take $B \equiv n/M$ so that $BM = bm = n$, the total sample size from which $\mathcal{V}^{\mathrm{O}}(B, M)$ is computed. To construct an OLCOVE using different variance estimators of the same type as $\mathcal{V}^{\mathrm{O}}$, we proceed as follows:

- For $j = 1, 2, \ldots, k$, let $r_j \in [1, b)$ and take $M_j \equiv \lfloor r_j m \rfloor$ as the jth batch size, with the corresponding quantity $B_j \equiv \lfloor bm \rfloor / M_j$ so that the total sample size is $B_j M_j = \lfloor bm \rfloor$, and B_j is the associated sample-size-to-batch-size ratio. Calculate k overlapping estimators, $\mathcal{V}^{\mathrm{O}}(B_1, M_1)$, $\mathcal{V}^{\mathrm{O}}(B_2, M_2), \ldots, \mathcal{V}^{\mathrm{O}}(B_k, M_k)$, where we use the same data set $\{Y_i : i = 1, 2, \ldots, n\}$ with a variety of different integer batch sizes, $M_1, M_2, \ldots, M_k$.
- Form a linear combination of these k estimators and scale appropriately. Use standard regression techniques underlying the method of control variates (see Lavenberg and Welch 1981) to determine the corresponding coefficients (multipliers, scaling factors) that preserve low bias and minimize variance.

Let $\boldsymbol{M} \equiv [M_1, M_2, \ldots, M_k]$ and $\boldsymbol{B} \equiv [B_1, B_2, \ldots, B_k]$. Further, let $\boldsymbol{\alpha} \equiv [\alpha_1, \alpha_2, \ldots, \alpha_{k-1}]$ denote the vector of coefficients in the linear combination. Corresponding to the variance estimator type $\mathcal{V}^{\mathrm{O}}$ and the vectors $\boldsymbol{M}$, $\boldsymbol{B}$, and $\boldsymbol{\alpha}$ is the new estimator of σ^2 given by the linear combination

$$\mathcal{V}^{\mathrm{LO}}(\boldsymbol{B},\boldsymbol{M},\boldsymbol{\alpha}) \equiv \sum_{i=1}^{k-1} \alpha_i \mathcal{V}^{\mathrm{O}}(B_i, M_i) + \left[1 - \sum_{i=1}^{k-1} \alpha_i\right] \mathcal{V}^{\mathrm{O}}(B_k, M_k)\,. \tag{11}$$

For $j = 1, \ldots, k$, we have $M_j \geq m$ because $r_j \geq 1$; hence, the expected values of the overlapping estimators in Table 1 suggest that $\left|\mathrm{Bias}\left[\mathcal{V}^{\mathrm{LO}}(\boldsymbol{B},\,\boldsymbol{M},\,\boldsymbol{\alpha})\right]\right|$ might be smaller than $\left|\mathrm{Bias}\left[\mathcal{V}^{\mathrm{O}}(b,m)\right]\right|$. In addition, if the k estimators composing $\mathcal{V}^{\mathrm{LO}}(\boldsymbol{B},\boldsymbol{M},\boldsymbol{\alpha})$ are all first-order unbiased (i.e., their bias is of the order $o(1/m)$), then $\mathcal{V}^{\mathrm{LO}}(\boldsymbol{B},\boldsymbol{M},\boldsymbol{\alpha})$ will also be first-order unbiased. Furthermore, $\mathcal{V}^{\mathrm{LO}}(\boldsymbol{B},\boldsymbol{M},\boldsymbol{\alpha})$ will likely have lower variance, even if its constituent estimators are correlated.

Remark 1 In the subsequent development of OLCOVE properties, the various definitions and arguments will be formulated for the general case in which b is real (not necessarily integer) and $b > 1$ so that the total sample size $n = \lfloor bm \rfloor$. This is done because the arguments must be applied to each $\mathcal{V}^{\mathrm{O}}(B_i, M_i)$, in which case b and m are replaced by B_i and M_i, respectively, for $i = 1, \ldots, k$; and in general B_i is a real number (not necessarily an integer) satisfying $B_i > 1$. This usage should not cause any confusion.

Example 1 As a simple motivational example, we combine two overlapping estimators—one based on batches of size m and the other on batches of size $2m$ so that $\boldsymbol{M} = [m, 2m]$, $\boldsymbol{B} = [b, b/2]$, and $\boldsymbol{\alpha} = [\alpha] = \alpha$, a scalar constant; and then we have

$$\mathcal{V}^{\mathrm{LO}}([b, b/2], [m, 2m], \alpha) = \alpha\mathcal{V}^{\mathrm{O}}(b,m) + (1-\alpha)\mathcal{V}^{\mathrm{O}}(b/2, 2m)\,.$$

If $\mathrm{Bias}\left[\mathcal{V}^{\mathrm{O}}(b,m)\right] \doteq \delta/m$ for some constant δ, then

$$\begin{aligned}\mathrm{E}\left[\mathcal{V}^{\mathrm{LO}}([b, b/2], [m, 2m], \alpha)\right] &= \alpha\mathrm{E}\left[\mathcal{V}^{\mathrm{O}}(b,m)\right] + (1-\alpha)\mathrm{E}\left[\mathcal{V}^{\mathrm{O}}(b/2, 2m)\right] \\ &\doteq \sigma^2 + \frac{(1+\alpha)\delta}{2m}\,. \end{aligned} \tag{12}$$

If $\alpha \in [-3, 1]$ and we ignore small-order terms, then we see that the bias of the linear combination $\mathcal{V}^{\mathrm{LO}}([b, b/2], [m, 2m], \alpha)$ is lower than that of the original overlapping estimator with batch size m. In fact, taking $\alpha = -1$, we see the first-order bias in (12) disappears entirely. The choice of α also affects the variance of $\mathcal{V}^{\mathrm{LO}}([b, b/2], [m, 2m], \alpha)$. Specifically,

$$\begin{aligned}&\mathrm{Var}\left[\mathcal{V}^{\mathrm{LO}}([b, b/2], [m, 2m], \alpha)\right] \\ &\quad = \alpha^2\,\mathrm{Var}\left[\mathcal{V}^{\mathrm{O}}(b,m)\right] + (1-\alpha)^2\,\mathrm{Var}\left[\mathcal{V}^{\mathrm{O}}(b/2, 2m)\right] \\ &\quad\quad + 2\alpha(1-\alpha)\,\mathrm{Cov}\left[\mathcal{V}^{\mathrm{O}}(b,m), \mathcal{V}^{\mathrm{O}}(b/2, 2m)\right]\,. \end{aligned} \tag{13}$$

Our goal is to find $\alpha^\star$, the value of the coefficient α that will minimize the variance (13). To do this, we need to calculate the two variance terms on the right-hand side of (13) as well as the covariance term. The covariance results for overlapping area estimators are derived in Aktaran-Kalaycı et al. (2007a). In the following subsections, we tackle this task for overlapping CvM and OBM estimators, respectively.

3.2 Covariance between Overlapping CvM Estimators

We define the estimator arising from a linear combination of overlapping CvM estimators as

$$\mathcal{C}^{\mathrm{LO}}(g;\boldsymbol{B},\boldsymbol{M},\boldsymbol{\alpha}) \equiv \sum_{i=1}^{k-1} \alpha_i \mathcal{C}^{\mathrm{O}}(g;B_i,M_i) + \left[1-\sum_{i=1}^{k-1}\alpha_i\right]\mathcal{C}^{\mathrm{O}}(g;B_k,M_k)\,,$$

where the overlapping CvM estimator is defined by Equation (10). Motivated by the discussion in Section 3.1, we will find an approximate expression for

$$\begin{aligned} &c_C(g;b,m;r_1,r_2)\\ &\equiv \mathrm{Cov}\big[\mathcal{C}^{\mathrm{O}}(g;\lfloor bm\rfloor\,/\,\lfloor r_1 m\rfloor\,,\lfloor r_1 m\rfloor),\mathcal{C}^{\mathrm{O}}(g;\lfloor bm\rfloor/\lfloor r_2 m\rfloor,\lfloor r_2 m\rfloor)\big]\\ &\quad \text{for } r_1,r_2\in[1,b)\,. \end{aligned}$$

Definition 1. Let $\boldsymbol{r} \equiv [r_1,\ldots,r_k]$ denote the vector of batch-size multipliers, and let $\boldsymbol{\mathcal{C}}^{\mathrm{O}}(g;\boldsymbol{B},\boldsymbol{r},\boldsymbol{M})$ denote the $k\times 1$ vector whose jth component is $\mathcal{C}^{\mathrm{O}}(g;B_j,M_j)$ for $j=1,\ldots,k$. Similarly, let $\boldsymbol{\mathcal{C}}^{\mathrm{O}}(g;b,\boldsymbol{r})$ denote the $k\times 1$ vector whose jth component is

$$\mathcal{C}^{\mathrm{O}}_j(g;b,r_j) \equiv (b-r_j)^{-1}\int_0^{b-r_j}\sigma^2\left[\int_0^1 g(u)\mathcal{B}^2_{s,r_j}(u)\,\mathrm{d}u\right]\mathrm{d}s$$

for $j=1,\ldots,k$, where the Brownian bridge $\mathcal{B}_{s,r_j}(u)$ is defined in Equation (6).

Theorem 1 *If Assumption FCLT and Equation (9) hold, then*

$$\boldsymbol{\mathcal{C}}^{\mathrm{O}}(g;\boldsymbol{B},\boldsymbol{r},\boldsymbol{M}) \underset{m\to\infty}{\Longrightarrow} \boldsymbol{\mathcal{C}}^{\mathrm{O}}(g;b,\boldsymbol{r})\,.$$

To prove Theorem 1, we first need the following definition to establish the asymptotic distribution of the random vector $\boldsymbol{\mathcal{C}}^{\mathrm{O}}(g;\boldsymbol{B},\boldsymbol{r},\boldsymbol{M})$ as $m\to\infty$.

Definition 2 Let $D[0,b]$ denote the space of functions on $[0,b]$ that are right-continuous and have left-hand limits. Let Λ_b denote the class of strictly increasing, continuous mappings of $[0,b]$ onto itself such that for every $\lambda\in\Lambda_b$, we have $\lambda(0)=0$ and $\lambda(b)=b$. If $X,Z\in D[0,b]$, then the Skorohod metric $\rho_b(X,Z)$ defining the "distance" between X and Z in $D[0,b]$ is the infimum of those positive ξ for which there exists a $\lambda\in\Lambda_b$ such that $\sup_{t\in[0,b]}|\lambda(t)-t|\le\xi$ and $\sup_{t\in[0,b]}|X(t)-Z[\lambda(t)]|\le\xi$. (See Billingsley 1968 for more particulars.)

Definition 3 Suppose that $r \in [1, b)$ and $g(\cdot)$ satisfies (9). Associated with $g(\cdot)$ is the overlapping CvM map $\vartheta^r : Z \in D[0, b] \mapsto \vartheta^r(Z) \in \mathbb{R}$ given by

$$\vartheta^r(Z) \equiv \frac{1}{b-r} \int_0^{b-r} \int_0^1 g(u) \left[\sigma \Omega^r_{Z,s}(u)\right]^2 du \, ds \,, \tag{14}$$

where the bridging map $\Omega^r : (Z, s) \in D[0, b] \times [0, b - r] \mapsto \Omega^r_{Z,s} \in D[0, 1]$ is defined by

$$\Omega^r_{Z,s}(t) \equiv \frac{t[Z(s+r) - Z(s)] - [Z(s+tr) - Z(s)]}{\sqrt{r}} \quad \text{for } t \in [0, 1] \,.$$

Moreover, for $m = 1, 2, \ldots$ with $M(r) \equiv \lfloor rm \rfloor$ and $B(r) \equiv \lfloor bm \rfloor / M(r)$, the approximate (discrete) bridging map $\Omega^{r,m} : (Z, s) \in D[0, b] \times [0, b-r] \mapsto \Omega^{r,m}_{Z,s} \in D[0, 1]$ is defined by

$$\begin{aligned}\Omega^{r,m}_{Z,s}(t) \equiv \frac{\lfloor M(r)t \rfloor}{M(r)} \sqrt{\frac{m}{M(r)}} \left\{ Z\left(\frac{\lceil sm \rceil - 1 + M(r)}{m}\right) - Z\left(\frac{\lceil sm \rceil - 1}{m}\right) \right\} \\ - \sqrt{\frac{m}{M(r)}} \left\{ Z\left(\frac{\lceil sm \rceil - 1 + \lfloor M(r)t \rfloor}{m}\right) - Z\left(\frac{\lceil sm \rceil - 1}{m}\right) \right\}\end{aligned}$$

for $t \in [0, 1]$, where $\lceil x \rceil$ denotes the smallest integer not less than x (that is, the ceiling of x). The associated discrete overlapping CvM map $\vartheta^r_m : Z \in D[0, b] \mapsto \vartheta^r_m(Z) \in \mathbb{R}$ is given by

$$\vartheta^r_m(Z) \equiv \tag{15}$$

$$\frac{1}{[B(r)-1]M(r)+1} \sum_{i=1}^{[B(r)-1]M(r)+1} \frac{1}{M(r)} \sum_{\ell=1}^{M(r)} g\left(\tfrac{\ell}{M(r)}\right) \left[\sigma \Omega^{r,m}_{Z,i/m}\left(\tfrac{\ell}{M(r)}\right)\right]^2 .$$

Remark 2 Using Equation (10), we see from the definition of the overlapping CvM estimator that

$$\vartheta^r_m(X_m) = \mathcal{C}^{\mathrm{O}}(g; B(r), M(r)) \,.$$

Therefore, Theorem 1 is equivalent to

$$\left[\vartheta^{r_1}_m(X_m), \ldots, \vartheta^{r_k}_m(X_m)\right] \underset{m\to\infty}{\Longrightarrow} \left[\vartheta^{r_1}(\mathcal{W}), \ldots, \vartheta^{r_k}(\mathcal{W})\right] . \tag{16}$$

To prove Equation (16), we first justify the assumptions of the generalized continuous mapping theorem (CMT) with the following proposition.

Proposition 1 *If $\vartheta^r(\cdot)$ and $\vartheta^r_m(\cdot)$ are defined by (14) and (15), respectively, then*

$$\Pr\left\{\mathcal{W} \in D[0, b] - \mathbb{D}_{\vartheta^r}\right\} = 1 \,,$$

where $\Pr\{\cdot\}$ *denotes Wiener measure and*

$$\mathbb{D}_{\vartheta^r} \equiv \Big\{x \in D[0,b] : \textit{for some sequence } \{x_m\} \subset D[0,b] \textit{ with } \lim_{m\to\infty} \rho_b(x_m, x) = 0, \textit{the sequence } \{\vartheta^r_m(x_m)\} \textit{ does not converge to } \vartheta^r(x)\Big\}\,.$$

Sketch of Proof of Proposition 1 Although the proof of Proposition 1 is detailed in Appendix A.2 of Aktaran-Kalaycı (2006), it can be summarized as follows. First, we exploit the almost-sure continuity of $\mathcal{W}(u)$ at every $u \in [0,b]$ and the convergence of $\{x_n\}$ to $\mathcal{W}$ in $D[0,b]$ so as to establish the following properties, each of which holds with probability one:

$$\lim_{m\to\infty} \left|\Omega^{r,m}_{x_m,s}(u) - \Omega^{r,m}_{\mathcal{W},s}(u)\right| = 0 \text{ for every } (s,u) \in [0,b-r]\times[0,1]\,; \tag{17}$$

$$\Omega^{r,m}_{\mathcal{W},s}(u) \text{ is continuous at every } (s,u) \in [0,b-r]\times[0,1] \text{ for } m = 1,2,\ldots\,; \tag{18}$$

$$\lim_{m\to\infty} \left|\Omega^{r,m}_{\mathcal{W},s}(u) - \Omega^{r}_{\mathcal{W},s}(u)\right| = 0 \text{ for every } (s,u) \in [0,b-r]\times[0,1]\,; \tag{19}$$

and

$$\Omega^{r}_{\mathcal{W},s}(u) \text{ is continuous at every } (s,u) \in [0,b-r]\times[0,1]\,. \tag{20}$$

Properties (18)–(20) imply that with probability one, $\left\{g(u)\Omega^{r,m}_{\mathcal{W},s}(u) : m = 1,2,\ldots\right\}$ is a uniformly bounded sequence of continuous functions of $(s,u) \in [0,b-r]\times[0,1]$. Using the latter property and (17), we can prove that

$$\lim_{m\to\infty} \left|\vartheta^r_m(x_m) - \vartheta^r_m(\mathcal{W})\right| = 0 \text{ with probability one.} \tag{21}$$

A similar analysis shows that with probability one, $g(u)\big[\Omega^{r}_{\mathcal{W},s}(u)\big]^2$ is continuous for all $(s,u) \in [0,b-r]\times[0,1]$ and hence is Riemann integrable on $[0,b-r]\times[0,1]$. Almost-sure Riemann integrability of $g(u)\big[\Omega^{r}_{\mathcal{W},s}(u)\big]^2$ immediately implies that

$$\lim_{m\to\infty} \left|\vartheta^r_m(\mathcal{W}) - \vartheta^r(\mathcal{W})\right| = 0 \text{ with probability one.} \tag{22}$$

Combining (21) and (22) and applying the triangle inequality, we see that

$$\lim_{m\to\infty} \left|\vartheta^r_m(x_m) - \vartheta^r(\mathcal{W})\right| = 0 \text{ with probability one.}$$ □

Proof of Theorem 1 Proposition 1 can be applied to $\vartheta^{r_j}(\cdot)$ and $\vartheta^{r_j}_m(\cdot)$ for each $j \in \{1,2,\ldots,k\}$ to conclude that if the sequence $\{x_m\}$ converges to $\mathcal{W}$ in $D[0,b]$, then the vector-valued sequence $\{[\vartheta_m^{r_1}(x_m), \ldots, \vartheta_m^{r_k}(x_m)] : m = 1, 2, \ldots\}$ converges to $[\vartheta^{r_1}(\mathcal{W}), \ldots, \vartheta^{r_k}(\mathcal{W})]$ in $\mathbb{R}^k$ with probability one. Thus (16) follows from the generalized CMT since its assumptions are satisfied. □

Now if we also assume that corresponding to the sequence (4), the vector-valued process $\{([\vartheta_m^{r_1}(X_m)]^2, \ldots, [\vartheta_m^{r_k}(X_m)]^2) : m = 1, 2, \ldots\}$ is uniformly integrable, then by our Theorem 1 and Theorem 5.4 of Billingsley (1968), we have

$$\begin{aligned}\lim_{m\to\infty} c_C(g; b, m; r_1, r_2) &= c_C(g; b; r_1, r_2) \qquad (23)\\ &\equiv \mathrm{Cov}\left[\mathcal{C}_1^{\mathrm{O}}(g; b, r_1), \mathcal{C}_2^{\mathrm{O}}(g; b, r_2)\right]\\ &= \frac{2\sigma^4}{(b-r_1)(b-r_2)} \int_0^{b-r_1}\int_0^{b-r_2}\int_0^1\int_0^1 g(t)g(t')\mathrm{Cov}^2\left[\mathcal{B}_{u,r_1}(t), \mathcal{B}_{v,r_2}(t')\right]\\ &\qquad \times \mathrm{d}t\,\mathrm{d}t'\,\mathrm{d}v\,\mathrm{d}u\;.\end{aligned}$$

The theorem, along with a great deal of algebra involving (23), lets us make the necessary covariance calculations (see Appendix A.6 of Aktaran-Kalaycı 2006). We illustrate the covariance calculations with several examples.

Example 2 For $1 \le r_2 \le r_1$ and $r_1 + r_2 \le b$, we have

$$\begin{aligned}&c_C(g_0; b; r_1, r_2)\\ &\quad = \frac{r_2^2(-280r_1^3 + 196r_1^2 r_2 - 12r_1r_2^2 - 19r_2^3 + 4b(70r_1^2 - 63r_1r_2 + 15r_2^2))}{210(b-r_1)(b-r_2)r_1^3}\\ &\quad \doteq \frac{2r_2^2(70r_1^2 - 63r_1r_2 + 15r_2^2)\sigma^4}{105br_1^3} \qquad (\text{for } b \gg r_1, r_2)\;.\end{aligned}$$

Example 3 For $1 \le r_2 \le r_1 \ll b$, we have

$$c_C(g_2^\star; b; r_1, r_2) \doteq \frac{2r_2^3(11{,}286r_1^3 - 19{,}140r_1^2r_2 + 11{,}825r_1r_2^2 - 2{,}625r_2^3)\sigma^4}{3{,}465br_1^5}\;.$$

Remark 3 Following a similar analysis, we can obtain the covariance result analogous to (23) for overlapping area estimators,

$$\begin{aligned}&c_A(f; b; r_1, r_2)\\ &\quad \equiv \lim_{m\to\infty} \mathrm{Cov}\left[\mathcal{A}^{\mathrm{O}}(f; \lfloor bm\rfloor / \lfloor r_1 m\rfloor, \lfloor r_1 m\rfloor), \mathcal{A}^{\mathrm{O}}(f; \lfloor bm\rfloor / \lfloor r_2 m\rfloor, \lfloor r_2 m\rfloor)\right]\\ &\quad = \frac{2\sigma^4}{(b-r_1)(b-r_2)} \int_0^{b-r_1}\int_0^{b-r_2}\left\{\int_0^1\int_0^1 f(t)f(t')\mathrm{Cov}\left[\mathcal{B}_{u,r_1}(t), \mathcal{B}_{v,r_2}(t')\right]\right.\\ &\qquad \left.\times\, \mathrm{d}t\,\mathrm{d}t'\right\}^2 \mathrm{d}v\,\mathrm{d}u\;;\end{aligned}$$

see Section 3.2.3 in Aktaran-Kalaycı (2006). Then for $1 \leq r_2 \leq r_1 \ll b$, we can compute that

$$c_A(f_0; b; r_1, r_2) \doteq \frac{6r_2^2(7r_1^2 - 3r_2^2)\sigma^4}{35r_1^3 b}$$

and

$$c_A(f_2; b; r_1, r_2) \doteq \frac{7(3{,}003r_1^5 - 3{,}250r_1^4 r_2 + 875r_1^2 r_2^3 - 126r_2^5)r_2^3\sigma^4}{4{,}290\, r_1^7\, b} .$$

Example 4 We use results from Table 1 and Example 2 to obtain

$$\begin{aligned}
&\mathrm{Var}\big[\mathcal{C}^{\mathrm{LO}}(g_0; [b, b/3], [m, 3m], \alpha)\big] \\
&\quad = \alpha^2\, \mathrm{Var}\big[\mathcal{C}^{\mathrm{O}}(g_0; b, m)\big] + (1-\alpha)^2\, \mathrm{Var}\big[\mathcal{C}^{\mathrm{O}}(g_0; b/3, 3m)\big] \\
&\quad\quad + 2\alpha(1-\alpha)\, c_C(g_0; b, m; 3, 1) \\
&\quad \doteq \alpha^2 \frac{44\sigma^4}{105b} + (1-\alpha)^2 \frac{44\sigma^4}{35b} + 2\alpha(1-\alpha) \frac{304\sigma^4}{945b}
\end{aligned}$$

for large b as $m \to \infty$. This quantity is minimized by taking $\alpha^\star = 0.9057$, whence $\mathrm{Var}\big[\mathcal{C}^{\mathrm{LO}}(g_0; [b, b/3], [m, 3m], \alpha^\star)\big] \doteq 0.410\sigma^4/b$, which is a modest improvement compared to $\mathrm{Var}\big[\mathcal{C}^{\mathrm{O}}(g_0; b, m)\big] \doteq 0.419\sigma^4/b$ from Table 1.

Example 5 From Example 3, we have

$$\begin{aligned}
&\mathrm{Var}\big[\mathcal{C}^{\mathrm{LO}}(g_2^\star; [b, b/3], [m, 3m], \alpha)\big] \\
&\quad = \alpha^2\, \mathrm{Var}\big[\mathcal{C}^{\mathrm{O}}(g_2^\star; b, m)\big] + (1-\alpha)^2\, \mathrm{Var}\big[\mathcal{C}^{\mathrm{O}}(g_2^\star; b/3, 3m)\big] \\
&\quad\quad + 2\alpha(1-\alpha)\, c_C(g_2^\star; b, m, 3, 1) \\
&\quad \doteq \alpha^2 \frac{2{,}692\sigma^4}{3{,}465b} + (1-\alpha)^2 \frac{2{,}692\sigma^4}{1{,}155b} + 2\alpha(1-\alpha) \frac{5{,}248\sigma^4}{13{,}365b}
\end{aligned}$$

for large b as $m \to \infty$. This variance is minimized if $\alpha^\star = 0.8345$, and then $\mathrm{Var}\big[\mathcal{C}^{\mathrm{LO}}(g_2^\star; [b, b/3], [m, 3m], \alpha^\star)\big] \doteq 0.713\sigma^4/b$, which compares favorably to $\mathrm{Var}\big[\mathcal{C}^{\mathrm{O}}(g_2^\star; b, m)\big] \doteq 0.777\sigma^4/b$ from Table 1.

3.3 Covariance between OBM Estimators

We define the estimator arising from the linear combination of OBM estimators,

$$\mathcal{O}^{\mathrm{LO}}(\boldsymbol{B}, \boldsymbol{M}, \boldsymbol{\alpha}) \equiv \sum_{i=1}^{k-1} \alpha_i \mathcal{O}(B_i, M_i) + \left[1 - \sum_{i=1}^{k-1} \alpha_i\right] \mathcal{O}(B_k, M_k) ,$$

where the OBM estimator is defined by Equation (3) (see also Pedrosa and Schmeiser 1993). In this section, we aim to find an approximate expression for

$$c_O(b, m; r_1, r_2) \equiv \operatorname{Cov}\big[\mathcal{O}(\lfloor bm \rfloor / \lfloor r_1 m \rfloor, \lfloor r_1 m \rfloor), \mathcal{O}(\lfloor bm \rfloor / \lfloor r_2 m \rfloor, \lfloor r_2 m \rfloor)\big] .$$

Definition 4 Recall $\boldsymbol{r}$, $\boldsymbol{M}$, and $\boldsymbol{B}$ from Definition 1. Let $\boldsymbol{\mathcal{O}}(\boldsymbol{B}, \boldsymbol{r}, \boldsymbol{M})$ denote the $k \times 1$ vector whose jth component is $\mathcal{O}(B_j, M_j)$ for $j = 1, \ldots, k$. Similarly, let $\boldsymbol{\mathcal{O}}(b, \boldsymbol{r})$ denote the $k \times 1$ vector whose jth component is

$$\mathcal{O}_j(b, r_j) \equiv \frac{b\sigma^2}{r_j(b - r_j)^2} \int_0^{b - r_j} \left[\mathcal{W}(t + r_j) - \mathcal{W}(t) - r_j \mathcal{W}(b)/b\right]^2 \mathrm{d}t$$

for $j = 1, \ldots, k$.

Theorem 2 *If Assumption FCLT holds, then*

$$\boldsymbol{\mathcal{O}}(\boldsymbol{B}, \boldsymbol{r}, \boldsymbol{M}) \underset{m \to \infty}{\Longrightarrow} \boldsymbol{\mathcal{O}}(b, \boldsymbol{r}) .$$

To establish the asymptotic distribution of $\boldsymbol{\mathcal{O}}(\boldsymbol{B}, \boldsymbol{r}, \boldsymbol{M})$ as $m \to \infty$, we need the following definitions.

Definition 5 Let $r \in [1, b)$ and define the differencing map $\mathcal{Q}^r : Z \in D[0, b] \mapsto \mathcal{Q}_Z^r \in D[0, b - r]$ by

$$\mathcal{Q}_Z^r(s) \equiv \frac{Z(s + r) - Z(s) - rZ(b)/b}{\sqrt{r}} \quad \text{for } s \in [0, b - r] .$$

Definition 6 We define the simple overlapping map $\Upsilon^r : Z \in D[0, b] \mapsto \Upsilon^r(Z) \in \mathbb{R}$ as follows:

$$\Upsilon^r(Z) \equiv \frac{b}{(b - r)^2} \int_0^{b - r} \left[\sigma\, \mathcal{Q}_Z^r(s)\right]^2 \mathrm{d}s . \tag{24}$$

Moreover, for $m = 1, 2, \ldots$ with $M(r) \equiv \lfloor rm \rfloor$ and $B(r) \equiv \lfloor bm \rfloor / M(r)$, we also define the approximate (discrete) differencing map $\mathcal{Q}^{r,m} : Z \in D[0, b] \mapsto \mathcal{Q}_Z^{r,m} \in D[0, b - r]$ by

$$\mathcal{Q}_Z^{r,m}(s) \equiv \sqrt{\frac{m}{M(r)}} \left\{ Z\left(\frac{\lceil sm \rceil - 1 + M(r)}{m}\right) - Z\left(\frac{\lceil sm \rceil - 1}{m}\right) \right\} - \frac{\sqrt{mM(r)}}{\lfloor bm \rfloor} Z\left(\frac{\lfloor bm \rfloor}{m}\right) \tag{25}$$

for $s \in [0, b - r]$ and the associated discrete overlapping map $\Upsilon_m^r : Z \in D[0, b] \mapsto \Upsilon_m^r(Z) \in \mathbb{R}$ by

$$\Upsilon_m^r(Z) \equiv \frac{\lfloor bm \rfloor \, \sigma^2}{\big[\, \lfloor bm \rfloor - M(r) + 1\big]\big[\, \lfloor bm \rfloor - M(r)\big]} \sum_{i=1}^{\lfloor bm \rfloor - M(r)+1} \left\{ \mathcal{Q}_Z^{r,m}(i/m) \right\}^2 , \tag{26}$$

if $\lfloor bm \rfloor > M(r)$, and $\Upsilon_m^r(Z) \equiv 0$ if $\lfloor bm \rfloor = M(r)$.

Remark 4 In view of Equations (4) and (25), we have

$$\mathcal{Q}_{X_m}^{r,m}(i/m) \;=\; \sigma^{-1}\sqrt{M(r)}\,\big[\bar{Y}_{i,M(r)}^{\,\mathrm{O}} - \bar{Y}_{\lfloor bm \rfloor}\big]$$

for $i = 1, \ldots, \lfloor bm \rfloor - M(r) + 1$ and $m = 1, 2, \ldots$. In this case, it follows from the definition of the OBM estimator (Equation 3) that

$$\Upsilon_m^r(X_m) = \mathcal{O}[B(r), M(r)] \quad (\text{provided } \lfloor bm \rfloor > M(r))\,.$$

In this case, Theorem 2 is equivalent to

$$\big[\Upsilon_m^{r_1}(X_m), \ldots, \Upsilon_m^{r_k}(X_m)\big] \underset{m\to\infty}{\Longrightarrow} \big[\Upsilon^{r_1}(\mathcal{W}), \ldots, \Upsilon^{r_k}(\mathcal{W})\big]\,. \tag{27}$$

We seek to apply the generalized CMT to prove Equation (27). Similar to the development in Section 3.2, we first justify the hypotheses of the generalized CMT with Proposition 2.

Proposition 2 *If* $\Upsilon^r(\cdot)$ *and* $\Upsilon_m^r(\cdot)$ *are defined by Equations (24) and (26), respectively, then*

$$\Pr\big\{\mathcal{W} \in D[0,b] - \mathbb{D}_{\Upsilon^r}\big\} \;=\; 1\,, \tag{28}$$

where $\Pr\{\cdot\}$ *denotes Wiener measure and*

$$\mathbb{D}_{\Upsilon^r} \equiv \Big\{x \in D[0,b] : \textit{for some sequence } \{x_m\} \subset D[0,b] \textit{ with } \lim_{m\to\infty} \rho_b(x_m, x) = 0, \textit{ the sequence } \{\Upsilon_m^r(x_m)\} \textit{ does not converge to } \Upsilon^r(x)\Big\}\,.$$

Sketch of Proof of Proposition 2 Although the proof of Proposition 2 is detailed in Appendix A.3 of Aktaran-Kalaycı (2006), it can be summarized as follows. First we exploit the almost-sure continuity of $\mathcal{W}(u)$ at every $u \in [0,b]$ and the convergence of $\{x_n\}$ to $\mathcal{W}$ in $D[0,b]$ so as to establish the following properties, each of which holds with probability one:

$$\lim_{m\to\infty} \left|\mathcal{Q}_{x_m}^{r,m}(s) - \mathcal{Q}_{\mathcal{W}}^{r,m}(s)\right| = 0 \text{ for every } s \in [0, b-r]\,; \tag{29}$$

$$\mathcal{Q}_{\mathcal{W}}^{r,m}(s) \text{ is continuous at every } s \in [0, b-r] \text{ for } m = 1, 2, \ldots\,; \tag{30}$$

$$\lim_{m\to\infty} \left| \mathcal{Q}_{\mathcal{W}}^{r,m}(s) - \mathcal{Q}_{\mathcal{W}}^{r}(s) \right| = 0 \text{ for every } s \in [0, b-r]\,; \tag{31}$$

and

$$\mathcal{Q}_{\mathcal{W}}^{r}(s) \text{ is continuous at every } s \in [0, b-r]\,. \tag{32}$$

The almost-sure properties (30)–(32) and the uniform continuity of $\mathcal{W}$ on $[0, b]$ imply that with probability one, $\left\{\mathcal{Q}_{\mathcal{W}}^{r,m}(s) : m = 1, 2, \ldots\right\}$ is a uniformly bounded sequence of functions on $[0, b-r]$. The latter almost-sure uniform-boundedness property and (29) imply that

$$\lim_{m\to\infty} \left| \Upsilon_m^r(x_m) - \Upsilon_m^r(\mathcal{W}) \right| = 0 \text{ with probability one.} \tag{33}$$

From (32) we see that with probability one, $\left[\mathcal{Q}_{\mathcal{W}}^{r}(s)\right]^2$ is Riemann integrable on $[0, b-r]$. It follows immediately that

$$\lim_{m\to\infty} \left| \Upsilon_m^r(\mathcal{W}) - \Upsilon^r(\mathcal{W}) \right| = 0 \text{ with probability one.} \tag{34}$$

Combining (33) and (34) and applying the triangle inequality, we see that

$$\lim_{m\to\infty} \left| \Upsilon_m^r(x_m) - \Upsilon^r(\mathcal{W}) \right| = 0 \text{ with probability one.}$$ □

Proof of Theorem 2 Assume that the sequence $\{x_m\}$ converges to $\mathcal{W}$ in $D[0, b]$. Combining Equations (27) and (28), we see that for each $j \in \{1, 2, \ldots, k\}$, the random variable $\Upsilon_m^{r_j}(x_m)$ converges to $\Upsilon^{r_j}(\mathcal{W})$ with probability one as $m \to \infty$ by the generalized CMT. It follows that the corresponding vector-valued sequence $\left\{\left[\Upsilon_m^{r_1}(x_m), \ldots, \Upsilon_m^{r_k}(x_m)\right] : m = 1, 2, \ldots\right\}$ converges to $\left[\Upsilon^{r_1}(\mathcal{W}), \ldots, \Upsilon^{r_k}(\mathcal{W})\right]$ in $\mathbb{R}^k$ with probability one. Thus the generalized CMT yields the result. □

If the vector-valued process $\left\{ \left([\Upsilon_m^{r_1}(X_m)]^2, \ldots, [\Upsilon_m^{r_k}(X_m)]^2\right) : m = 1, 2, \ldots\right\}$ is uniformly integrable, then our Theorem 2 and Theorem 5.4 of Billingsley (1968) imply

$$\begin{aligned} \lim_{m\to\infty} c_O(b, m; r_1, r_2) &= c_O(b; r_1, r_2) \\ &\equiv \frac{2b^2 r_1 r_2 \sigma^4}{(b-r_1)^2(b-r_2)^2} \int_0^{b-r_2} \int_0^{b-r_1} \mathcal{G}(x, y)\, \mathrm{d}x\, \mathrm{d}y\,, \end{aligned} \tag{35}$$

where

$$\mathcal{G}(x, y) \equiv \left[\frac{\min(x + r_1, y + r_2) - \min(x + r_1, y) - \min(x, y + r_2) + \min(x, y)}{r_1 r_2} - \frac{1}{b} \right]^2$$

for all $(x, y) \in \mathbb{R}^2$. After a series of calculations involving Equation (35), detailed in Appendix A.7 of Aktaran-Kalaycı (2006), we can show that for $r_1, r_2 \in [1, b-1]$, such that $r_2 \le r_1$, $r_1 + r_2 \le b$, and $2r_1 - r_2 \le b$,

$$\begin{aligned} &\frac{c_O(b;r_1,r_2)}{\sigma^4} \\ &= \frac{2r_2[b^3(3r_1-r_2) + b^2(7r_1r_2 - 9r_1^2 - 5r_2^2) + 3br_1(3r_1^2 - 5r_1r_2 + 4r_2^2)]}{3(b-r_1)^2(b-r_2)^2 r_1} \\ &\quad - \frac{3r_1^2(r_1-r_2)(r_1-2r_2)}{3(b-r_1)^2(b-r_2)^2 r_1} . \end{aligned} \tag{36}$$

Remark 5 When b is large, Equation (36) gives

$$c_O(b;r_1,r_2) \doteq \frac{2(3r_1 - r_2)r_2\sigma^4}{3r_1 b} ,$$

which is the same as Equation (24) in Pedrosa and Schmeiser (1993).

Example 6 From Equation (13) and Remark 5, we have

$$\begin{aligned} &\mathrm{Var}\big[\mathcal{O}^{\mathrm{LO}}([b, b/2], [m, 2m], \alpha)\big] \\ &= \alpha^2\,\mathrm{Var}\big[\mathcal{O}(b, m)\big] + (1-\alpha)^2\,\mathrm{Var}\big[\mathcal{O}(b/2, 2m)\big] + 2\alpha(1-\alpha)\,c_O(b, m; 2, 1) \\ &\doteq \alpha^2\frac{4\sigma^4}{3b} + (1-\alpha)^2\frac{8\sigma^4}{3b} + 2\alpha(1-\alpha)\frac{5\sigma^4}{3b} , \end{aligned}$$

for large b and $m \to \infty$. This quantity is minimized by taking $\alpha^\star = 1.5$; and then we obtain $\mathrm{Var}\big[\mathcal{O}^{\mathrm{LO}}([b, b/2], [m, 2m], \alpha^\star)\big] \doteq 7\sigma^4/(6b)$, which compares well to $\mathrm{Var}\big[\mathcal{O}(b, m)\big] \doteq 4\sigma^4/(3b)$.

Example 7 To enable exact calculations for the covariance results of the previous sections, suppose that the Y_i's are i.i.d. Nor(0, 1). Of course, the sample variance $S_Y^2(n)$ is the best estimator for σ^2 in this situation; but we ignore this fact temporarily in order to check if the OLCOVE estimators perform as advertised. Specifically, we calculate $c_A(f;b;r_1,r_2)$, $c_C(g;b;r_1,r_2)$, and $c_O(b;r_1,r_2)$ analytically and find that the exact calculations match up perfectly with the corresponding results from Remark 3, Sections 3.2 and 3.3. For complete details, see Appendices A.8, A.9, and A.10 in Aktaran-Kalaycı (2006).

3.4 General Linear Combinations of Overlapping Variance Estimators

We consider the general linear combination (11) of k overlapping estimators with $r_1, r_2, \ldots, r_k \ge 1$; and we take $\alpha_k \equiv 1 - \sum_{i=1}^{k-1}\alpha_i$ so that $\sum_{i=1}^{k}\alpha_i = 1$. As in

Section 3.1, suppose $\text{Bias}\big[\mathcal{V}^{\text{O}}(b, m)\big] = \delta/m + o(1/m)$ for some constant δ. Then

$$\text{E}\big[\mathcal{V}^{\text{LO}}(\boldsymbol{B}, \boldsymbol{M}, \boldsymbol{\alpha})\big] = \sigma + \delta \sum_{i=1}^{k} \frac{\alpha_i}{M_i} + o(1/m) \tag{37}$$

and

$$\begin{aligned}\text{Var}\big[\mathcal{V}^{\text{LO}}(\boldsymbol{B}, \boldsymbol{M}, \boldsymbol{\alpha})\big] &= \sum_{i=1}^{k} \alpha_i^2 \text{Var}\big[\mathcal{V}^{\text{O}}(B_i, M_i)\big] \\ &\quad + 2 \sum_{i=2}^{k} \sum_{j=1}^{i-1} \alpha_i \alpha_j \text{Cov}\left[\mathcal{V}^{\text{O}}(B_i, M_i), \mathcal{V}^{\text{O}}(B_j, M_j)\right] .\end{aligned}$$

As discussed in Section 3.1, one can choose the coefficients $\{\alpha_i\}$ in (37) to produce an estimator $\mathcal{V}^{\text{LO}}(\boldsymbol{B}, \boldsymbol{M}, \boldsymbol{\alpha})$ whose first-order bias is less than that of the original overlapping estimator using batches of size m; but this may cause increased estimator variance (see Examples 4 and 6). One can also select a weight function so that the OLCOVE is first-order unbiased. In any case, we henceforth assume that the base batch size m is large enough to ensure negligible bias for the estimator $\mathcal{V}^{\text{LO}}(\boldsymbol{B}, \boldsymbol{M}, \boldsymbol{\alpha})$; and then we proceed to find the value of the vector $\boldsymbol{\alpha}$ that minimizes variance.

For large b and $m \to \infty$, we can minimize $\text{Var}\big[\mathcal{V}^{\text{LO}}(\boldsymbol{B}, \boldsymbol{M}, \boldsymbol{\alpha})\big]$ subject to the constraint that $\sum_{i=1}^{k} \alpha_i = 1$ by adopting the perspective of the method of control variables (see Lavenberg and Welch 1981). We have the "response" $Z \equiv \mathcal{V}^{\text{O}}(B_k, M_k)$ and the "control vector" $\boldsymbol{C} \equiv [C_1, C_2, \ldots, C_{k-1}]$ with jth element $C_j \equiv \mathcal{V}^{\text{O}}(B_k, M_k) - \mathcal{V}^{\text{O}}(B_j, M_j)$ for $j = 1, 2, \ldots, k-1$. Let $\mathbf{0}_{k-1}$ denote the $1 \times (k-1)$ vector of zeros. Because $\text{E}[\boldsymbol{C}] \to \mathbf{0}_{k-1}$ as $m \to \infty$, we see that

$$\mathcal{V}^{\text{LO}}(\boldsymbol{B}, \boldsymbol{M}, \boldsymbol{\alpha}) = Z - \boldsymbol{\alpha}\boldsymbol{C}^{\text{T}}$$

has the form of a control-variate estimator when the base batch size m is sufficiently large; and thus we can use standard regression techniques to determine the value $\boldsymbol{\alpha}^{\star}$ that minimizes $\text{Var}[Z - \boldsymbol{\alpha}\boldsymbol{C}]$ asymptotically as $m \to \infty$. In particular, if $\boldsymbol{\Sigma}_C$ denotes the asymptotic variance-covariance matrix of $\boldsymbol{C}$ and if $\boldsymbol{\sigma}_{ZC}$ denotes the $1 \times (k-1)$ vector whose jth element is the asymptotic covariance $\text{Cov}[Z, C_j]$ as $m \to \infty$, then the asymptotically optimal control-coefficient vector is given by

$$\boldsymbol{\alpha}^{\star} = \boldsymbol{\sigma}_{ZC} \boldsymbol{\Sigma}_C^{-1} .$$

Now that we have the machinery to obtain the optimal coefficients $\{\alpha_i^{\star}\}$ in (11), we compute the variance-optimal estimators from linear combinations of overlapping CvM estimators or OBM estimators. The results are presented in Tables 2 and 3; the corresponding results for OLCOVEs formed from overlapping area estimators are presented in Aktaran-Kalaycı et al. (2007a).

Table 2 Approximate asymptotic performance of variance-optimal estimators of the form $\mathcal{C}^{\mathrm{LO}}(g; \boldsymbol{B}, \boldsymbol{M}, \boldsymbol{\alpha}^\star)$

		g_0		$g_2^\star$	
k	$\boldsymbol{M}$	$\frac{m}{\gamma}$Bias	$\frac{b}{\sigma^4}$Var	$\frac{m}{\gamma}$Bias	$\frac{b}{\sigma^4}$Var
1	$[m]$	5.00	0.419	$o(1)$	0.777
2	$[m, 2m]$	4.90	0.418	$o(1)$	0.756
3	$[m, 2m, 3m]$	4.86	0.399	$o(1)$	0.701
4	$[m, 2m, 3m, 4m]$	4.82	0.397	$o(1)$	0.696
5	$[m, 2m, \ldots, 5m]$	4.78	0.394	$o(1)$	0.685
10	$[m, 2m, \ldots, 10m]$	4.74	0.390	$o(1)$	0.673
20	$[m, 2m, \ldots, 20m]$	4.72	0.388	$o(1)$	0.667

Table 3 Approximate asymptotic performance of variance-optimal estimators of the form $\mathcal{O}^{\mathrm{LO}}(\boldsymbol{B}, \boldsymbol{M}, \boldsymbol{\alpha}^\star)$

k	$\boldsymbol{M}$	$\frac{m}{\gamma}$Bias	$\frac{b}{\sigma^4}$Var
1	$[m]$	1	1.333
2	$[m, 2m]$	1.25	1.167
3	$[m, 2m, 3m]$	1.267	1.156
5	$[m, 2m, \ldots, 5m]$	1.268	1.155
20	$[m, 2m, \ldots, 20m]$	1.279	1.155
3	$[m, 1.5m, 2m]$	1.167	1.156
7	$[m, 1.5m, \ldots, 4m]$	1.168	1.155
39	$[m, 1.5m, \ldots, 20m]$	1.168	1.155
41	$[m, 1.05m, \ldots, 2m]$	1.021	1.155
381	$[m, 1.05m, \ldots, 20m]$	1.021	1.155

The estimator $\mathcal{C}^{\mathrm{LO}}(g_0; \boldsymbol{B}, \boldsymbol{M}, \boldsymbol{\alpha}^\star)$ has bias of the form $\psi\gamma/m$, where the constant ψ decreases from 5.00 to 4.72 as we add up to 20 terms to the linear combination. By contrast, $\mathcal{C}^{\mathrm{LO}}(g_2^\star; \boldsymbol{B}, \boldsymbol{M}, \boldsymbol{\alpha}^\star)$ has bias of the form $o(1/m)$. Furthermore, as we add more terms to these OLCOVEs, the "standardized" variance $b\sigma^{-4}\mathrm{Var}\big[\mathcal{C}^{\mathrm{LO}}(g_0; \boldsymbol{B}, \boldsymbol{M}, \boldsymbol{\alpha}^\star)\big]$ decreases from 0.419 to 0.388 (about a 9% reduction), while that of $\mathcal{C}^{\mathrm{LO}}(g_2^\star; \boldsymbol{B}, \boldsymbol{M}, \boldsymbol{\alpha}^\star)$ decreases from 0.777 to 0.667 (about a 15% reduction). Thus, we have the familiar bias-variance trade-off: $\mathcal{C}^{\mathrm{LO}}(g_0; \boldsymbol{B}, \boldsymbol{M}, \boldsymbol{\alpha}^\star)$ has higher bias but lower variance than $\mathcal{C}^{\mathrm{LO}}(g_2^\star; \boldsymbol{B}, \boldsymbol{M}, \boldsymbol{\alpha}^\star)$.

We also see that as we augment $\boldsymbol{M}$ from the scalar $[m]$ to the 20-dimensional vector $[m, \ldots, 20m]$, the "standardized" variance $b\sigma^{-4}\mathrm{Var}\big[\mathcal{O}^{\mathrm{LO}}(\boldsymbol{B}, \boldsymbol{M}, \boldsymbol{\alpha}^\star)\big]$ decreases from 1.333 to 1.155 (about a 14% reduction) while the corresponding "standardized" bias $m\gamma^{-1}\mathrm{Bias}\big[\mathcal{O}^{\mathrm{LO}}(\boldsymbol{B}, \boldsymbol{M}, \boldsymbol{\alpha}^\star)\big]$ *increases* from 1 to 1.279. This bias increase is caused by the coefficients used in the linear combination, which is undesirable. For this reason, we try different batch-size sets in an attempt to reduce the bias. For example, when $\boldsymbol{M}$ changes from the three-dimensional vector $[m, 1.5m, 2m]$ to the 39-dimensional vector $[m, 1.5m, \ldots, 20m]$, the bias actually increases slightly from 1.167 to 1.168; and the variance decreases slightly from 1.156 to 1.155. When $\boldsymbol{M}$ changes from the 41-dimensional vector $[m, 1.05m, \ldots, 2m]$ to the 381-dimensional vector $[m, 1.05m, \ldots, 20m]$, the bias and variance of $\mathcal{O}^{\mathrm{LO}}(\boldsymbol{B}, \boldsymbol{M}, \boldsymbol{\alpha}^\star)$

stay fixed around 1.021 and 1.155, respectively. In comparison with the conventional OBM estimator, the latter results represent a 14% decrease in variance and a 2% increase in bias.

3.5 Monte Carlo Study

In the previous sections, we established some (asymptotic) analytical properties of OLCOVEs. In this section, we summarize the results of Monte Carlo studies on these estimators to see if they perform as advertised in practice. More specifically, we estimate the expected values and variances of various OLCOVEs; and then we compare the empirical results with our analytical results when those OLCOVEs are applied to the following test processes:

- The first test process is a first-order autoregressive (AR(1)) process, $Y_i = \phi Y_{i-1} + \varepsilon_i$, for $i = 1, 2, \ldots$, where: $|\phi| < 1$; the ε_i's are i.i.d. Nor(0, $1 - \phi^2$); and $Y_0 \sim$ Nor(0, 1) so that $\{Y_i : i = 1, 2, \ldots\}$ is stationary.
- The second test process is an $M/M/1$ queue-waiting-time process, where: the interarrival times are i.i.d. exponential with mean $1/\lambda$; the service times are i.i.d. exponential with mean $1/\omega$; and the traffic intensity $\rho \equiv \lambda/\omega < 1$. To ensure that $\{Y_i : i = 1, 2, \ldots\}$ is stationary, we take $Y_1 = 0$ with probability $1 - \rho$, and we sample Y_1 from an exponential distribution having mean $1/(\omega - \lambda)$ with probability ρ.

For the AR(1) process, we take $\phi = 0.9$, and for the $M/M/1$ queue-waiting-time process, we take $\lambda = 0.8$ and $\omega = 1.0$ so that $\rho = 0.8$. It can be shown that the steady-state variance parameters for these AR(1) and $M/M/1$ processes are $\sigma^2 = 19$ and 1,976, respectively; see, for example, Alexopoulos et al. (2007b).

Table 4 summarizes the Monte Carlo performance results for the variance estimators under consideration when those estimators are applied to the selected AR(1) and $M/M/1$ processes. All entries in the table are based on 10,000 independent replications using $b = 20$, various values of m, and batch-size vectors $\boldsymbol{M}_i \equiv [M(1), \ldots, M(i)]$, where $M(i) \equiv im$ for $i = 1, 2, 3, 5$. Common random numbers were employed whenever possible. The last two columns in Table 4 provide the exact asymptotic expected values and variances of the selected variance estimators as $m \to \infty$. In Table 4, we let $\widehat{\mathrm{E}}$ and $\widehat{\mathrm{Var}}$ respectively denote the sample mean and variance of the 10,000 independent replications of each selected estimator of σ^2.

From the experimental results displayed in Table 4, we concluded that the expected values of all selected variance estimators converged to σ^2 as m increased, in accordance with the theory. Moreover for the AR(1) process, the estimated variance of each selected estimator of σ^2 converged to its theoretical (asymptotic) value relatively quickly as m became large. The analogous results for the $M/M/1$ queue-waiting-time process exhibited slower convergence to their asymptotic values. This behavior was observed for all the variance estimators under consideration—OLCOVEs as well as the classical OBM estimator.

Table 4 Estimated performance of OLCOVEs based on area, CvM, or OBM variance estimators

(a) AR(1) process with $\phi = 0.9$ and $\sigma^2 = 19$

	$m = 256$		$m = 512$		$m = 1{,}024$		$m \to \infty$	
	$\widehat{\mathrm{E}}$	$\widehat{\mathrm{Var}}$	$\widehat{\mathrm{E}}$	$\widehat{\mathrm{Var}}$	$\widehat{\mathrm{E}}$	$\widehat{\mathrm{Var}}$	E	Var
$\mathcal{A}^{\mathrm{O}}(f_0; b, m)$	16.92	11.90	17.95	12.76	18.44	12.73	19.00	12.81
$\mathcal{A}^{\mathrm{LO}}(f_0; \boldsymbol{B}, \boldsymbol{M}_2, \boldsymbol{\alpha}^\star)$	17.05	11.58	18.02	12.31	18.47	12.28	19.00	12.29
$\mathcal{A}^{\mathrm{LO}}(f_0; \boldsymbol{B}, \boldsymbol{M}_3, \boldsymbol{\alpha}^\star)$	17.08	11.21	18.04	11.95	18.48	11.93	19.00	12.05
$\mathcal{A}^{\mathrm{LO}}(f_0; \boldsymbol{B}, \boldsymbol{M}_5, \boldsymbol{\alpha}^\star)$	17.15	11.04	18.08	11.71	18.49	11.65	19.00	11.88
$\mathcal{A}^{\mathrm{O}}(f_2; b, m)$	18.15	14.21	18.75	15.27	18.89	15.22	19.00	15.38
$\mathcal{A}^{\mathrm{LO}}(f_2; \boldsymbol{B}, \boldsymbol{M}_2, \boldsymbol{\alpha}^\star)$	18.23	13.72	18.79	14.60	18.90	15.59	19.00	14.73
$\mathcal{A}^{\mathrm{LO}}(f_2; \boldsymbol{B}, \boldsymbol{M}_3, \boldsymbol{\alpha}^\star)$	18.25	12.95	18.79	13.75	18.89	13.67	19.00	13.86
$\mathcal{A}^{\mathrm{LO}}(f_2; \boldsymbol{B}, \boldsymbol{M}_5, \boldsymbol{\alpha}^\star)$	18.27	12.67	18.79	13.44	18.89	13.31	19.00	13.53
$\mathcal{C}^{\mathrm{O}}(g_0; b, m)$	15.76	7.30	17.27	7.58	18.12	7.77	19.00	7.56
$\mathcal{C}^{\mathrm{LO}}(g_0; \boldsymbol{B}, \boldsymbol{M}_2, \boldsymbol{\alpha}^\star)$	15.81	7.30	17.30	7.57	18.13	7.77	19.00	7.55
$\mathcal{C}^{\mathrm{LO}}(g_0; \boldsymbol{B}, \boldsymbol{M}_3, \boldsymbol{\alpha}^\star)$	15.85	7.00	17.33	7.24	18.15	7.49	19.00	7.20
$\mathcal{C}^{\mathrm{LO}}(g_0; \boldsymbol{B}, \boldsymbol{M}_5, \boldsymbol{\alpha}^\star)$	15.90	6.94	17.35	7.19	18.15	7.42	19.00	7.11
$\mathcal{C}^{\mathrm{O}}(g_2^\star; b, m)$	18.06	13.36	18.68	13.85	18.88	14.38	19.00	14.02
$\mathcal{C}^{\mathrm{LO}}(g_2^\star; \boldsymbol{B}, \boldsymbol{M}_2, \boldsymbol{\alpha}^\star)$	18.16	13.18	18.71	13.57	18.88	14.02	19.00	13.65
$\mathcal{C}^{\mathrm{LO}}(g_2^\star; \boldsymbol{B}, \boldsymbol{M}_3, \boldsymbol{\alpha}^\star)$	18.17	12.17	18.72	12.72	18.88	13.00	19.00	12.54
$\mathcal{C}^{\mathrm{LO}}(g_2^\star; \boldsymbol{B}, \boldsymbol{M}_5, \boldsymbol{\alpha}^\star)$	18.18	12.01	18.73	12.51	18.89	12.84	19.00	12.36
$\mathcal{O}(b, m)$	18.22	25.14	18.61	26.31	18.82	26.60	19.00	25.56
$\mathcal{O}^{\mathrm{LO}}(\boldsymbol{B}, \boldsymbol{M}_2, \boldsymbol{\alpha}^\star)$	18.09	22.47	18.52	23.09	18.77	23.01	19.00	21.48
$\mathcal{O}^{\mathrm{LO}}(\boldsymbol{B}, \boldsymbol{M}_5, \boldsymbol{\alpha}^\star)$	18.08	22.27	18.53	22.95	18.77	22.87	19.00	21.28

(b) $M/M/1$ waiting-time process with $\rho = 0.8$ and $\sigma = 1{,}976$; all variance entries $\times 1{,}000$

	$m = 4{,}096$		$m = 8{,}192$		$m = 16{,}278$		$m = 32{,}768$		$m \to \infty$	
	$\widehat{\mathrm{E}}$	$\widehat{\mathrm{Var}}$	$\widehat{\mathrm{E}}$	$\widehat{\mathrm{Var}}$	$\widehat{\mathrm{E}}$	$\widehat{\mathrm{Var}}$	$\widehat{\mathrm{E}}$	$\widehat{\mathrm{Var}}$	E	Var
$\mathcal{A}^{\mathrm{O}}(f_0; b, m)$	1,931	642	1,935	381	1,958	267	1,971	206	1,976	139
$\mathcal{A}^{\mathrm{LO}}(f_0; \boldsymbol{B}, \boldsymbol{M}_2, \boldsymbol{\alpha}^\star)$	1,935	645	1,939	378	1,960	264	1,972	202	1,976	135
$\mathcal{A}^{\mathrm{LO}}(f_0; \boldsymbol{B}, \boldsymbol{M}_3, \boldsymbol{\alpha}^\star)$	1,936	640	1,939	365	1,961	260	1,971	198	1,976	130
$\mathcal{A}^{\mathrm{LO}}(f_0; \boldsymbol{B}, \boldsymbol{M}_5, \boldsymbol{\alpha}^\star)$	1,937	642	1,940	375	1,962	259	1,972	195	1,976	128
$\mathcal{A}^{\mathrm{O}}(f_2; b, m)$	1,993	738	1,967	428	1,975	298	1,979	235	1,976	166
$\mathcal{A}^{\mathrm{LO}}(f_2; \boldsymbol{B}, \boldsymbol{M}_2, \boldsymbol{\alpha}^\star)$	1,993	740	1,970	421	1,975	297	1,979	229	1,976	159
$\mathcal{A}^{\mathrm{LO}}(f_2; \boldsymbol{B}, \boldsymbol{M}_3, \boldsymbol{\alpha}^\star)$	1,993	724	1,970	413	1,975	281	1,979	220	1,976	150
$\mathcal{A}^{\mathrm{LO}}(f_2; \boldsymbol{B}, \boldsymbol{M}_5, \boldsymbol{\alpha}^\star)$	1,993	724	1,970	412	1,976	282	1,979	216	1,976	146
$\mathcal{C}^{\mathrm{O}}(g_0; b, m)$	1,859	451	1,913	287	1,954	214	1,967	149	1,976	82
$\mathcal{C}^{\mathrm{LO}}(g_0; \boldsymbol{B}, \boldsymbol{M}_2, \boldsymbol{\alpha}^\star)$	1,861	453	1,914	287	1,954	214	1,967	149	1,976	82
$\mathcal{C}^{\mathrm{LO}}(g_0; \boldsymbol{B}, \boldsymbol{M}_3, \boldsymbol{\alpha}^\star)$	1,862	453	1,915	284	1,955	211	1,968	146	1,976	78
$\mathcal{C}^{\mathrm{LO}}(g_0; \boldsymbol{B}, \boldsymbol{M}_5, \boldsymbol{\alpha}^\star)$	1,864	454	1,916	285	1,955	210	1,968	145	1,976	77
$\mathcal{C}^{\mathrm{O}}(g_2^\star; b, m)$	1,946	620	1,967	417	1,984	299	1,981	224	1,976	152
$\mathcal{C}^{\mathrm{LO}}(g_2^\star; \boldsymbol{B}, \boldsymbol{M}_2, \boldsymbol{\alpha}^\star)$	1,942	679	1,967	413	1,983	293	1,981	222	1,976	148
$\mathcal{C}^{\mathrm{LO}}(g_2^\star; \boldsymbol{B}, \boldsymbol{M}_3, \boldsymbol{\alpha}^\star)$	1,941	679	1,967	404	1,985	285	1,982	212	1,976	137
$\mathcal{C}^{\mathrm{LO}}(g_2^\star; \boldsymbol{B}, \boldsymbol{M}_5, \boldsymbol{\alpha}^\star)$	1,941	678	1,967	404	1,984	282	1,982	210	1,976	134
$\mathcal{O}(b, m)$	1,971	820	1,968	533	1,978	415	1,973	342	1,976	277
$\mathcal{O}^{\mathrm{LO}}(\boldsymbol{B}, \boldsymbol{M}_2, \boldsymbol{\alpha}^\star)$	1,967	786	1,964	495	1,975	382	1,973	312	1,976	245
$\mathcal{O}^{\mathrm{LO}}(\boldsymbol{B}, \boldsymbol{M}_5, \boldsymbol{\alpha}^\star)$	1,967	786	1,964	492	1,978	379	1,973	310	1,976	242

From Table 4, we also observed that the bias of each STS-based OLCOVE had magnitude no larger than that of the individual estimators composing the OLCOVE. On the other hand, the bias of OBM-based OLCOVEs sometimes appeared to be worse than that of some of the individual estimators composing the OLCOVE. This seemingly anomalous behavior is also evident in Tables 2 and 3. Finally from Table 4, we found that the expected values of the various area estimators with weight function $f_2(\cdot)$ and the various CvM estimators with weight function $g_2^\star(\cdot)$ converged to σ^2 much faster than did the expected values of the other estimators of σ^2. Overall, the various CvM estimators with weight function $g_0(\cdot)$ had the poorest performance in terms of bias. For both test processes, as m became large, each OLCOVE eventually achieved lower variance than was achieved by any of the estimators composing the OLCOVE. All in all, the observed performance of the OLCOVEs agreed with the asymptotic results developed in Section 3.

4 Efficiency Analysis

In the previous sections, we introduced new linear combinations of overlapping estimators for the variance parameter of a stationary simulation output process. We have shown analytically and experimentally that linear combinations improve the performance of the estimators. Natural concerns that arise at this point are: What is the cost of this improved performance? And is the improvement actually worth such a cost?

The purpose of this section is to address the costs (i.e., computational requirements) associated with OLCOVEs. Alexopoulos et al. (2007a) find that for a simulation-generated time series of length n, all overlapping estimators considered in this article have $O(n)$ computational complexity, which is the same order of magnitude as that of NBM, computationally the simplest estimator. Then it follows that OLCOVEs also have $O(n)$ computational complexity; and so linear combinations do not cause a substantial increase in the computational requirements.

In practice, an OLCOVE will require greater computation time than the sum of the computation times of its constituent estimators. Here, we are interested in the increase in the actual computation times, as opposed to theoretical computational complexity. Using a Monte Carlo example, we show that linear combinations of overlapping STS estimators improve the computational *efficiency* when compared with their constituent overlapping STS estimators. The Monte Carlo study we use is from Alexopoulos et al. (2007c), where $n = 160{,}000$ observations are generated from the waiting-time process of an $M/M/1$ queue with 80% traffic intensity.

We choose a base batch size $m = 4{,}000$ and consider OLCOVEs with three different batch-size vectors: $\boldsymbol{M}_{(1)} \equiv [m]$, $\boldsymbol{M}_{(2)} \equiv [m, 2m]$, and $\boldsymbol{M}_{(4)} \equiv [m, 2m, 4m]$. For these batch-size vectors, we find the bias, variance, and MSE of overlapping area and CvM OLCOVEs using several weight functions (see Section 3.4 of Aktaran-Kalaycı 2006 for complete details). Using Table 1 from Alexopoulos et al. (2007c), we also find the computation times for the selected OLCOVEs. The total computation time $\mathbb{C}$ for each estimator is the sum of two parts,

$$\mathbb{C} \equiv \mathbb{C}_{\text{sim}} + \mathbb{C}_{\text{est}} ,$$

as follows:

(i) We let $\mathbb{C}_{\text{sim}}$ denote the time to generate 160,000 observations from the $M/M/1$ queue-waiting-time process via an Arena simulation model (Kelton et al. 2007). Using an IBM T43p laptop with an Intel Pentium M processor having 1.00 GB of memory and a processor speed of 2.13 GHz, Alexopoulos et al. (2007c) find the average execution time for this model is $\mathbb{C}_{\text{sim}} = 4.22$ seconds.
(ii) We let $\mathbb{C}_{\text{est}}$ denote the time to compute (on the same laptop computer) the variance estimator from the 160,000 observations of the $M/M/1$ queue-waiting-time process that are already stored in computer memory.

We present our findings for the OLCOVEs based on CvM estimators in Table 5; similar results are reported for OLCOVEs based on overlapping area estimators in Aktaran-Kalaycı et al. (2007c). Specifically, we report bias, variance, MSE, and average computation times of these OLCOVEs. We also give two relative computational efficiency indices:

$$\mathbb{Q}_{\text{V}} \equiv (\mathbb{C}_{\text{LO}}\text{Var}_{\text{LO}})/(\mathbb{C}_{\text{O}}\text{Var}_{\text{O}})$$

and

$$\mathbb{Q}_{\text{M}} \equiv (\mathbb{C}_{\text{LO}}\text{MSE}_{\text{LO}})/(\mathbb{C}_{\text{O}}\text{MSE}_{\text{O}}) ,$$

where $\mathbb{C}_{\text{O}}$, Var_{O}, and MSE_{O} denote the performance characteristics of an overlapping estimator and $\mathbb{C}_{\text{LO}}$, Var_{LO}, and MSE_{LO} denote the performance characteristics of the corresponding OLCOVE.

From Table 5, we concluded that OLCOVEs yielded $\mathbb{Q}_{\text{V}}$ and $\mathbb{Q}_{\text{M}}$ values smaller than 1, except for those involving $\mathcal{C}^{\text{LO}}(g_0; \boldsymbol{B}, \boldsymbol{M}_{(2)}, \boldsymbol{\alpha}^\star)$. Thus, the increase in cost (computation time) for the linear-combination estimators was offset by the improved performance (the decrease in variance and bias). This performance-evaluation experiment showed that net improvements in computational efficiency were achieved

Table 5 Performance and computation times of various OLCOVEs for the $M/M/1$ queue-waiting-time process with $\rho = 0.8$

	$\mathcal{C}^{\text{LO}}(g_0; \boldsymbol{B}, \boldsymbol{M}, \boldsymbol{\alpha}^\star)$			$\mathcal{C}^{\text{LO}}(g_2^\star; \boldsymbol{B}, \boldsymbol{M}, \boldsymbol{\alpha}^\star)$		
	$\boldsymbol{M}_{(1)}$	$\boldsymbol{M}_{(2)}$	$\boldsymbol{M}_{(4)}$	$\boldsymbol{M}_{(1)}$	$\boldsymbol{M}_{(2)}$	$\boldsymbol{M}_{(4)}$
$\mathbb{C}_{\text{est}}$	0.053	0.104	0.168	0.057	0.117	0.204
$\mathbb{C}$	4.273	4.324	4.388	4.277	4.337	4.424
|Bias|	120.8	118.6	116.7	15.2	13.6	13.4
Var	40,900	40,803	38,753	75,846	73,796	67,940
MSE	55,493	54,872	52,359	76,077	73,982	68,118
$\mathbb{Q}_{\text{V}}$	1.000	1.010	0.973	1.000	0.987	0.926
$\mathbb{Q}_{\text{M}}$	1.000	1.001	0.969	1.000	0.986	0.926

with the proposed OLCOVEs in a test problem whose characteristics are arguably typical of many practical applications.

5 OLCOVE Density and Confidence Intervals for μ and σ^2

In this section, we develop an approximation to the p.d.f. of the limiting distribution of an OLCOVE as the base batch size $m \to \infty$. Our ultimate objective is to compute approximate confidence intervals for μ and σ^2 that are sufficiently precise and reliable for routine use in practice.

5.1 Approximation to Asymptotic Density of an OLCOVE

To estimate the probability density function of selected OLCOVEs, we generated 1,000,000 independent sample paths from the following processes: (a) an i.i.d. standard normal process, where each path consists of 10,240 observations with $m = 512$ and $b = 20$; and (b) the AR(1) process with $\phi = 0.9$, as specified in Section 3.5, where each path consists of 20,480 observations with $m = 1{,}024$ and $b = 20$. Note that these settings are large enough for approximate convergence of the estimators to their limiting distributions. From these sample paths, we computed the following estimators, all for various choices of weight functions: (i) the nonoverlapping variance estimators, $\mathcal{N}(b, m)$, $\mathcal{A}(f; b, m)$, and $\mathcal{C}(g; b, m)$; (ii) the overlapping variance estimators, $\mathcal{O}(b, m)$, $\mathcal{A}^{\mathrm{O}}(f; b, m)$, and $\mathcal{C}^{\mathrm{O}}(g; b, m)$; and (iii) the OLCOVEs, $\mathcal{O}^{\mathrm{LO}}(\boldsymbol{B}, \boldsymbol{M}_5, \boldsymbol{\alpha}^\star)$, $\mathcal{A}^{\mathrm{LO}}(f; \boldsymbol{B}, \boldsymbol{M}_5, \boldsymbol{\alpha}^\star)$, and $\mathcal{C}^{\mathrm{LO}}(g; \boldsymbol{B}, \boldsymbol{M}_5, \boldsymbol{\alpha}^\star)$, where $\boldsymbol{M}_5 = [m, \ldots, 5m]$.

From the computed estimates, we obtained the empirical p.d.f.'s (e.p.d.f.'s) of the corresponding variance estimators, and these were plotted in MATLAB. We used frequency polygons with equal class intervals (bins); and for each plotted point (vertex of the frequency polygon) representing a bin, the abscissa is the bin's midpoint and the ordinate is the bin's relative frequency divided by the bin width (Hald 1952). Superimposing the e.p.d.f.'s of the various estimators, we obtained Figs. 1 and 2 for the i.i.d. normal and AR(1) processes, respectively.

The e.p.d.f.'s in Figs. 1 and 2 conformed to the theoretical properties discussed in previous sections. For each OLCOVE, its p.d.f. appeared to have lighter tails (lower variance) than did the individual overlapping estimators composing the OLCOVE; and in turn, the overlapping estimators appeared to have lighter tails than did their nonoverlapping counterparts. The moment properties of these variance estimators were not obvious from the figures, even though the subplots of Fig. 2 suggested that OLCOVEs performed the best in terms of bias.

We adopt the methodology of Alexopoulos et al. (2007b) to approximate the p.d.f.'s for these estimators by scaled chi-squared distributions. In Alexopoulos et al. (2007b), the approximation technique of Satterthwaite (1941) is applied to obtain

$$\mathcal{V}(b, m) \;\dot{\sim}\; \mathrm{E}[\mathcal{V}(b, m)]\chi^2_{\nu_{\mathrm{eff}}}/\nu_{\mathrm{eff}} ,$$

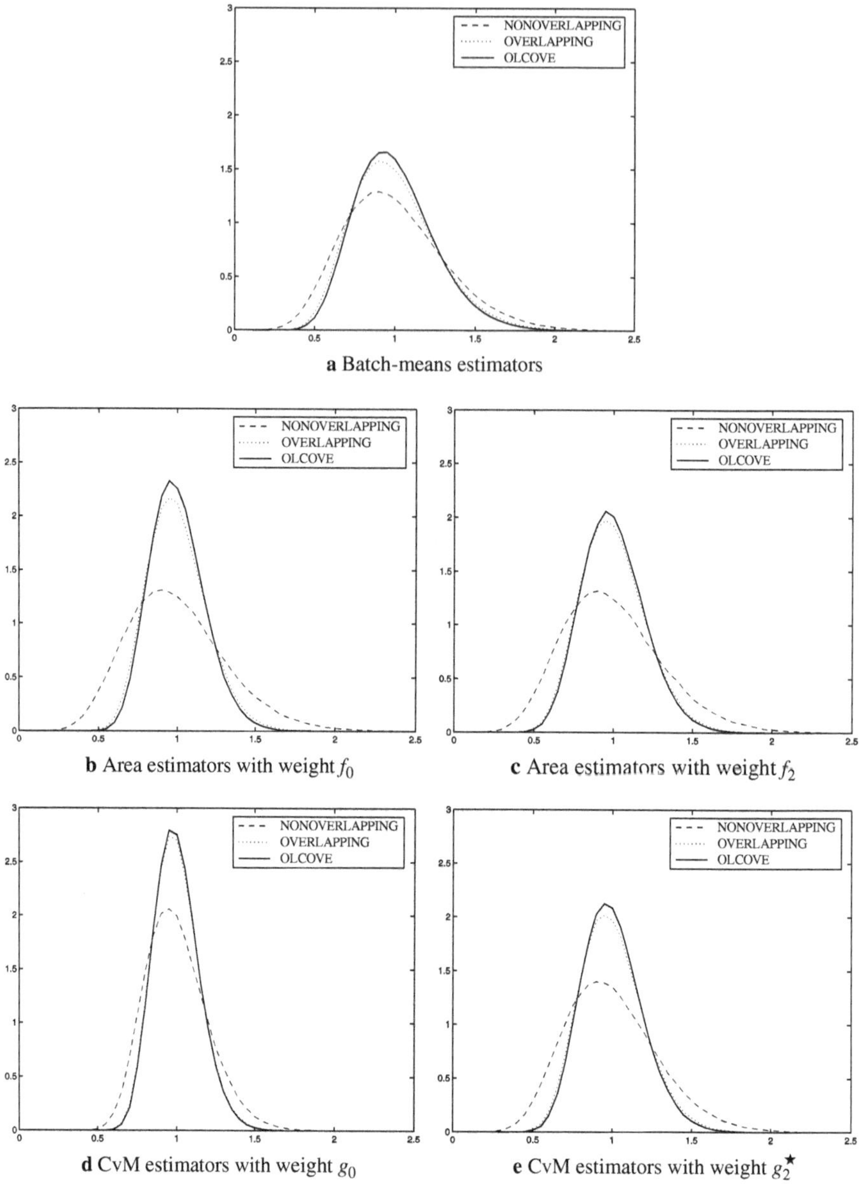

a Batch-means estimators

b Area estimators with weight f_0

c Area estimators with weight f_2

d CvM estimators with weight g_0

e CvM estimators with weight $g_2^{\star}$

Fig. 1 Estimated p.d.f.'s for various variance estimators based on 1,000,000 sample paths of the i.i.d. standard normal process with $m = 512$ and $b = 20$. (**a**) Batch-means estimators. (**b**) Area estimators with weight function f_0. (**c**) Area estimators with weight function f_2. (**d**) CvM estimators with weight function g_0. (**e**) CvM estimators with weight function $g_2^{\star}$

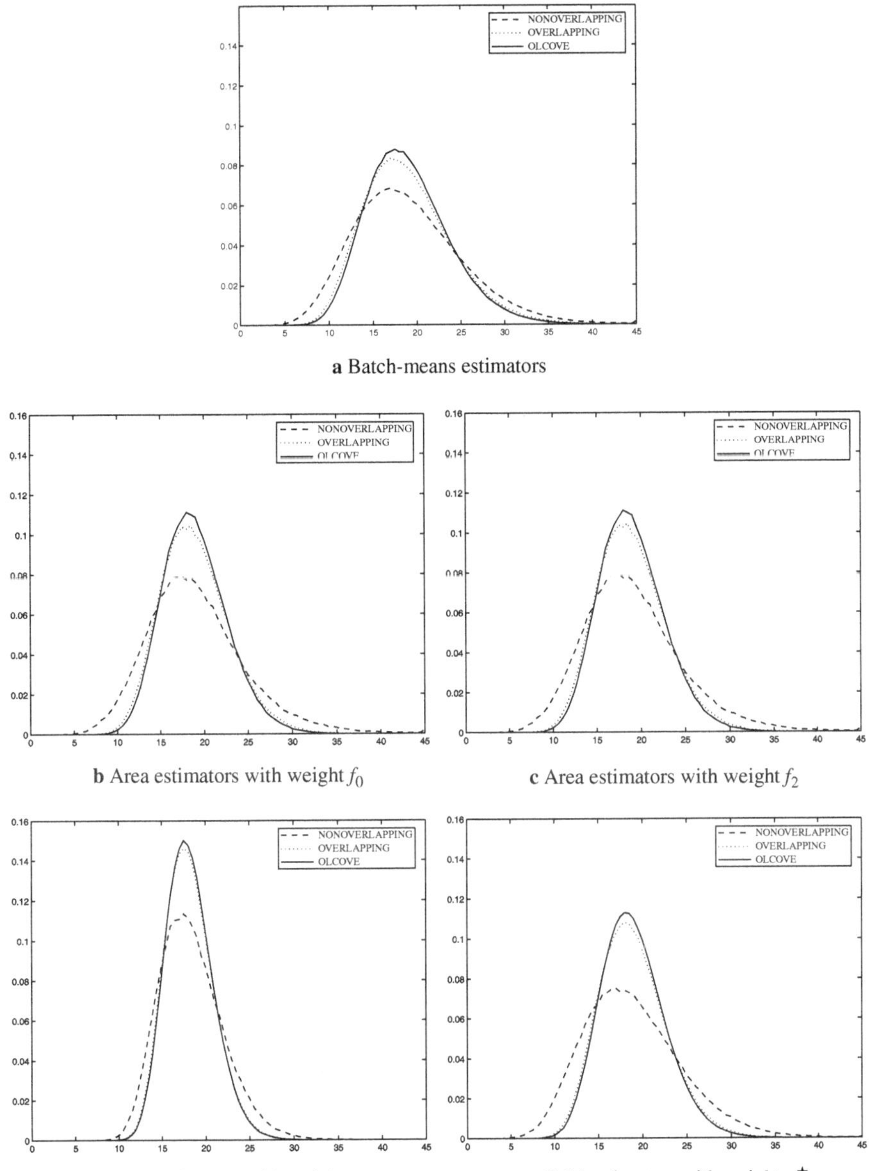

Fig. 2 Estimated p.d.f.'s for various variance estimators based on 1,000,000 sample paths of the AR(1) process with $\phi = 0.9$, $m = 1{,}024$, and $b = 20$ ($\sigma^2 = 19$ for this process). (**a**) Batch-means estimators. (**b**) Area estimators with weight function f_0. (**c**) Area estimators with weight function f_2. (**d**) CvM estimators with weight function g_0. (**e**) CvM estimators with weight function $g_2^\star$

where $\mathcal{V}(b, m)$ is a generic overlapping variance estimator, χ^2_ν is a chi-squared random variable with ν degrees of freedom, and the quantity ν_{eff} is called the effective degrees of freedom (e.d.o.f.) and is given by

$$\nu_{\text{eff}} = \left[\!\left[2\text{E}^2[\mathcal{V}(b, m)]/\text{Var}[\mathcal{V}(b, m)] \right]\!\right], \tag{38}$$

where $[\![z]\!]$ denotes the rounding of z to the nearest integer.

Using Equation (38), we calculate ν_{eff} for various OLCOVEs, with $\boldsymbol{M}_i \equiv [m, \ldots, im]$ and $\boldsymbol{B}_i \equiv [b/1, \ldots, b/i]$ for $i = 1, \ldots, 5$. For example, in a process consisting of i.i.d. standard normal variates with $b = 20$, we obtain $\text{E}\big[\mathcal{A}^{\text{LO}}(f_0; \boldsymbol{B}, \boldsymbol{M}_5, \boldsymbol{\alpha}^\star)\big] = 1$ and $\text{Var}\big[\mathcal{A}^{\text{LO}}(f_0; \boldsymbol{B}, \boldsymbol{M}_5, \boldsymbol{\alpha}^\star)\big] = 0.0329$, which gives $\nu_{\text{eff}} = [\![2/0.0329]\!] = 61$. Similarly, without loss of generality, we use the i.i.d. standard normal process to calculate the e.d.o.f.'s for other batch size sets and other types of estimators. The results are given in Table 6 for the various OLCOVEs under consideration here.

To show how well we can approximate the limiting p.d.f.'s of the variance estimators, we superimpose the empirical p.d.f.'s of the OLCOVEs and the corresponding fitted p.d.f.'s based on the approximation

$$\mathcal{V}^{\text{LO}}(\boldsymbol{B}, \boldsymbol{M}, \boldsymbol{\alpha}) \mathrel{\dot\sim} \sigma^2 \chi^2_{\nu_{\text{eff}}} / \nu_{\text{eff}}, \tag{39}$$

where ν_{eff} was taken from Table 6. From Fig. 3, where we used OLCOVEs with $\boldsymbol{M}_5$ and $b = 20$, we concluded that we obtained very good approximations to the target p.d.f.'s. This finding immediately suggested that using OLCOVEs, we could construct approximately valid confidence intervals for the parameters μ and σ^2.

Table 6 Estimated effective degrees of freedom ν_{eff} for various OLCOVEs with $b = 20$

	E.d.o.f. ν_{eff}				
Estimator	$\boldsymbol{M}_1$	$\boldsymbol{M}_2$	$\boldsymbol{M}_3$	$\boldsymbol{M}_4$	$\boldsymbol{M}_5$
$\mathcal{A}^{\text{LO}}(f_0; \boldsymbol{B}, \boldsymbol{M}, \boldsymbol{\alpha}^\star)$	56	58	60	60	61
$\mathcal{A}^{\text{LO}}(f_2; \boldsymbol{B}, \boldsymbol{M}, \boldsymbol{\alpha}^\star)$	47	49	52	53	53
$\mathcal{C}^{\text{LO}}(g_0; \boldsymbol{B}, \boldsymbol{M}, \boldsymbol{\alpha}^\star)$	95	96	100	101	102
$\mathcal{C}^{\text{LO}}(g_2^\star; \boldsymbol{B}, \boldsymbol{M}, \boldsymbol{\alpha}^\star)$	51	53	57	58	58
$\mathcal{O}^{\text{LO}}(\boldsymbol{B}, \boldsymbol{M}, \boldsymbol{\alpha}^\star)$	28	32	32	32	32

5.2 Confidence Intervals for μ

It follows from the properties given in Equations (7), (8), and (39) that

$$\frac{\bar{Y}_n - \mu}{\sqrt{\mathcal{V}^{\text{LO}}(\boldsymbol{B}, \boldsymbol{M}, \boldsymbol{\alpha})/n}} \mathrel{\dot\sim} t_{\nu_{\text{eff}}}$$

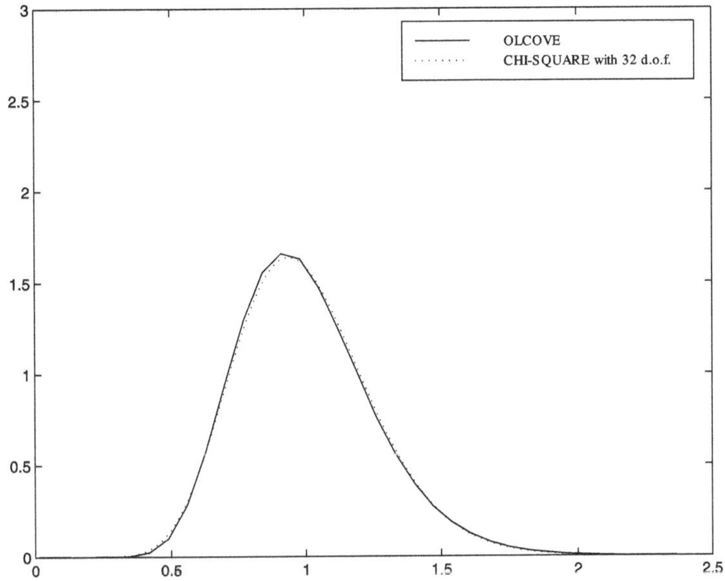

a Overlapping batch-means estimators

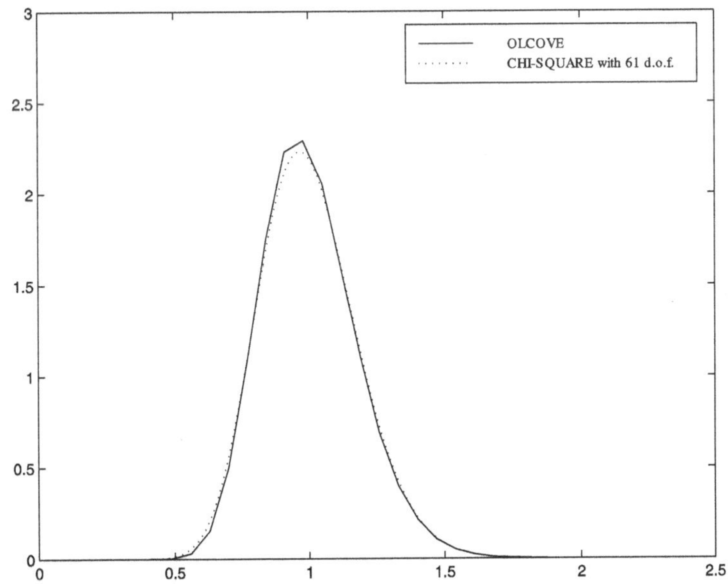

b Area estimators with weight function f_0

Fig. 3 Estimated and fitted p.d.f.'s for various OLCOVEs based on 1,000,000 sample paths of the i.i.d. standard normal process with $\boldsymbol{M}_5$, $m = 512$, and $b = 20$. (**a**) Overlapping batch-means estimators. (**b**) Area estimators with weight function f_0. (**c**) Area estimators with weight function f_2. (**d**) CvM estimators with weight function g_0. (**e**) CvM estimators with weight function $g_2^\star$

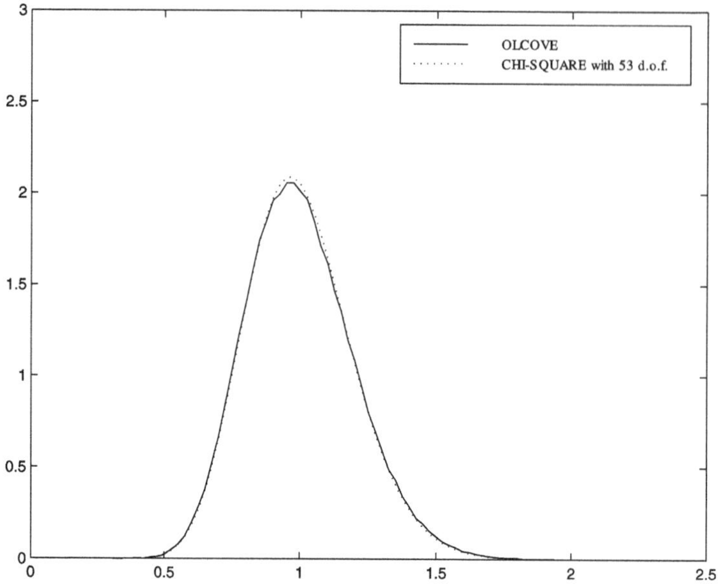

c Area estimators with weight function f_2

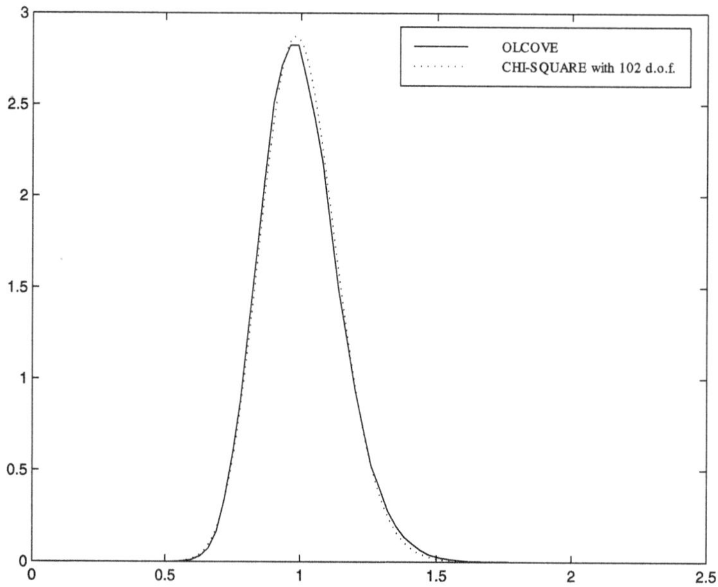

d CvM estimators with weight function g_0

Fig. 3 (continued)

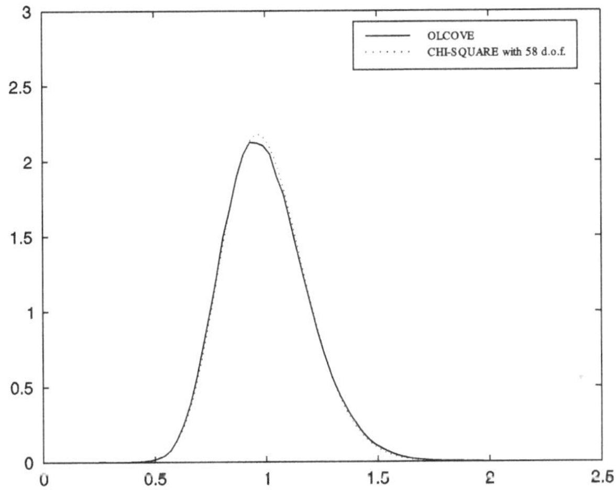

e CvM estimators with weight function $g_2^\star$

Fig. 3 (continued)

for sufficiently large m, where t_ν denotes a Student's t random variable with ν degrees of freedom. Then an approximate $100(1-\beta)\%$ confidence interval for μ is given by

$$\bar{Y}_n - t_{1-\beta/2,\nu_{\text{eff}}}\sqrt{\mathcal{V}^{\text{LO}}(\boldsymbol{B},\boldsymbol{M},\boldsymbol{\alpha})/n} \;\le\; \mu \;\le\; \bar{Y}_n + t_{1-\beta/2,\nu_{\text{eff}}}\sqrt{\mathcal{V}^{\text{LO}}(\boldsymbol{B},\boldsymbol{M},\boldsymbol{\alpha})/n}\,, \tag{40}$$

where $t_{\omega,\nu}$ denotes the ω-quantile of a Student's t random variable with ν degrees of freedom.

Example 8 We use the 1,000,000 independent sample paths of the specified AR(1) process with $\phi = 0.9$ as described in Section 5.1 to compute 1,000,000 realizations of $\bar{Y}_n$ used together with various OLCOVEs. We construct two-sided 90% confidence intervals for μ from Equation (40), where the corresponding ν_{eff}'s are given in Table 6. Then we estimate the empirical coverage probabilities for the two-sided 90% confidence intervals—that is, the proportion of the confidence intervals containing the steady-state mean μ of the process. These results are presented in Table 7. From Table 7, we observed that the achieved empirical coverage probabilities did not differ substantially from the targeted coverage probability, 0.90, thus indicating that the confidence interval procedure worked reasonably well for sufficiently large base batch size m.

5.3 Confidence Intervals for σ^2

Assuming $\text{E}\big[\mathcal{V}^{\text{LO}}(\boldsymbol{B},\boldsymbol{M},\boldsymbol{\alpha})\big] = \sigma^2$ and using Equation (39), we see that an approximate $100(1-\beta)\%$ confidence interval for σ^2 is given by

Table 7 Estimated coverage probabilities of two-sided 90% confidence intervals for μ from various OLCOVEs based on 1,000,000 sample paths of the AR(1) process with $\phi = 0.9$, $m = 1{,}024$, and $b = 20$

	Estimated coverage probability				
Estimator	$\boldsymbol{M}_1$	$\boldsymbol{M}_2$	$\boldsymbol{M}_3$	$\boldsymbol{M}_4$	$\boldsymbol{M}_5$
$\mathcal{A}^{\text{LO}}(f_0; \boldsymbol{B}, \boldsymbol{M}, \boldsymbol{\alpha}^\star)$	0.8958	0.8961	0.8960	0.8962	0.8962
$\mathcal{A}^{\text{LO}}(f_2; \boldsymbol{B}, \boldsymbol{M}, \boldsymbol{\alpha}^\star)$	0.8997	0.8997	0.8999	0.8998	0.8999
$\mathcal{C}^{\text{LO}}(g_0; \boldsymbol{B}, \boldsymbol{M}, \boldsymbol{\alpha}^\star)$	0.8916	0.8917	0.8918	0.8918	0.8919
$\mathcal{C}^{\text{LO}}(g_2^\star; \boldsymbol{B}, \boldsymbol{M}, \boldsymbol{\alpha}^\star)$	0.8990	0.8989	0.8989	0.8988	0.8989
$\mathcal{O}^{\text{LO}}(\boldsymbol{B}, \boldsymbol{M}, \boldsymbol{\alpha}^\star)$	0.8992	0.8986	0.8986	0.8987	0.8987

$$\frac{\nu_{\text{eff}}\, \mathcal{V}^{\text{LO}}(\boldsymbol{B}, \boldsymbol{M}, \boldsymbol{\alpha})}{\chi^2_{1-\beta/2,\nu_{\text{eff}}}} \le \sigma^2 \le \frac{\nu_{\text{eff}}\, \mathcal{V}^{\text{LO}}(\boldsymbol{B}, \boldsymbol{M}, \boldsymbol{\alpha})}{\chi^2_{\beta/2,\nu_{\text{eff}}}}, \tag{41}$$

where $\chi^2_{\omega,\nu}$ denotes the ω-quantile of the chi-squared distribution with ν degrees of freedom.

Example 9 We use the OLCOVEs computed from the 1,000,000 independent sample paths of the AR(1) process of Section 5.1 with the intent to construct two-sided 90% confidence intervals for σ^2. The two-sided confidence intervals for σ^2 are given by Equation (41), where the corresponding ν_{eff}'s are from Table 6. We obtain the estimated coverage probabilities for the two-sided 90% confidence intervals as presented in Table 8.

The confidence intervals using Equation (41) are based on the assumption that all the estimators are unbiased for σ^2. However, we know that $\mathcal{A}^{\text{LO}}(f_0; \boldsymbol{B}, \boldsymbol{M}, \boldsymbol{\alpha}^\star)$ and $\mathcal{O}^{\text{LO}}(\boldsymbol{B}, \boldsymbol{M}, \boldsymbol{\alpha}^\star)$ are moderately biased, whereas $\mathcal{C}^{\text{LO}}(g_0; \boldsymbol{B}, \boldsymbol{M}, \boldsymbol{\alpha}^\star)$ has significant bias for σ^2. Thus, our unbiasedness assumption caused the coverage probabilities to be a bit off for the $\mathcal{A}^{\text{LO}}(f_0; \boldsymbol{B}, \boldsymbol{M}, \boldsymbol{\alpha}^\star)$ and $\mathcal{O}^{\text{LO}}(\boldsymbol{B}, \boldsymbol{M}, \boldsymbol{\alpha}^\star)$ estimators and significantly below the nominal coverage level for the $\mathcal{C}^{\text{LO}}(g_0; \boldsymbol{B}, \boldsymbol{M}, \boldsymbol{\alpha}^\star)$ estimator in Table 8. On the other hand, the empirical coverage probabilities did not differ substantially from the targeted coverage probability for the $\mathcal{A}^{\text{LO}}(f_2; \boldsymbol{B}, \boldsymbol{M}, \boldsymbol{\alpha}^\star)$ and $\mathcal{C}^{\text{LO}}(g_2^\star; \boldsymbol{B}, \boldsymbol{M}, \boldsymbol{\alpha}^\star)$ estimators. This finding makes sense in light of the fact that these estimators are first-order unbiased for σ^2.

Table 8 Estimated coverage probabilities of two-sided 90% confidence intervals for σ^2 from various OLCOVEs based on 1,000,000 sample paths of the AR(1) process with $\phi = 0.9$, $m = 1{,}024$, and $b = 20$

	Estimated coverage probability				
Estimator	$\boldsymbol{M}_1$	$\boldsymbol{M}_2$	$\boldsymbol{M}_3$	$\boldsymbol{M}_4$	$\boldsymbol{M}_5$
$\mathcal{A}^{\text{LO}}(f_0; \boldsymbol{B}, \boldsymbol{M}, \boldsymbol{\alpha}^\star)$	0.8697	0.8708	0.8706	0.8730	0.8719
$\mathcal{A}^{\text{LO}}(f_2; \boldsymbol{B}, \boldsymbol{M}, \boldsymbol{\alpha}^\star)$	0.8891	0.8895	0.8902	0.8887	0.8906
$\mathcal{C}^{\text{LO}}(g_0; \boldsymbol{B}, \boldsymbol{M}, \boldsymbol{\alpha}^\star)$	0.8396	0.8391	0.8393	0.8385	0.8380
$\mathcal{C}^{\text{LO}}(g_2^\star; \boldsymbol{B}, \boldsymbol{M}, \boldsymbol{\alpha}^\star)$	0.8855	0.8835	0.8826	0.8807	0.8825
$\mathcal{O}^{\text{LO}}(\boldsymbol{B}, \boldsymbol{M}, \boldsymbol{\alpha}^\star)$	0.8802	0.8781	0.8793	0.8794	0.8794

6 Limitations and Evaluation of the OLCOVE Technique

In this section, we discuss some of the limitations of the OLCOVE technique and point out some challenges that might arise when the linear combinations of overlapping estimators are to be used in practice. First, only modest improvements in precision are achieved by using the proposed linear combinations of variance estimators, at least in comparison with the improvements typically achieved by using the method of control variables in some other well-known applications (Lavenberg and Welch 1981).

Second, the ability to use different batch sizes in one estimator with linear combinations relaxes the problem of choosing the batch size to some extent. Nevertheless, the performance of an OLCOVE still depends considerably on the base batch size m. Therefore, choosing the base batch size remains an important decision with respect to implementation of OLCOVEs. Another important decision is that of choosing the batch-size vector $\boldsymbol{M}$. For a given sample size n, there are many possible choices of $\boldsymbol{M}$; and in general it is difficult to recommend guidelines for choosing $\boldsymbol{M}$ that will work reasonably well for all steady-state simulation scenarios and still provide good computational performance.

Third, even though the idea behind OLCOVEs is very simple, the resulting estimator structure is not intuitive; and the estimators might seem prohibitively complex for practitioners. To facilitate use of these estimators in practice, we must develop portable, robust software that automates the technique. We are currently working on the development of such software.

Lastly, our Monte Carlo results indicate that net increases in computational efficiency can be achieved with OLCOVEs—at least for the $M/M/1$ queue-waiting-time process considered in the experiment. But the actual computational efficiency is unclear for other specific practical applications. In addition, when used in iterative routines, such as ranking-and-selection procedures, an OLCOVE may slow down the computations somewhat. Thus, some further development to increase the computational efficiency of OLCOVEs (for example, computing all overlapping area estimators in the linear combination simultaneously) may be required before using OLCOVEs in such procedures.

7 Summary and Recommendations

Our goal has been to study a new class of estimators for the variance parameter of a stationary simulation output process. The new estimators are simply linear combinations of overlapping estimators, where each constituent of the linear combination uses a different batch size. We consider OLCOVEs based on overlapping area, overlapping CvM, and overlapping batch means estimators.

We established the theoretical convergence properties of the joint distribution of overlapping estimators based on different batch sizes, with the aim of calculating the underlying covariances between these estimators. Using these covariance results, we found the variance-optimal coefficients to use in constructing the OLCOVEs.

For the variance-optimal STS estimators, we found that the magnitude of the bias was also smaller than that of the original overlapping estimators. Then it follows that the MSE properties for the OLCOVEs were better than those of the underlying components. We have shown that the new estimators perform as expected (in terms of bias and variance) on the exact and empirical examples presented.

We have also shown that the distribution of an OLCOVE can be accurately approximated by a scaled chi-squared distribution with the appropriate effective degrees of freedom (at least for the specific processes we considered). We have applied this approximation to construct confidence intervals for the parameters μ and σ^2 of the underlying steady-state stochastic process based on various OLCOVEs; and we have conducted Monte Carlo studies to see how these confidence intervals perform when applied to simple stochastic processes. Regarding the confidence intervals for μ, we found that the achieved coverage probability was practically the same as the targeted coverage probability (for large enough batch sizes). Regarding the confidence intervals for σ^2, we found that actual coverage probabilities based on first-order unbiased estimators of σ^2 did not differ substantially from the targeted coverage probabilities, at least for batch sizes that were sufficiently large. In particular, we found that the confidence intervals based on $\mathcal{A}^{\mathrm{LO}}(f_2; \boldsymbol{B}, \boldsymbol{M}, \boldsymbol{\alpha}^\star)$ and $\mathcal{C}^{\mathrm{LO}}(g_2^\star; \boldsymbol{B}, \boldsymbol{M}, \boldsymbol{\alpha}^\star)$ performed comparatively well.

As an immediate consequence of Theorem 2, we have the following characterization of the asymptotic distribution of the classical OBM estimator $\mathcal{O}(b, m)$ for the fixed integer $b > 1$ as $m \to \infty$:

$$\mathcal{O}(b, m) \underset{m \to \infty}{\Longrightarrow} \frac{b\sigma^2}{(b-1)^2} \int_0^{b-1} \left[\mathcal{W}(t+1) - \mathcal{W}(t) - \mathcal{W}(b)/b\right]^2 \mathrm{d}t \,. \tag{42}$$

To the best of our knowledge, this is the first formulation of the exact asymptotic distribution of $\mathcal{O}(b, m)$; and we believe that (42) is a result of independent interest beyond our discussion of OLCOVEs.

When we combine (42) with the scaled chi-squared approximation (38)–(39) to the asymptotic distribution of $\mathcal{O}(b, m)$ and with the experimental results in Tables 6 and 7 for $\mathcal{O}(b, m)$ with $b = 20$ and $\boldsymbol{M}_1$ (that is, a single batch size m with overall sample size $n = 20m$), we have some indication of the performance of the classical OBM variance estimator $\mathcal{O}(b, m)$ that can be expected in practical applications—provided the target output process $\{Y_i : i = 1, 2, \ldots\}$ is observed beyond the end of the warm-up period, and the batch size m is sufficiently large to ensure adequate convergence of $\mathcal{O}(b, m)$ to its limiting distribution (42). Specifically in the AR(1) process with lag-one correlation 0.9, we found that by taking $b = 20$ and $m = 1{,}024$ so that the total sample size $n = bm = 20{,}480$, we obtained an OBM variance estimator with 28 degrees of freedom; and the corresponding 90% confidence intervals for μ and σ^2 had empirical coverage probabilities of 89.9% and 88.0%, respectively. We hope that these insights into the performance of $\mathcal{O}(b, m)$ will stimulate more interest in overlapping variance estimators of all types, including the OLCOVEs developed in this article.

References and Selected Bibliography

Aktaran-Kalaycı, T. 2006. Steady-state analyses: variance estimation in simulations and dynamic pricing in service systems. PhD diss., H. Milton Stewart School of Industrial and Systems Engineering, Georgia Institute of Technology. Available online via `hdl.handle.net/1853/13993` [accessed November 11, 2007].

Aktaran-Kalaycı, T., and D. Goldsman. 2005. Linear combinations of overlapping variance estimators for simulations. In *Proceedings of the 2005 Winter Simulation Conference*, eds. M. E. Kuhl, N. M. Steiger, F. B. Armstrong, and J. A. Joines, pp. 756–762. Piscataway, New Jersey: Institute of Electrical and Electronics Engineers. Available online via `http://www.informs-sim.org/wsc05papers/087.pdf` [accessed November 28, 2008].

Aktaran-Kalaycı, T., D. Goldsman, and J. R. Wilson. 2007a. Linear combinations of overlapping variance estimators for simulation. *Operations Research Letters* 35:439–447.

Aktaran-Kalaycı T., D. Goldsman, and J. R. Wilson. 2007b. Online supplement to "Linear combinations of overlapping variance estimators for simulation." Available online via `www.ise.ncsu.edu/jwilson/files/taorlos13.pdf` [accessed November 11, 2007].

Aktaran-Kalaycı, T., D. Goldsman, and J. R. Wilson. 2007c. Confidence interval estimation using linear combinations of overlapping variance estimators. In *Proceedings of the 2007 Winter Simulation Conference*, eds. S. G. Henderson, B. Biller, M.-H. Hsieh, J. Shortle, J. D. Tew, and R. R. Barton, pp. 448–454. Piscataway, New Jersey: Institute of Electrical and Electronics Engineers. Available online via `www.informs-sim.org/wsc07papers/053.pdf` [accessed January 26, 2008].

Alexopoulos, C., and D. Goldsman. 2004. To batch or not to batch? *ACM Transactions on Modeling and Computer Simulation* 14(1):76–114.

Alexopoulos C., N. T. Argon, D. Goldsman, G. Tokol, and J. R. Wilson. 2007a. Overlapping variance estimators for simulation. *Operations Research* 55(6):1090–1103.

Alexopoulos C., N. T. Argon, D. Goldsman, N. M. Steiger, G. Tokol, and J. R. Wilson. 2007b. Efficient computation of overlapping variance estimators for simulation. *INFORMS Journal on Computing* 19(3):314–327.

Alexopoulos C., N. T. Argon, D. Goldsman, N. M. Steiger, G. Tokol, and J. R. Wilson. 2007c. Online supplement to "Efficient computation of overlapping variance estimators for simulation." Available online via `http://www.ise.ncsu.edu/jwilson/files/oveosv14.pdf` [accessed November 11, 2007].

Billingsley, P. 1968. *Convergence of Probability Measures*. New York: Wiley.

Fishman, G. S. 2001. *Discrete-Event Simulation: Modeling, Programming, and Analysis*. New York: Springer.

Fishman, G. S., and L. S. Yarberry. 1997. An implementation of the batch means method. *INFORMS Journal on Computing* 9:296–310.

Glynn, P. W., and D. L. Iglehart. 1990. Simulation output analysis using standardized time series. *Mathematics of Operations Research* 15:1–16.

Goldsman, D., K. Kang, and A. F. Seila. 1999. Cramér–von Mises variance estimators for simulations. *Operations Research* 47:299–309.

Goldsman, D., M. S. Meketon, and L. W. Schruben. 1990. Properties of standardized time series weighted area variance estimators. *Management Science* 36:602–612.

Hald, A. 1952. *Statistical Theory with Engineering Applications*. John Wiley & Sons, New York.

Kelton, W. D., R. P. Sadowski, and D. T. Sturrock. 2007. *Simulation with Arena*, 4th edition Boston: McGraw-Hill.

Lavenberg, S. S., and P. D. Welch. 1981. A perspective on the use of control variables to increase the efficiency of Monte Carlo simulations. *Management Science* 27:322–335.

Meketon, M. S., and B. W. Schmeiser. 1984. Overlapping batch means: something for nothing? In *Proceedings of the 1984 Winter Simulation Conference*, eds. S. Sheppard, U. W. Pooch, and C. D. Pegden, pp. 227–230. Piscataway, New Jersey: Institute of Electrical and Electronics Engineers.

Pedrosa, A., and B. W. Schmeiser. 1993. Asymptotic and finite-sample correlations between OBM estimators. In *Proceedings of the 1993 Winter Simulation Conference*, eds. G. W. Evans, M. Mollaghassemi, E. C. Russell, and W. E. Biles, pp. 481–488. Piscataway, New Jersey: Institute of Electrical and Electronics Engineers.

Satterthwaite, F. E. 1941. Synthesis of variance. *Psychometrika* 6:309–316.

Schruben, L. W. 1983. Confidence interval estimation using standardized time series. *Operations Research* 31:1090–1108.

Song, W.-M. T., and B. W. Schmeiser. 1995. Optimal mean-squared-error batch sizes. *Management Science* 41:110–123.

Author Index

C. Alexopoulos et al. (eds.), *Advancing the Frontiers of Simulation: A Festschrift in Honor of George Samuel Fishman*, International Series in Operations Research & Management Science 133, DOI 10.1007/b110059, © Springer Science+Business Media, LLC 2009

FSC
www.fsc.org
MIX
Papier | Fördert gute Waldnutzung
FSC® C083411

Kolibri 360 GmbH
Ferdinand-Jühlke-Straße 7
99095 Erfurt, Deutschland
produktsicherheit@kolibri360.de